# RADIATION AND COMBINED HEAT TRANSFER IN CHANNELS

**Experimental and Applied Heat Transfer Guide Books**

A. Žukauskas, *Editor*

*A. Žukauskas* and *J. Žiugžda,* Heat Transfer of a Cylinder in Crossflow

*J. Vilemas, B. Česna,* and *V. Survila,* Heat Transfer in Gas-Cooled Annular Channels

*M. Tamonis,* Radiation and Combined Heat Transfer in Channels

IN PREPARATION

*J. Stasiulevičius* and *A. Skrinska,* Heat Transfer of Bundles of Finned Tubes in Crossflow

*A. Žukauskas* and *A. Šlančiauskas,* Heat Transfer in Turbulent Fluid Flows

*A. Žukauskas, V. Katinas,* and *R. Ulinskas,* Vibration and Fluid Dynamics of Tube Bundles in Crossflow

*A. Žukauskas* and *R. Ulinskas,* Heat Transfer of Tube Bundles in Crossflow

# RADIATION AND COMBINED HEAT TRANSFER IN CHANNELS

**M. Tamonis**

*Institute of Physical and Technical Problems of Energetics*
*Kaunas, Lithuanian SSR*

Edited by

**A. Žukauskas**

*Academy of Sciences of the Lithuanian SSR, Vilnius*

**HEMISPHERE PUBLISHING CORPORATION**

A subsidiary of Harper & Row, Publishers, Inc.

Washington New York London

**DISTRIBUTION OUTSIDE NORTH AMERICA**

**SPRINGER-VERLAG**

Berlin Heidelberg New York London Paris Tokyo

**RADIATION AND COMBINED HEAT TRANSFER IN CHANNELS**



1 2 3 4 5 6 7 8 9 0    B C B C    8 9 8 7 6

This book was set in Press Roman. The editors were Barbara A. Bodling and Elizabeth Dugger. BookCrafters, Inc. was printer and binder.
Originally published by Mokslas, Vilnius, as *Radiatsionnyĭ i slozhnyĭ teploobmen v kanalakh* in the series Teplofizika, vol. 12.

**Library of Congress Cataloging in Publication Data**

Tamonis, M. (Matas)
Radiation and combined heat transfer in channels.

(Experimental and applied heat transfer guide books)
Translation of: Radiatsionnyĭ i slozhnyĭ teploobmen v kanalakh.
Bibliography: p.
Includes index.
1. Heat—Transmission. 2. Gas flow. 3. Pipe—Fluid dynamics. I. Zhukauskas, A., date. II. Title. III. Series.

**DISTRIBUTION OUTSIDE NORTH AMERICA:**
ISBN 3-540-17552-0 Springer-Verlag Berlin

# CONTENTS

Preface vii
Nomenclature xi

1 OPTICAL PROPERTIES OF MOLECULAR GASES 1

1.1 Physical Processes Attending the Radiation of Gases 2
1.2 Models of Spectral Line Groups 8
1.3 Radiation in Spectral Line Bands 11
1.4 Total Emmissivity of Gases 22
1.5 Radiation of Nonuniform Gases 29
1.6 Results and Recommendations 41

2 TRANSFER EQUATIONS FOR COMBINED HEAT TRANSFER 43

2.1 Principal Equations 43
2.2 Boundary Conditions 47
2.3 Boundary-Layer Equations in a Weakly Radiating Medium 49
2.4 Equations of Combined Heat Transfer of a Medium in Channels 57
2.5 Equation of Radiative Heat Flux in Fundamental Formulation of the Problem 63

3 EXPERIMENTAL TECHNIQUE 75

3.1 Experimental Stands 75
3.2 Test Channels 80
3.3 Determination of Flow Temperature in the Channel 83
3.4 Radiative Heat Flux Sensors 84
3.5 Calibration of Radiometers 88

4 CONVECTIVE HEAT TRANSFER IN HEATED GAS FLOWS 91

4.1 Features of Heat Transfer in Laminar Flows 91
4.2 Features of Heat Transfer in Turbulent Flows 95
4.3 Determination of Turbulence Parameters of Flows 99
4.4 Experimental Results on Convective Heat Transfer in the Inlet Region of Channels 103
4.5 Experimental Results on Convective Heat Transfer in Stabilized Flow 107

5 RADIATIVE HEAT TRANSFER IN GASEOUS MEDIA 111

5.1 Effect of the Geometry of the Isothermal Medium 112
5.2 Radiation in a Nonisothermal Planar Layer of Gas 115
5.3 Radiation of a Nonisothermal Medium in a Cylindrical Channel 121
5.4 Experimental Study of Radiative Heat Flux 129

6 COMBINED HEAT TRANSFER 137

6.1 Similitude Parameters for Combined Heat Transfer 138
6.2 Conductive-Radiative Heat Transfer in a Planar Layer 144
6.3 Laminar Convective-Radiative Heat Transfer in a Planar Duct 153
6.4 Turbulent Convective-Radiative Heat Transfer in a Cylindrical Channel 160
6.5 Experimental Study of Combined Heat Transfer in Channels 166

7 RADIATIVE AND COMBINED HEAT TRANSFER IN APPLIED PROBLEMS 173

7.1 On the Calculation of Radiative Heat Transfer in Heat Engineering 173
7.2 Heat Transfer in MHD Generator Channels 177
7.3 Heat Transfer in Indirectly Heated Furnaces 180
7.4 Heat Transfer in a Hydrogen Plasma Arc 183

Appendices

1 Nomograns of the Values of $Nu_T$ in High-Temperature Turbulent Flows 193
2 Experimental Data on Convective Heat Transfer in Test Channels 1 and 2 199
3 Experimental Data on Convective Heat Transfer in Test Channel 4 212
4 Experimental Data on Combined and Radiative Heat Transfer in Test Channels 3 and 4 219

References 225

Index 237

# PREFACE

The present volume in the Experimental and Applied Heat Transfer Guide Books series is concerned with the theoretical and experimental study of radiation and combined (with convection and conduction) heat transfer in channels flows of high-temeprature gases performed at the Institute of Physical and Technical Problems of Energetics, Academy of Sciences of the Lithuanian SSR.

Work in this field was begun when interest in the study of radiative heat transfer in gray gases started lessening and investigators were faced with a new problem of practical importance—finding the capabilities and methods for making allowances for the selectivity of optical properties of various gases. In 1971, on the initiative of Professor A. Žukauskas, a scientific team was organized at the Institute for investigating the governing relationships of radiative and combined heat transfer, under the leadership of the present author. Professor Žukauskas provided constant attention to the formation of the scientific direction taken by the team, and a number of problems were solved with his direct participation.

The scientific direction of studies of this team was formed while maintaining contact with leading specialists on radiative and combined heat transfer, such as Professor S. Shorin, Doctor of Technical Sciences A. Blokh; Corresponding Member of the U.S.S.R. Academy of Sciences L. Biberman; professor I. Mikk; Doctors of Technical Sciences A. Nevskiy, S. Detkov, S. Filimonov, and B. Khrustalev; and Professor R. Viskanta (U.S.A.), to whom I express my heartfelt thanks. I also feel obligated to Corresponding Member of the U.S.S.R. Academy of Sciences B. Petukhov and Professor Ye. Dyban for useful advice on a number of problems of convective heat transfer in turbulent flows.

The present study is a compilation of the results of investigations of individual problems of radiative and combined heat transfer, performed at the Institute during the past several years with the direct participation of the author. In performing these studies we encountered various aspects of the problem, starting with calculations and determination of optical properties of gases, solution of rather complex sets of integrodifferential equations, allowance for the turbulent flow structure, and ending with problems of thermal engineering measurements in high-temperature flows.

The first chapter presents a survey of the methods for calculating optical properties of molecular gases in the infrared spectrum, analyzes them, and presents recommendations for determining these properties when using a multistage wide-band model. A great deal of attention is paid to radiation in nonisothermal media. The circle of problems considered in this book is limited to assumption of the existence of local thermodynamic equilibrium in the media under study.

The second chapter gives a mathematical description of convective and radiative heat transfer in turbulent flows. Iterative schemes of their numerical calculation are based on linearizing the starting integrodifferential equations.

The experimental arrangements and techniques form the subject matter of Chapter 3. The experimental facilities permitted investigation of radiative and combined heat transfer under various conditions of flow for combustion products of natural gas in cooled channels.

The theoretical study of convective heat transfer, the results of which are presented in Chapter 4, served as a mathematical model in analyzing experimental results obtained for high-temperature turbulized flows. It was found that the turbulent Prandtl number for high-temperature flows in cooled channels is a function of temperature conditions of the flow and other factors.

The fifth chapter is concerned with relationships governing the radiative transport of energy in nonisothermal media in the fundamental formulation of the problem. Results of calculations and measurements of radiation under various temperature and geometric conditions of the medium are presented.

The results of convective and radiative heat transfer studies were employed in the analysis of governing relationships of combined heat transfer presented in Chapter 6. Results of analytic and experimental studies of complex heat transfer are correlated in nondimensional variables.

Chapter 7 describes the results of calculation of radiative and combined heat transfer in MHD channels, in electric-arc heaters, and in other high-temperature devices.

The appendixes give nomograms for determining the convective heat transfer in turbulized flows as well as the main data of experimental results of the study.

In conclusion, I wish to thank Professor A. Žukauskas, editor of the Experimental and Applied Heat Transfer Guide Books series; Doctors of Technical Sciences S. Detkov and F. Yurevich; Candidate of Physical and Mathematical

Sciences Ye. Nogotov for useful advice and recommendations expressed in reviewing the book; Candidates of Technical Sciences L. Dagys and V. Šid-lauskas for active participation in present studies; Candidate of Technical Sciences A. Tamulionis and L. L. Burkoy for assistance in preparing the manuscript of the present book; and all who assisted in the preparation of the manuscript.

*M. Tamonis*

# NOMENCLATURE

| | |
|---|---|
| $a$ | parameter of the Voigt broadening of spectral lines, Eq. (1.3) |
| $a_\omega$ | spectral hemispherical absorptivity of surface |
| $a_0$, $a_{-1}$, $a_{-2}$ | parameters of the Lagrange approximation, Eqs. (2.81)–(2.86) |
| $A$ | spectral absorptivity, $cm^{-1}$ |
| $\hat{A}$ | total absorption of band, $cm^{-1}$ |
| $B$ | spectral intensity of blackbody radiation, Eq. (2.17), W/cm·ster |
| $B_e$ | rotational constant, $cm^{-1}$ |
| $c_0 = 2.9979 \times 10^{10}$ | speed of light, cm/s |
| $c_p$ | specific heat of gas, kJ/kg·K |
| $C \equiv S/d$ | first parameter of the rotational structure of a group of lines, $cm^{-1}$ |
| $d$ | distance between lines within a band, $cm^{-1}$ |
| $D$ | spectral transmissivity; proportionality factor for kinetic energy transfer in a compressible flow with convective energy transfer |
| $D_n$ | $n$th-order cylindrical integral function, Eq. (2.125) |
| $D_x$ | integral of the flux density over the thickness of the velocity boundary layer, Eq. (2.43) |
| $e_\omega$ | spectral hemispherical emissivity of surface |
| $E$ | electric field strength, V/m |
| $E_n$ | $n$th-order exponential integral function, Eq. (2.108) |

| | |
|---|---|
| $f$ | stream function for laminar or turbulent flow, Eq. (2.30) or (2.41) |
| $g$ | relative wave number for spectral lines or relative thermophysical properties of flow, defined by Eqs. (1.5), (1.7), (2.35)–(2.38), and (2.64) |
| $G$ | flowrate of gas, defined by Eq. (2.9), kg/s |
| $h = 6.6256 \times 10^{-34}$ | Planck's constant, J·s |
| $h$ | enthalpy of gas, kJ/kg |
| $h_r$ | radiative enthalpy of gas, kJ/kg |
| $h_{\text{kin}}$ | kinetic energy of gas, kJ/kg |
| $h_{\text{chem}}$ | dissociation energy of gas, kJ/kg |
| $H$ | height of duct in the case of a planar layer or duct diameter for an axisymmetric layer, j |
| $H_h$ | hydraulic diameter of duct, m |
| $I$ | spectral intensity of radiation, W/cm·ster |
| $\hat{I}$ | intensity of radiation, integral over spectrum, W/cm$^2$·ster |
| $J_n$ | $n$th-order Bessel function of the second kind |
| $k = 1.38054 \times 10^{-23}$ | Boltzmann's constant, J/K |
| $K_1, K_2, K_3$ | temperature functions for the wide-band Edwards model, Eqs. (1.49)–(1.51) |
| $K_T$ | proportionality factor in the mixing length in Eq. (2.6) |
| $K(a, g_D)$ | Voigt function, Eq. (1.9) |
| $l_T$ | mixing length, m |
| Le | Lewis–Semenov number |
| $M$ | molecular weight, kg/mol |
| $n_\omega$ | spectral index of refraction of the medium into which the body radiates |
| $N_T$ | damping factor, allowing for the decay of the eddy viscosity near the wall |
| $\text{Nu} \equiv \alpha x/\lambda$ | Nusselt number |
| $\text{Nu}^* \equiv q_k x \bar{C}_p / \lambda(h_f - h_w)$ | Nusselt number based on the total enthalpy difference |
| $p$ | pressure, atm |
| $p_i$ | partial pressure of gas, atm |
| $P(\bar{S}, S)$ | line intensity distribution function within a band |
| $\text{Pr} \equiv \mu c_p/\lambda$ | Prandtl number |
| $\text{Pr}_T$ | turbulent Prandtl number, Eq. (2.13) |
| $q$ | heat flux density, kW/m$^2$ |
| $q_{\text{conv}}$ | convective heat flux density, kW/m$^2$ |
| $q_{\text{chem}}$ | heat flux density due to transfer of chemical energy, kW/m$^2$ |

| | |
|---|---|
| $q_r$ | radiative heat flux density, $kW/m^2$ |
| $q_{es}$ | strength of internal energy source, $kW/m^3$ |
| $r_\omega$ | spectral hemispherical reflectivity of surface |
| $r$, $R$ | current and total length of beam, cm |
| $R = 8.3143$ | universal gas constant, $J/mol \cdot K$ |
| $Re \equiv U_f x/\nu$ | Reynolds number |
| $Re_\delta \equiv U_f \delta_x/\nu$ | Reynolds number for the thickness of the velocity boundary layer |
| $s$ | nondimensional contour of spectral line, Eqs. (1.4), (1.6), or (1.8) |
| $S$ | spectral line intensity, $cm^{-2}$ |
| $S_u$, $S_\Theta$ | auxiliary functions, Eqs. (2.89) and (2.98) |
| $t_\omega$ | spectral hemispherical transmissivity of surface |
| $T$ | temperature, K |
| $T_*$ | characteristic temperature of turbulent flow, Eq. (2.63), K |
| $Tu \equiv (u'^2)^{1/2}/U_f$ | turbulence factor |
| $u$ | longitudinal velocity component, m/s |
| $u_* \equiv (\tau_w/\rho_w)^{1/2}$ | friction velocity |
| $U$ | parameter of spectral line strength, longitudinal flow velocity, m/s |
| $v$ | transverse component of flow velocity, m/s |
| $V_d$ | diffusion transport velocity, m/s |
| $w$ | transverse velocity coordinate in three-dimensional flow, m/s |
| $W$ | equivalent line width, $cm^{-1}$ |
| $x$ | longitudinal coordinate, m |
| $y$ | transverse coordinate, m |
| $Y$ | half height of planar duct or radius of axisymmetric channel, m |
| $Y_n$ | first kind, $n$th order Bessel function |
| $\alpha$ | excess oxidant factor; heat transfer coefficient, $kW/m^2 \cdot K$ |
| $\beta$ | second parameter of the rotation structure of a group of lines, Eq. (1.20) |
| $\gamma$ | half width of spectral lines, broadened in different fashions, Eq. (1.1) or Eq. (1.2), $cm^{-1}$; configuration factor for cylindrical geometry in calculating the radiant heat flux |
| $\Gamma$ | gamma function |
| $\delta_x$ | thickness of velocity boundary layer, m |
| $\Delta_k$, $\Delta_a$ | rms and arithmetic mean errors, % |
| $\epsilon$ | total emissivity of gases |
| $\epsilon_\tau$ | eddy viscosity coefficient, $m^2/s$ |

| | |
|---|---|
| $\epsilon_q$ | eddy thermal diffusivity, $m^2/s$ |
| $\zeta$ | nondimensional longitudinal coordinate, Eq. (2.75) |
| $\eta$ | nondimensional transverse coordinate, defined by Eqs. (1.96), (2.75), (2.33), or (2.42) |
| $\theta$ | angle between the opposite direction normal to the surface element and the incident beam |
| $\Theta$ | nondimensional temperature, defined by Eqs. (2.32), (2.60), or (2.94) |
| $\kappa$ | spectral absorption coefficient, $cm^{-1}$ |
| $\kappa_T$ | von Karman constant |
| $\lambda$ | thermal conductivity, $kW/m \cdot K$ |
| $\Lambda$ | proportionality factor between diffusive and convective energy transports |
| $\mu$ | dynamic viscosity, $kg/m \cdot s$ |
| $\nu$ | kinematic viscosity, $m^2/s$ |
| $\rho$ | density, $kg/m^3$ |
| $\sigma = 5.6697 \times 10^{-12}$ | Stefan–Boltzmann constant, $W/cm^2 \cdot K^4$ |
| $\sigma$ | electrical conductivity, $ohm^{-1} \cdot m^{-1}$ |
| $\tau$ | spectral optical thickness of layer; shear stress |
| $\varphi$ | azimuthal angle; relative flow velocity; factor allowing for the variability of physical properties as a function of temperature |
| $\varphi_K$ | factor allowing for the effect of free-stream turbulence on heat transfer |
| $\chi$ | parameter characterizing the interaction between different kinds of energy transfer in combined heat transfer |
| $\omega$ | wave number, $cm^{-1}$ |
| $\Omega$ | solid angle, steradians |

## Subscripts

| | |
|---|---|
| $w$ | at the wall |
| $f$ | in the flow |
| $l$ | laminar |
| $T$ | turbulent |
| $L$ | Lorentz |
| $D$ | Doppler |
| $V$ | Voigt |
| exp | experimental |
| hem | hemispherical |

| | |
|---|---|
| pl | planar |
| $c$ | cylindrical |
| iso | isothermal |
| m | bulk |
| Σ | combined |

CHAPTER

# ONE

# OPTICAL PROPERTIES OF MOLECULAR GASES

The study of relationships governing the thermal radiation of hot gases has for a long time attracted the attention of investigators, since it is important to the economic development of countries. In conventional boilers and furnaces, use is made basically of gases heated to 1500–2000 K. Advances in technology continuously widen the range of temperatures encountered in practice. For example, in missile engineering one has to deal with gas temperatures of the order of 3500–4500 K, whereas the use of nuclear energy for peaceful purposes requires the use of engineering devices that operate at even higher temperatures. The successes attained at present in rocket and missile technology are extensively used in industries and in power production, where various processes with high-temperature gaseous working fluids are introduced. In conjunction with this it becomes extremely important to investigate transfer of energy in gases by radiation; here the computational accuracy depends to a large degree on reliability of the optical properties of gases, as used in the calculations.

The main tenets of the theory of infrared (IR) radiation of molecular gases are presented in [1–11], which give data describing the spectral and integral optical properties of various gases. Up to now, due to the extreme complexity of the physical processes of radiation, the dependence of optical properties of gases on temperature, pressure, layer thickness, and other factors has been insufficiently explored. In practice the increasingly high temperatures and pressures have forced investigators to find methods for sufficiently reliable approximation of the optical properties of gases under conditions for which no experimental data exist, or to perform additional experiments. Data on optical properties of gases (primarily for combustion product components), conforming to modern requirements, are given by Ludwig et al. [10].

In this chapter we consider aspects of the radiation theory of gases, which must be clarified to allow the description and calculation of the parameters of the optical properties of molecular gases at the modern level. Subsequently these parameters are used for investigating the governing relationship of radiative and combined heat transfer under various conditions.

## 1.1. PHYSICAL PROCESSES ATTENDING THE RADIATION OF GASES

In the majority of engineering applications, with the exception of cases of very high temperatures, radiation of gases occurs upon transitions of bound atomic or molecular states from one energy level to another (i.e., in bound–bound transitions). These transitions are accompanied by absorption or emission of a photon by the gas atom or molecule. The absorption coefficient in such a process is a peak-shaped function of the photon energy, i.e., it corresponds to a linear spectrum. Since bound–bound transitions in molecular gases are accompanied by changes in the rotational and vibrational states of the molecules, there form as a result vibrational–rotational bands, consisting of a large number of overlapping lines. These bands lie in the spectral region from 1.5 to 20 μm. The bands of purely rotational lines are situated in the far-IR spectral region (from 20 to 1000 μm.).

A gas is capable not only of absorbing and emitting photons, but also of scattering them; however, in the majority of cases the molecular scatter effect is negligible [12].

Radiation of gases in the ultraviolet (UV), visible, and near-IR spectral regions is due to electronic transitions. At moderate temperatures ($T < 3000$ K), the contribution of these transitions to the thermal radiation of gases is insignificant, for which reason they are not considered in this study.

Since the microscopic mechanism of the molecular spectrum is based on concepts of the rotational and vibrational motions of a molecule, quantitative description of the molecular spectrum involves the use of various molecular models (the rigid rotator, anharmonic oscillator, etc. models). At the current level of computer technology it is possible to calculate the spectra also of polyatomic molecules; however, these calculations are extremely work-consuming [13]. Hence the optical properties of polyatomic gases are determined by band modeling.

Calculations of the radiation of gases within the limits of vibrational-rotational bands are based on relationships governing radiation in spectral lines, which are characterized by the line intensity $S$ and shape $s$. The Einstein coefficients are used for relating the spectral line intensity $S$ to the microscopic description of radiative processes, which are already the subject of study of quantum mechanics [14]. The spectral line shape can be used to judge the effect of pressure or temperature and also the effect of other properties of the radiating gas on emission or absorption of radiation. The principal parameter describing the contour is the spectral line half-width $\gamma$, defined at a line intensity one half of its maximum value.

The contour of the spectral line is affected by various broadening mechanisms: the Lorentz (shock), Doppler, natural, Stark, etc. The main line-broadening mechanism in the IR spectrum is Lorentz broadening, due to the pressure of the gas. In certain cases, in conjunction with increasing interest in the radiation of gases at increasingly higher temperatures, it may become necessary to calculate the spectral line contour with allowance for the Voigt broadening, when the Lorentz broadening is joined by Doppler broadening. Expressions for the half-width and contour of the spectral line with different broadening mechanisms are listed in Table 1.1

The half-width $\gamma_L$ of the Lorentz line, according to the molecular theory of gases [15], is in direct proportion to the total pressure $p$ and approximately inversely proportional to the square root of temperature $T$ (Table

**TABLE 1.1** Expressions for Half-width and Contour of Spectral According Tien [9]

| | Characteristic of half-width of lines | | Characteristic of nondimensional line contours | |
|---|---|---|---|---|
| Lorentz broadening | $\gamma_L=\gamma_{L,0}\left(\frac{p}{p_0}\right)\left(\frac{T}{T_0}\right)^n$ | (1.1) | $s_L\equiv\frac{\varkappa\gamma_L}{S}=\frac{1}{\pi}\,\frac{1}{1+g_L^2}$ | (1.4) |
| | where $n=-0.5$ | | where $g_L=(\omega-\omega_0)/\gamma_L$ | (1.5) |
| Doppler broadening | $\gamma_D=\frac{\omega_0}{c_0}\left(\frac{2kT}{M}\ln 2\right)^{1/2}$ | (1.2) | $s_D\equiv\frac{\varkappa\gamma_D}{S}=\left(\frac{\ln 2}{\pi}\right)^{1/2}\cdot\exp(-g_D^2)$ | (1.6) |
| | | | where $g_D=(\ln 2)^{1/2}\cdot(\omega-\omega_0)/\gamma_D$ | (1.7) |
| Voigt broadening | $a=(\ln 2)^{1/2}\cdot\frac{\gamma_L}{\gamma_D}$ | (1.3) | $s_V\equiv\frac{\varkappa\gamma_L}{S}=\frac{a}{\pi^{1/2}}K(a,\ g_D)$ | (1.8) |
| | | | where the Voigt function is $K(a,\ g_D)\equiv \equiv\frac{1}{\pi}\int_0^\infty \exp\left(-az-\frac{z^2}{4}\right)\cos(g_D\cdot z)\,dz$ | (1.9) |

1.1). Actually the value of $\gamma_L$ is a more complex function both of the overall pressure and of the partial pressures of the individual constituents of the gas. More detailed quantum mechanical calculations shown that the exponent of the temperature dependence of half line width $n$ is controlled by the location of the line within the band. According to Yamamoto et al. [16], at low values of quantum numbers $j$ the value of $n$ approaches 0.75, but with increasing $j$ it decreases to 0.3, whereupon it again rises to 0.5.

The half-width $\gamma_D$ of the line is a function of temperature to a power of ½. At high temperatures and low pressures, Doppler broadening may become predominant.

Contours (1.4) and (1.6) are limiting. The real spectral line contour is always produced both by Lorentz and Doppler broadening and is described by the complex expressions (1.8) and (1.9), which are a function of parameter $a$. The difficulty in using such a contour arises from the fact that the integral Voigt function (1.9), which is part of Eq. (1.8), does not easily lend itself to analytic transformations.

However, in the case of the pressure and temperature ranges most frequently encountered in practice, the Lorentz broadening is predominant, and hence the need to allow for Doppler broadening may arise only at high wave numbers $\omega$. Here one must have a simpler expression for the spectral line contour with Voigt broadening. The published approximations of the Voigt contour [1] are too complex for use over the range of $a$ of $0.5 \leq a < \infty$.

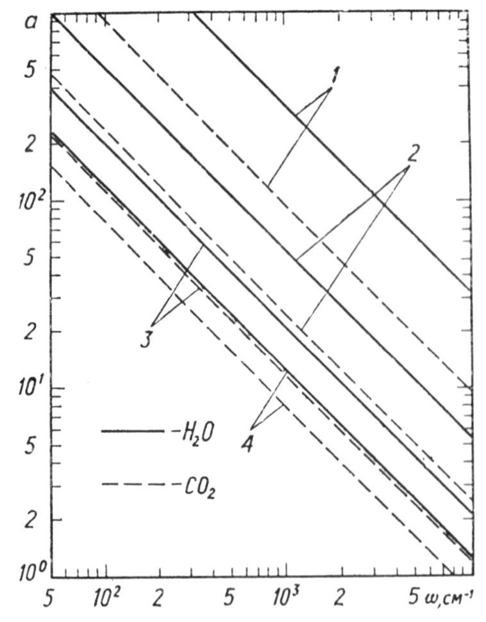


**FIG. 1.1** Parameter $a$ as a function of wave number $\omega$ and temperature $T$ for water vapor and gaseous carbon dioxide at atmospheric pressure, calculated from the data of Ludwig et al. [10]. $T$ = (1) 273 K; (2) 1000 K; (3) 2000 K; (4) 3000 K.

Numerical analysis of the Voigt contour showed that satisfactory results can be obtained from the empirical expression

$$s_V \equiv \frac{\varkappa \gamma_L}{S} \simeq \frac{1}{\pi(1+g_L^2)} \left[1 + a^{v_0} \sum_{i=1}^{4} v_i \frac{(2i-1)g_L^2 - 1}{(1+g_L^2)^i}\right] \tag{1.10}$$

The coefficients in Eq. (1.10) were determined by the method of minimizing the root mean square (rms) error. It was found that $v_0 = -1.172$, $v_1 = -0.0078$, $v_2 = 0.85$, $v_3 = -1.09$ and $v_4 = 0.46$.

The rms error of Eq. (1.10) at $a$ from 0.50 to $\infty$ is 0.017%, whereas the mean arithmetic deviation is 0.93%; here in the case of $a \geqslant 0.75$ the maximum deviation does not exceed $\pm 4\%$, and when $a$ is reduced to 0.5 it becomes as high as $\pm 10\%$.

At any values of $v_i$, Eq. (1.10) satisfies the norming condition

$$\int_{-\infty}^{+\infty} s(\omega - \omega_0, a, \gamma_L)\, d\omega = 1 \tag{1.11}$$

Attempts were made also at a simpler allowance for Voigt broadening. It should be noted, in particular, that the incorporation in Eq. (1.4) of Voigt broadening in the form of the additive sum of the Lorentz and Doppler half-width

$$\gamma = \gamma_L + \gamma_D \tag{1.12}$$

as done by Popov and Shvartsblat [17] can result in a significant underestimate of the values of radiation intensity in the vicinity of the center of the line and overestimation of it in line wings (Fig. 1.2).

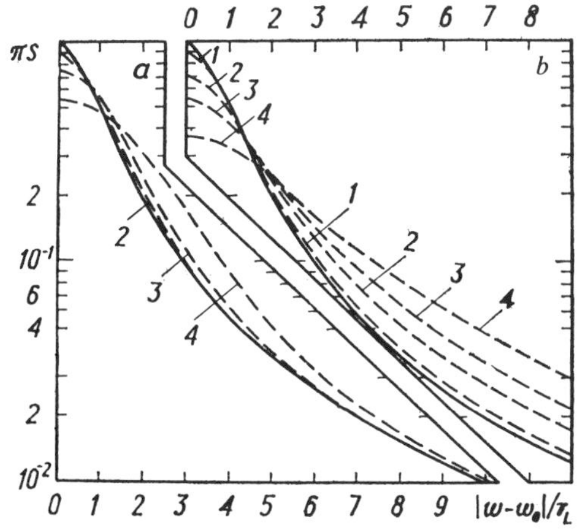


**FIG. 1.2** Comparison of the Lorentz (solid curves) and Voigt (dashed curves) spectral line contours at different values of $a$.
$a$= (1) 10; (2) 2; (3) 1; (4) 0.5. ($a$) Exact calculation of the Voigt contour, Eq. (1.8). (b) Calculation from Eqs. (1.4) and (1.12).

A quantity of great importance in calculating radiation both of isolated and overlapping spectral lines is the so-called equivalent line width:

$$W = \int_{-\infty}^{+\infty} [1 - \exp(-\varkappa R)]\, d\omega \qquad (1.13)$$

The equivalent spectral line width of the Lorentz contour is expressed in terms of the Ladenburg–Reiche function [1, 9, 10], and of the Voigt contour, by the formula

$$W = \frac{2\gamma_L}{a} \int_0^{\infty} \{1 - \exp[-2\pi^{1/2} aUK(a,\ g_D)]\}\, dg_L \qquad (1.14)$$

where the spectral line strength parameter is

$$U \equiv \frac{SR}{2\pi\gamma_L} \qquad (1.15)$$

Calculations show (Fig. 1.3) that the values of $W/2\pi\gamma_L$ for weak ($U \ll 1$) and strong ($U \gg 1$) spectral lines of the Voigt and Lorentz contours coincide. At $U \simeq 1$ Doppler broadening results in some increase in $W/2\pi\gamma_L$, which does not exceed 15% at $0.5 \leqslant a < \infty$.

Study of the equivalent line width can yield useful information on the applicability of the multiplicity property to gas mixtures, which consists in the fact that the spectral transmissivity of the mixture is equal to the product of the transmissivities of its constituents. This property, rigorously speaking, is valid only in the case of monochromatic radiation. For wider spectral intervals, for example within the limits of one spectral line, this property is not obvious [6].

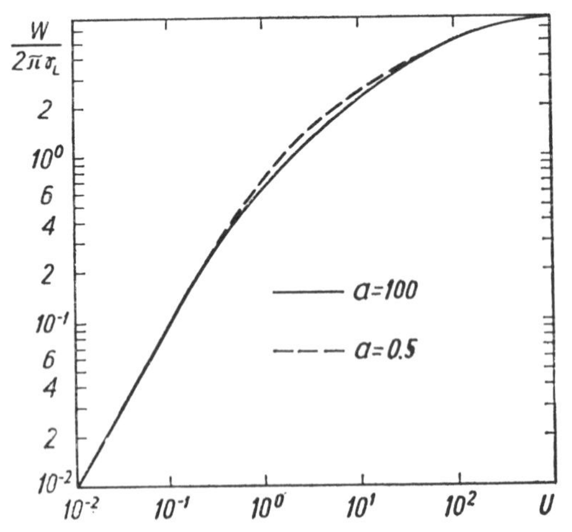


**FIG. 1.3** Curve of growth of spectral lines with a Voigt contour as a function of the line strength parameter $U$ and parameter $a$.

Let us consider isolated spectral lines with Lorentz contour of a gas mixture containing $M$ components, located near wave number $\omega_0$. Due to applicability of the multiplicity property to monochromatic radiation, the equivalent width of such a group of lines can be represented in the form

$$W = \int_{-\infty}^{+\infty} \left[1 - \exp\left(-\sum_{m=1}^{M} \varkappa_m R\right)\right] d\omega \tag{1.16}$$

or using Eqs. (1.4), (1.5), and (1.15),

$$W = \int_{-\infty}^{+\infty} \left[1 - \exp\left(-\sum_{m=1}^{M} \frac{2U_m \gamma_{L,m}^2}{(\omega-\omega_0)^2 + \gamma_{L,m}^2}\right)\right] d\omega \tag{1.17}$$

We shall restrict ourselves to analysis of this expression for weak ($U_m << 1$) and strong ($U_m >> 1$) lines. Since by its physical meaning, the value of $U_m$ is equal to one half of the optical path in the center of the line, then at $U_m << 1$ the optical path is found to be small for all the frequencies. Then the exponent in Eq. (1.17) can be replaced by the first two terms of its series expansion:

$$W = \sum_{m=1}^{M} \int_{-\infty}^{\infty} \frac{2U_m \gamma_{L,m}^2}{(\omega-\omega_0)^2 + \gamma_{L,m}^2} d\omega = 2\pi \sum_{m=1}^{M} U_m \gamma_{L,m} = \sum_{m=1}^{M} S_m R \tag{1.18}$$

In the case of strong lines ($U_m >> 1$) the absorption in the center of the line is complete. Then if the quantity $\gamma_{L,m}^2$ in the denominator of Eq. (1.17) is neglected, the absorption increases for all the frequencies. However, for $(\omega - \omega_0) >> \gamma_{L,m}^2$ this increase in absorption will be negligible and at moderate values of $(\omega - v_0)$ it will not affect the final result. For strong lines,

$$W = \sqrt{2 \sum_{m=1}^{M} U_m \gamma_{L,m}^2} \cdot \int_0^{\infty} \frac{1 - \exp(-\xi)}{\xi^{3/2}} d\xi = 2 \sqrt{\sum_{m=1}^{M} S_m \gamma_{L,m} R} \tag{1.19}$$

where $\xi = \dfrac{2 \sum_{m=1}^{M} U_m \gamma_{L,m}^2}{(\omega - \omega_0)^2}$

This means that the multiplicity property can be used only in the case of weak lines given by Eq. (1.18). On the other hand, in the case of strong lines the optical parameters should be summed under the square root sign. Note that this problem has been insufficiently explored in the literature, although it is quite important in calculating radiation in gas mixtures.

The above relationships for isolated lines can, strictly speaking, be used only for atomic spectra in the case of low pressures and short optical path lengths. In all other cases the lines in the spectra overlap quite arbitrarily. These overlapping lines are investigated in models of spectral line groups.

## 1.2. MODELS OF SPECTRAL LINE GROUPS

A spectral band with overlapping lines has an absorption which is always smaller than an analogous group of isolated lines. Absorption is a function of the distance $d$ between the spectral lines, the line intensities $S$ and half-widths $\gamma$. Exact quantum-mechanical calculations are possible only for bands with a relatively simple structure. At present it is easier to calculate the mean absorption for a narrow range of wave numbers inside a band on the basis of theoretical models of groups of spectral lines (or models of narrow bands).

There exists a large number of band models for a narrow interval of wave numbers with various spectral line profiles. In the first place models are classified on the basis of the intensity distributions and of regularity of lines within the band.

The best known models with regular distribution of spectral lines within the bands are the Elsasser model for Lorentz profiles [1, 9] and the Golden model for Doppler profiles [18].

The most frequently used models are those with random line arrangement. According to Table 1.2, the spectral transmissivity $D$ for such models of line groups depends in a complex functional manner on the parameters of the rotational structure of line groups

$$\beta_L \equiv \frac{\pi\gamma_L}{d} \quad \text{or} \quad \beta_D \equiv \frac{1}{(\ln 2)^{1/2}} \cdot \frac{\pi\gamma_D}{d} \tag{1.20}$$

and

$$C \equiv \frac{S}{d} \tag{1.21}$$

Comparison of models with regular distribution of spectral lines with models having a random line distribution shows that in the first case the absorption is always greater, when regularly distributed lines overlap less. As a rule, the difference between absorption calculated from these models is not too great and in most cases does not exceed 20%. Naturally, models with intermediate distributions of spectral lines, for example the King model [19], yield intermediate values of absorption.

In spite of the large variety of expressions for spectral transmission for a wide range of parameters, given by Eqs. (1.22)–(1.28), there exists a single

**TABLE 1.2** Brief description of the principal models of spectral line groups with random arrangement of lines within the bands

| Line intensity distribution | Line shape | Expression for spectral transmissivity |
|---|---|---|
| Equal strength lines | Lorentz | Goody model [1, 6, 10]<br>$\ln D_L = -2\beta_L \cdot f_{LR}(U)$ (1.22)<br>where<br>$f_{L-R}(U) = U \cdot \exp(-U) \cdot [Y_0(U) + Y_1(U)]$<br>is the Ladenburg–Reiche function [1,10]; $Y_0(U)$ and $Y_1(U)$ are modified Bessel functions. To within ±8% the actual expression is [10]:<br>$\ln D_L \simeq -CR \cdot \left(1 + \frac{\pi}{4} \cdot \frac{CR}{\beta_L}\right)^{-\frac{1}{2}}$ (1.23)<br>More exact approximations of spectral absorption for this model of line groups are given by Detkov [20] |
| | Doppler | According to Ludwig et al. [10]:<br>$\ln D_D = -\frac{\beta_D}{\pi^{1/2}} \cdot f_D\left(\frac{\pi^{1/2} CR}{\beta_D}\right)$ (1.24)<br>where<br>$f_D(x) = \frac{1}{\pi^{1/2}} \int_{-\infty}^{+\infty} \{1 - \exp[-x \exp(-x_1^2)]\} dx.$<br>To within ±9% one can use the expression[10]<br>$\ln D_D \simeq -\frac{2^{1/2}}{\pi} \beta_D \left\{\ln\left[1 + \frac{1}{2}\left(\frac{\pi CR}{\beta_D}\right)^2\right]\right\}^{1/2}$ (1.25) |
| | Voigt | The SLG model [10]<br>$\ln D_V \simeq -CR \cdot \left(1 - Y^{-\frac{1}{2}}\right)^{\frac{1}{2}}$<br>where<br>$Y = \left[1 - \left(\frac{\ln D_L}{CR}\right)^2\right]^{-2} + \left[1 - \left(\frac{\ln D_D}{CR}\right)^2\right]^{-2} - 1$ (1.26)<br>$\ln D_L$ is determined from Eq. (1.23) and $\ln D_D$ from Eq. (1.25) |
| Lines with random Lorentz intensity distribution | | Mayer–Goody model<br>$\ln D_L = -CR\left(1 + \frac{CR}{\beta_L}\right)^{-\frac{1}{2}}$ (1.27) |

**TABLE 1.2.** Brief Description of the Principal Models of Spectral line Groups with Random Arrangement of Lines within the Bands (continued)

| Line intensity distribution | Line shape | Expression for spectral transmissivity |
|---|---|---|
| Lines with random intensity distribution | Doppler | According to Ludwig et al. [10]<br>$\ln D_D = -\frac{\beta_D}{\pi^{1/2}} \cdot f_E \left( \frac{\pi^{1/2} CR}{\beta_D} \right)$ (1.28)<br>where<br>$f_E(x) = \frac{1}{\pi^{1/2}} \int_{-\infty}^{+\infty} \{x \exp(-x_1^2)/[x \exp(-x_1^2)+1]\} \, dx_1.$<br>To within ±10% one can use Eq. (1.25) |

expression, suitable for all band models in the weak-lines approximation, or in complete overlapping of lines, where Beer's law applies:

$$D = \exp(-CR) \tag{1.29}$$

On the other hand, the asymptotic expressions for strong lines are various functions of the same argument $(\beta_L CR)^{1/2}$[9].

Equation (1.26) for random distribution of lines with a Voigt profile is very cumbersome. At $a \geqslant 0.5$ one can use a simpler relationship.

It is known that for a group of lines with any profile the logarithm of the spectral transmissivity is given by the expression [1, 6]

$$\ln D = -\frac{1}{a} \int_{-\infty}^{+\infty} \frac{sSRd\,\omega_0}{\gamma + sSR} \tag{1.30}$$

Numerical analysis of Eq. (1.30) together with the Voigt line profile from Eq. (1.10) shows that in order to incorporate the Doppler broadening it suffices to add in Eq. (1.27) a multiplier in the form of some function of parameters $a$ and $U$. Thus, for $0.5 \leqslant a < \infty$ and $10^{-2} \leqslant U \leqslant 10^5$ one can use the expression

$$\ln D_V = -CR\,(1 + CR/\beta_L)^{-\frac{1}{2}} \left(1 + 0{,}033\, a^{-1.3}\, U^{1-\frac{\lg U}{2.4}}\right) \tag{1.31}$$

The results obtained from Eq. (1.31) differ from those given by the exact expression (1.30) by not more than ±1%.

Over the range of $a$ and $U$ under study, the Doppler broadening can increase the effective absorption coefficient only to 11% (Fig. 1.4). If it is remembered that the error in determining the remaining parameters is ±15% on the average, then it becomes meaningless to incorporate effects due to Doppler broadening. Hence in subsequent calculations we do not make allowance for the Voigt spectral line shape, and the subscript of the second parameter of rotation band structure β, which gives the line profile, is dropped.

## 1.3. RADIATION IN SPECTRAL LINE BANDS

In calculating the emissivity and absorption for a band from data on the spectral transmissivity of groups of lines, one must know the frequency dependence of the mean intensity of the lines $S(\omega)$, their half-width $(\omega)$, and the mean distance between the lines in the band $d(\omega)$. Such information is provided by the so-called wide-band models.

As a rule, wide-band models are used for describing a quantity known as the total absorption for a band:

$$\hat{A} = \int_{\Delta\omega} A d\omega = \int_{\Delta\omega} (1 - D)\, d\omega = \int_{\Delta\omega} [1 - \exp(-\tau)]\, d\omega \qquad (1.32)$$

A large number of various wide-band models is known (Table 1.3).

Of the simpler models with an arbitrarily selected frequency dependence of intensity, the most extensively used is the rectangular band model suggested by Penner and successfully used by him for calculating the emissivity of diatomic gases [1]. It consists of a rectangular model-line shape with

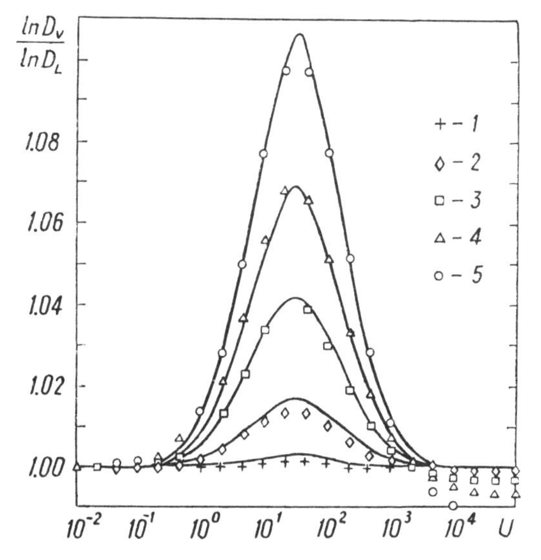


**FIG.1.4** The effect of Doppler broadening on the transmissivity of a group of line as a function of $U$ and $a$. The points were obtained analytically from Eq. (1.30) for the following values of $a$:(1) 100; (2) 2; (3) 1; (4) 0.7; (5) 0.5. The solid curves were calculated from Eq. (1.31).

**TABLE 1.3** Principal Features of Wide-Band Models in Use

| Reference | Definition of band absorption | Brief description of enveloping band |
|---|---|---|
| Penner [1] | $\hat{A}=[1-\exp(-\bar{\varkappa} R)]\cdot\Delta\omega_e$ (1.33) | Rectangular band model |
| Schack [8, 12] | $\hat{A}=A_0[1-\exp(-K_0 R)\,\mathrm{sh}\,(K_0 R)/(K_0 R)]$ (1.34)<br>where $A_0$ and $K_0$ are the correlation parameters | Triangular band model |
| Edwards and Menard [21] | At $\beta\leqslant 1$<br>$\hat{A}=\begin{cases} K_1 R, & \text{at } 0\leqslant\hat{A}\leqslant\beta K_3 \\ K_2\sqrt{p_e}\,R-\beta K_3, & \text{at } \beta K_3\leqslant\hat{A}\leqslant K_3(2-\beta) \\ K_3\left[\ln\left(\frac{K_2^2 p_e R}{4K_3}\right)+2-\beta\right], & \text{at } \hat{A}>K_3(2-\beta) \end{cases}$ (1.35)<br>At $\beta>1$<br>$\hat{A}=\begin{cases} K_1 R, & \text{at } 0\leqslant\hat{A}\leqslant K_3 \\ K_3[\ln(K_1 R/K_3)+1], & \text{at } \hat{A}>K_3 \end{cases}$ (1.36)<br>where $K_1$, $K_2$, and $K_3$ are correlation parameters | Exponential band model<br>The statistical Mayer–Goody model was used for line groups |
| Tien and Lowder [22] | $\hat{A}=K_3\ln\left\{\frac{K_1 R f(\beta)}{K_3}\cdot\left[\frac{\frac{K_1 R}{K_3}+2}{\frac{K_1 R}{K_3}+2f(\beta)}+1\right]\right\}$ (1.37)<br>where $K_1$, $K_2$, and $K_3$ are correlation parameters introduced by Edwards Menard [21] whereas $f(\beta)=2.94[1-\exp(-2.6\beta)]$ according to Tien and Lowder [22]<br>or $f(\beta)=\frac{1.78\cdot 2\beta}{1+2\beta}$ according to Detkov et al. [23] | Same as above |

**Table 1.3** Principal Features of Wide-Band Models in Use (Continued)

| Reference | Definition of band absorption | Brief description of enveloping band |
|---|---|---|
| Lukash [24] | $\hat{A} \simeq 4 \sqrt{\frac{kT}{hc_0}} \cdot B_e \cdot Y\left(\frac{K_1 R}{2} \cdot \sqrt{\frac{hc_0 B_e}{kT}}\right)$ (1.38)<br>where $Y(x) = -\frac{1}{2} \sum_{i=1}^{\infty} (-x)^i \cdot i^{-\frac{i+1}{2}} \frac{\Gamma\left(\frac{i+1}{2}\right)}{\Gamma(i+1)}$ (1.39) | Model of rigid rotator with band envelope<br>$f(g_L) \simeq \lvert g_L \rvert \exp(-g_L^2)$<br>in the approximation of total line overlapping |
| Popov and Shvartsblat [17] | $\hat{A} = 1 - \exp[-CR/(1+CR/\beta)]$ (1.40)<br>where $\beta = \frac{\pi(\gamma_L + \gamma_D)}{d}$ (1.41) | Model of nonrigid rotator with envelope<br>$f(g_L) \simeq \lvert g_L \rvert \cdot \exp\{-g_L^2 \times$<br>$\times [1 - f_2(g_L T)^2 + f_4(g_L T)^4]\}$<br>The statistical Mayer–Goody model was used for line groups |
| Hsieh and Greif [25] | $\hat{A} = 2K_3 \int_0^{\left[\frac{K_1 R\beta}{K_3} \exp\left(-\frac{2d}{K_3}\right)\right]^{1/2}} \operatorname{erf}(z) \frac{dz}{z}$ (1.42)<br>where $K_1$, and $K_3$ are correlation parameters introduced by Edwards Menard [21] | Exponential band envelope together with the regular Elsasser model |

**Table 1.3** Principal Features of Wide-Band Models in Use (Continued)

| Reference | Definition of band absorption | | Brief description of enveloping |
|---|---|---|---|
| Detkov el al. [23] | $\hat{A}=\gamma_e A_0 \left\{1-\exp\left[\frac{-SR}{\gamma_e A_0 \cdot \left(1+\frac{z_* \gamma_e SdR}{4A_0}\right)^{1/2}}\right]\right\}$ <br> where <br> $\gamma_e=B_* \cdot \sqrt{\ln\left(4.9+1,1\frac{SR}{\gamma}\right)}$ <br> $A_0$, $B_*$, $z_*$ — are correlation parameters | (1.43) <br> (1.44) | Bands with envelope according to the harmonioc oscillator model replaced by a model with a rectangular profile with some effective value of width parameter $\gamma_e$. The rotational structure is taken into account phenomenologically. The Mayer–Goody statistical model is used for line groups |
| Detkov el al. [23] | $\hat{A}=\gamma_e A_0 \sqrt{1-\exp\left[-\frac{S^2 R^2}{A_0^2 \gamma_e^2 \left(1+z_* \frac{SR}{4\gamma}\right)}\right]}$ <br> where $\gamma_e$ is defined by Eq. (1.44), $A_0$ and $z_*$ are correlation parameters | (1.45) | Same as above, but the regular Elsasser model was used for line groups |
| Detkov [20] | $\hat{A}=2\sqrt{[1-\exp(-U/2)]\cdot[0.5772+\ln(U/2)]+E_1(U/2)}$, <br> where $E_1(x)$ is an exponential integral function | (1.46) | The band envelope is assumed according to the rigid rotator model. The model of the line groups is given by Eq. (1.27) |
| Detkov and Tokmakov [26] | $\hat{A}=\left\{2m\left[1-\exp\left(-\frac{U}{2m\sqrt{1+zU/\beta}}\right)\right]\right\}^{\delta}\times$ <br> $\times\left[2\ln\left(1+\frac{U\cdot f}{2}\cdot\frac{U+4}{U+4f}\right)\right]^{1-\delta}$ <br> For the statistical model $f = 7.12\beta/(1 + 4\beta)$. For the regular model $f = 1.78\mathrm{th}(\beta/2)[1 - 0.88/\mathrm{ch}^2(\beta/2)]^{0.278}$; $m$, $z$ are correlation parameters. | (1.47) | The band envelope was expressed by the function <br> $f(g_L)=\frac{\exp(-g_L^n)}{2\Gamma\left(1+\frac{1}{n}\right)}$ <br> according to which the band envelope ranges from exponential ($n=1$) to rectangular ($n=\infty$); |

mean absorption coefficient $\bar{\varkappa}$, defined by the expression

$$\bar{\varkappa} = \frac{\alpha}{\Delta\omega_e} \tag{1.47}$$

where $\Delta\omega_e$ is the band width determined by the pertinent method.

The region of application of this model is limited to moderately high pressures and small to medium optical path lengths. These limitations stem from the fact that the rectangular band model does not make allowance for the structure of lines within the bands.

The Schack triangular model [8, 21] assumes that the spectral absorption coefficient $\varkappa$ is a linear function of the wave number $\omega$. The two-parametric equation (1.34) obtained by Schack does not satisfy the limiting case of large optical thicknesses, since at $R \rightarrow \infty$ the value of $\hat{A}$ approaches that of $A_0$, which contradicts the physical picture of the process. This means that the use of Eq. (1.34) is highly limited with respect to the range of optical paths. In conjunction with this, Howard suggested the use of a linear distribution of the absorption coefficient at small optical thicknesses, the quadratic at medium thicknesses, and the logarithmic at large optical thicknesses [21]. The main shortcoming of this correlation is the jump from one relationship to another, for which reason Goody suggested that the Howard equations be replaced by the expression obtained in the Mayer–Goody model for a line group (1.27); this still made it necessary to determine the functional dependence of the band parameters $C$ and $\beta$ on the wave number $\omega$. In general this relationship can be determined by analyzing the line structure in the vibrational–rotational bands, or by selecting arbitrarily, with subsequent correlation, the parameters contained in this expression from experimental measurements of absorption.

The values of $C$ and $\beta$ for diatomic gases can be determined analytically, by calculating the spectrum in the approximation of the rigid or nonrigid rotator model [27].

When the frequency dependence for determining the parameters of band structure is selected arbitrarily, it is best to assume that the mean distance $d$ between the lines depends little on wave number $\omega$. Then the remaining parameters can be obtained by dividing the band into individual segments, within which allowance is made for the rotational structure of the spectrum. This was done by Edwards and Menard [21], who expressed the band envelope by the function

$$C = C_0 \exp\left[-|\omega - \omega_0|/A_0\right] \tag{1.48}$$

The parameters in Eq. (1.48) are defined in terms of three temperature functions, $K_1$, $K_2$, and $K_3$, which basically reflect three regions of variation in the absorption as a function of the optical path length: the linear region, the square root region, and the logarithmic region. The temperature functions were selected in such a manner that

$$C_0 = \frac{K_1}{K_3} \rho_v \tag{1.49}$$

$$A_0 = K_3, \tag{1.50}$$

$$\beta = \frac{K_2^2}{4K_1 \cdot K_3} \cdot p_* \tag{1.51}$$

$$p_* = (p_y + b_p \cdot p_v)^{n_p} \tag{1.52}$$

where $b_p$ and $n_p$ are parameters making allowance for the partial pressure of the gas responsible for broadening of the spectral line ($p_b$) and of the absorbing gas ($p_{ab}$) in the Edwards wide-band model. The selected function $K_1$ was equated to the band intensity by

$$K_1 = \int_{\Delta\omega} C d\omega \tag{1.53}$$

The use of this functional relationship for finding the band parameters in the Mayer–Goody model (1.28) allows obtaining a number of approximate expressions for band absorption using Eqs. (1.35) and (1.36) (Table 1.3).

The Edwards wide-band model was transformed by Tien and Lowder [22], who suggested a single expression for band absorption (1.37) over the entire range of optical pathlengths. In spite of the fact that the Tien–Lowder approximation contains additional errors, it is extensively used by Western investigators [28–34], since it makes it possible to approximately integrate the absorption over the frequency.

The values of correlation parameters needed for calculating absorption for various gases from Eqs. (1.35)–(1.37) are listed in [4, 9, 28, 29, 35–40].

Real band spectra of gases are sometimes best described not by exponential band envelopes, which are proportional to $\exp(-|\omega-\omega_0|)$, but to band envelopes obtained in the approximation of the harmonic oscillator and the rigid rotator, proportional to $|\omega-\omega_0| \times \exp[-(\omega-\omega_0)^2]$. Such enveloping bands were used by Detkov et al. [23], Lukash [24], Leckner [41, 42], and others. The harmonic oscillator and rigid rotator model yields expressions for the band such as Eq. (1.38) (Table 1.3), which are independent of the total pressure, which does not agree with experimental data. To eliminate this shortcoming, Popov and Shvartsblat used a harmonic oscillator and nonrigid rotator model and obtained Eq. (1.40). By retaining three terms in the Taylor expansion for rotational energy they were able to obtain a band envelope in the form of $|\omega-\omega_0|\exp\{-(\omega-\omega_0)^2[1-f_2(\omega-\omega_0)^2 T^2+f_4(\omega-\omega_0)^4 T^4]\}$, where $f_2$ and $f_4$ are correlation parameters. In this way they obtained an analytic expression for the spectral transmissivity of a band with a small number of undefined parameters, which were determined for the individual gas bands from the available, albeit sparse, experimental data. One of the shortcomings of this model is the fact that it incorporates the Doppler line broadening in the form of an additive sum of half-widths of lines from Eq. (1.12), which may result in a significant distortion of the physical picture of the process (see Fig. 1.2).

The main shortcoming of most wide-band models is the fact that in the case of complex band envelopes it is difficult to obtain analytic expressions for the total band absorption. A way out of this may be selection of sufficiently reliable empirical relationships using the phenomenological approach. Such an approach is exemplified by Detkov and his co-workers [20,23,26] [Eqs. (1.43)–(1.47) in Table 1.3].

Calculation of band absorption on the basis of group models with regular line arrangement reduces to the use of no less complex expressions, such as Eq. (1.42). The correlation parameters for calculating the absorption of carbon dixide using the regular Elsasser model were determined by Hsieh and Greif [25].

A general shortcoming of the aforementioned models is the fact that all of them contain a number of correlation parameters determined from the best agreement between available experimental data, which means that their suitability has been validated only for the range of variables used in the experiments.

One of the studies that did not use arbitrary selection of the band envelope is that by Detkov [43], who obtained an expression for the absorptivity of $CO_2$ in the 4.3-μm band, expressed as an ensemble of a series of overlapping bands. His results are in satisfactory agreement with the available spectral measurements.

The large variety and complexity of functional relationships for $\hat{A}$ makes the calculation of other integral radiation parameters very difficult.

The most work-consuming operation in the numerical solution of radiation gas dynamics by computer is integration over the spectrum. To obtain optimal results one needs simple relationships. A successful way out of this situation is to tabulate the spectral parameters as a function of the wave number and temperature, as was done by Ludwig et al. [10]; however, the use of tabulated parameters requires a computer with a large operative memory.

Experience shows that an acceptable accuracy of results can be attained by integrating over the spectrum by the method of rectangles. Here it suffices to divide the spectrum into 15–20 parts within a given band, since a further increase in the number of subdivisions has little effect on the final result. Integration of the expression.

$$\hat{A} = \int_{\Delta\omega} A\,d\omega = \sum_{i=1}^{N-1} A_i \cdot \Delta\omega_i \tag{1.54}$$

by the method of rectangles with a moderate number of steps results in further error. For example, when $A$ is determined by the Edwards–Menard or Tien–Lowder equations, the error for individual bands may be greater than 50%. In this case it is necessary to correlate the parameters in the expressions for $A$, which, in substance, corresponds to replacing the smooth exponential band envelope by some multistep function. The correlation constants for such an envelope, determined by way of minimization of the function

$$\sum_{m=1}^{M} \left[ \frac{\hat{A}_{m,i}(K_1, K_2, K_3, b_p, n_p, \ldots)}{\hat{A}_{exp\,m,i}} - 1 \right]^2 = \min \tag{1.55}$$

are listed in the paper by the present author and his co-workers [44] (Table 1.4). Function (1.55) corresponds to summation of the rms deviations of the calculated values of total absorption $\hat{A}_{\exp} m,i$. In analyzing data in [44] the authors expressed $K_1$, $K_2$, and $K_3$ by functional relationships, as well as by others (Table 1.5). The temperature dependences of structural parameters $C$ and ß for certain gases determined in this manner are listed in Figs. 1.5 and 1.6.

The available experimental results on total band absorption $\hat{A}_{\exp}$ were

**TABLE 1.4** Correlation Constants for the Bands of Individual Gases

| Gas | Band, μm | Wave number $\omega_0$ of the centre of the band, cm$^{-1}$ | $b_p$ | $n_p$ | $k_1$ | $r_1$ | $r_2$ | $k_2$ | $s_1$ | $s_2$ | $k_3$ | $t_1$ |
|---|---|---|---|---|---|---|---|---|---|---|---|---|
| $H_2O$ | Rotational. | 370 | 3,62 | 0.98 | 6437 | 17.6 | 0,50 | 215 | 8.72 | 0.50 | 28.3 | 0,50 |
| ” | 6.3 | 1600 | 4.81 | 1.14 | 37.87 | 0.0 | – | 26.15 | 0.19 | – | 32.74 | 0.51 |
| ” | 2.7–3.2 | 3750 | 5.03 | 1.00 | 30.3 | 0.0 | – | 26.67 | 0.01 | – | 52.33 | 0.39 |
| ” | 1.87 | 5350 | 4.26 | 1.12 | 2.85 | 0.0 | – | 6.30 | 0.04 | – | 46.52 | 0.14 |
| ” | 1.38 | 7250 | 1.41 | 0.73 | 2.71 | 0.0 | – | 5.98 | 0.11 | – | 44.56 | 0.38 |
| $CO_2$ | 15 | 667 | 1.3* | 0.7* | 9.1 | 0.05 | – | 3.1 | 1.2 | – | 7.1 | 0.56 |
| ” | 10.4** | 960 | 1.3 | 0.8 | 0.76 | 0.0 | – | 1.6 | 0.5 | – | 12.4 | 0.50 |
| ” | 9.4** | 1060 | 1.3 | 0.8 | 0.76 | 0.0 | – | 1.6 | 0.5 | – | 12.4 | 0.50 |
| ” | 4.3 | 2350 | 1.3* | 0.80* | 86.2 | 0.40 | – | 8.7 | 1.25 | – | 7.0 | 0.40 |
| ” | 2.7 | 3715 | 1.3* | 0.65* | 4.5 | 0.09 | – | 4.4 | 0.80 | – | 12.5 | 0.51 |
| CO | 4.7** | 2143 | 0.153 | 1.0 | 21.21 | 0.0 | – | 0.182 | 1.57 | 26.54 | 15.62 | 0.53 |
| ” | 2.35 | 4260 | 0.807 | 1.33 | 0.14 | 0.0 | – | 0.22 | 1.50 | 15.15 | 22.0 | 0.50 |
| $CH_4$ | 7.66 | 1310 | 1.3** | 0.8** | 40.0 | –0.4 | – | 7.7 | 0.7 | – | 12.0 | 0.51 |
| ” | 3.31 | 3020 | 1.3** | 0.8** | 27.6 | 0.28 | – | 10.7 | 0.53 | – | 54.0 | 0.07 |

* Parameters not included in the present analysis.

** The correlation constants for the 10, 4, and 9.4 μm $CO_2$ bands and for the base band of CO were taken from Tien [9] and Edwards [38]

**TABLE 1.5** Listing of Expressions for Temperature Functions $K_1$, $K_2$, and $K_3$ ($T_0 = 100$)

| Description of band | $K_1$, $cm^{-1}/(g \cdot m^{-2})$ | $K_2$, $cm^{-1}/(g \cdot m^{-2} \cdot atm)$ | $K_3$, $cm^{-1}$ |
|---|---|---|---|
| Rotational band of $H_2O$ | $k_1 \exp[-r_1 \cdot (T/T_0)^{r_2}]$ | $k_2 \exp[-s_1 \cdot (T/T_0)^{-s_2}]$ | $k_3 \cdot (T/T_0)^{t_1}$ |
| 6.3 and 2.7 μm $H_2O$ band<br>7.6 and 3.3 μm $CH_4$ band<br>15, 4.3 and 2.7 μm $CO_2$ band | $k_1 \cdot (T/T_0)^{r_1}$ | $k_2 \cdot (T/T_0)^{s_1}$ | " |
| 1.87 μm $H_2O$ band | $k_1 \cdot (T/T_0)^{r_1} \cdot \Phi_{0,1,1}(T)$ | $k_2 \cdot (T/T_0)^{s_1} \cdot K_1^{1/2}$ | " |
| 1.38 μm $H_2O$ band | $k_1 \cdot (T/T_0)^{r_1} \cdot \Phi_{1,0,1}(T)$ | " | " |
| 10.4 and 9.4 μm $CO_2$ band | $k_1 \cdot (T/T_0)^{r_1} \cdot \varphi_2(T)$ | " | " |
| 4.7 μm CO band | $k_1 \cdot (T/T_0)^{r_1}$ | $k_2 \cdot (T/T_0)^{s_1} \cdot \varphi_1(T) + s_2$ | " |
| 2.35 μm CO band | " | $[k_2 \cdot (T/T_0)^{s_1} \cdot \varphi_1(T) + s_2] \cdot 0.08$ | " |

$$\varphi_1(T) = 1 - \exp(-2143\ m/T); \quad m = 1.438776;$$

$$\varphi_2(T) = \left[1 - \exp\left(\frac{-1045\ m}{T}\right)\right] \cdot \exp\left(\frac{-1351\ m}{T}\right) - \frac{\exp\left(\frac{-2702\ m}{T}\right)}{2 \cdot \left[1 - \exp\left(\frac{-1351\ m}{T}\right)\right] \cdot \left[1 - \exp\left(\frac{-2396\ m}{T}\right)\right]},$$

$$\Phi_{0,1,1} = \frac{1 - \exp\left(\frac{-5351\ m}{T}\right)}{\left[1 - \exp\left(\frac{-1595\ m}{T}\right)\right] \cdot \left[1 - \exp\left(\frac{-3756\ m}{T}\right)\right]}$$

$$\Phi_{1,0,1} = \frac{1 - \exp\left(\frac{-7408\ m}{T}\right)}{\left[1 - \exp\left(\frac{-3652\ m}{T}\right)\right] \cdot \left[1 - \exp\left(\frac{-3756\ m}{T}\right)\right]}$$

obtained only for relatively narrow temperature and pressure ranges (Table 1.6). Given the unavailability of published experimental results, we used the values of spectral parameters of bands suggested by others. Owing to the lack of complete absorption data for certain bands, certaim parameters were not analyzed (Table 1.4).

| Gas | | Bands, μm | | | | |
|---|---|---|---|---|---|---|
| | | 1 | 2 | 3 | 4 | 5 |
| $CO_2$ | ——— | 15 | 10.4 | 9.4 | 4.3 | 2.7 |
| $H_2O$ | - - - - | 20 | 6.3 | 2.7 | 1.87 | 1.38 |
| CO | — · — | 4.7 | 2.35 | – | – | – |
| NO | — ○ — | 5.3 | – | – | – | – |
| $CH_4$ | — • — | 7.6 | 3.3 | – | – | – |

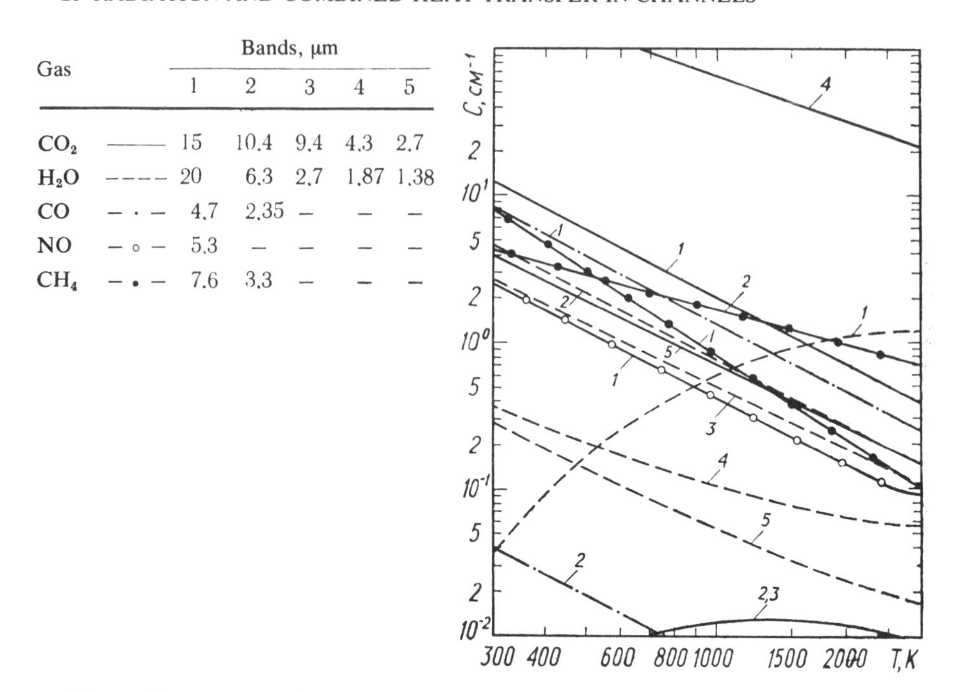


**FIG. 1.5** First parameter $C$ of the rotational structure of line groups for centers of bands of molecular gases as a function of temperature. The curve numbers correspond to the band numbers in the table.

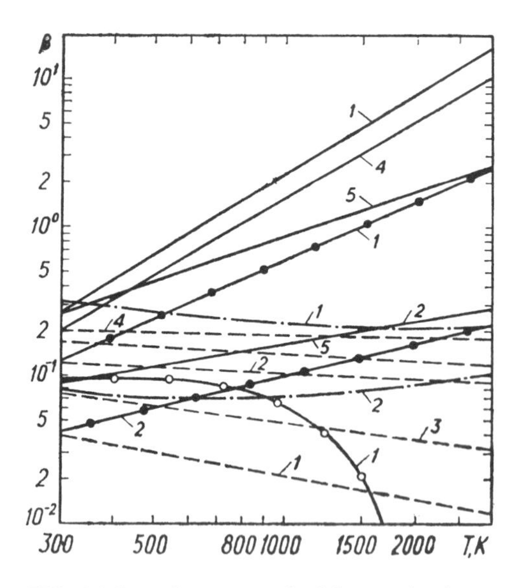


**FIG. 1.6** Second parameter β of the rotational structure of line groups as a function of temperature. For legend, see Fig. 1.5.

**TABLE 1.6** General Description of Experimental Results on Absorption $\hat{A}_{exp}$ Analyzed Here

| Gas | Band | No. of experimental runs | Ref. | Temperature range, K | Range of optical lengths | | Pressure range $p$, atm | | Range of partial pressures | $\Delta_k$ | $\Delta_a$ |
|---|---|---|---|---|---|---|---|---|---|---|---|
| | | | | | g/m$^2$ | cm·atm | Effective | Total | | % | % |
| $H_2O$ | Rotational | 25 | [45] | 555–1111 | 0,9 –310 | – | | 0,27 –2,75 | 0,01 –1 | 3,00 | 12,7 |
| $H_2O$ | 6,3 | 50 | [45] | 300–1111 | 0,90–14900 | – | | 0,003–2,75 | 0,004–1 | 1,13 | 8,1 |
| $H_2O$ | 2,7+3,2 | 50 | [45] | 300–1110 | 0,90–21000 | – | | 0,003–2,75 | 0,008–1 | 0,62 | 6,2 |
| $H_2O$ | 1,87 | 42 | [45] | 300–1111 | 3,5 –38500 | – | | 0,077–2,75 | 0,007–1 | 2,76 | 12,4 |
| $H_2O$ | 1,38 | 29 | [45] | 300–1111 | 8 –5310 | – | | 0,007–2,75 | 0,007–1 | 2,33 | 11,8 |
| $CO_2$ | 15 | 44 | [35] | 300–1390 | 0,1 –52200 | – | 0,04–13,5 | – | – | 3,77 | 12,7 |
| $CO_2$ | 4,3 | 42 | [35] | 300–1390 | 0,07–23400 | – | 0,03–13,1 | – | – | 1,95 | 10,6 |
| $CO_2$ | 2,7 | 33 | [35] | 300–1390 | 1,5 –23400 | – | 0,07–13,1 | – | – | 1,44 | 9,1 |
| CO | 4,7 | 130 | [37] | 300–1500 | – | 0,5–60 | | 0,25–3,0 | – | 2,16 | 10,4 |
| CO | 2,35 | 14 | [37] | 300–1500 | – | 20–60 | | 1–3 | – | 1,76 | 9,2 |
| $CH_4$ | 7,66 | 22 | [35] | 300–830 | 0,1 –1230 | – | 0,03– 7,1 | – | – | 1,49 | 9,5 |
| $CH_4$ | 3,31 | 22 | [35] | 300–830 | 0,1 –1230 | – | 0,03– 7,1 | – | – | 4,62 | 15,5 |

When using the best values of correlation constants, the accuracy in describing the experimental values of $\hat{A}_{exp}$ obtained by Edwards et al. [35,45] and Abu-Romia and Tien [37] for individual bands fluctuates within wide ranges; however, the value of the mean error does not exceed the experimental error (Table 1.6). The improvement of the accuracy of description is apparently limited both by errors of experimental results and by the nonconformance of the multistep band envelope, for which spectral parameters are determined from an experimental relationship, to actuality. The correlation parameters recommended for calculating absorption from Eq. (1.54) differ significantly from correlation constants listed by Edwards and his co-workers [29, 35, 45] and Abu-Romia and Tien [37]. The significant dependence of parameters $C$ and $\beta$ on the method of their determination is basically due to differences between the real band envelope and the assumed symmetrical multistage envelope. This means that parameters $C$ and $\beta$ lose their physical meaning and amount to fitting parameters for the selected wideband model. Nevertheless, the correlation constants obtained by the present author and his co-workers [44] yield satisfactory results in calculating the total emissivity of gases.

More real values of parameters $C$ and $\beta$, determined at low values of wave numbers over a wide range of temperatures, can be found in Ludwig et al. [10]. However, their use requires the use of computers with large operative memory. Incorporation of these results would allow us to significantly refine the calculation of radiation at high temperatures and high optical thicknesses of the radiating gas.

## 1.4 TOTAL EMISSIVITY OF GASES

The advances achieved in the study of band structure allow one to determine both analytically and experimentally the total emissivity of gases from their spectral parameters.

The total emissivity over the entire energy spectrum and the total band absorption are interrelated as

$$\varepsilon = \frac{\pi}{\sigma T^4} \int_0^\infty B(\omega) \cdot A(\omega)\, d\omega \tag{1.56}$$

Frequently this relationship is replaced by summation over bands,

$$\varepsilon \simeq \frac{\pi}{\sigma T^4} \sum_i B_i \cdot \hat{A}_i \tag{1.57}$$

neglecting the band overlap and assuming that the Planck function $B$ is independent of the wave number within the band [9, 23]. This may result in additional errors.

Using the relationship between the spectral absorption and spectral transmission of a gas

$$A = 1 - D \tag{1.58}$$

and introducing the effective spectral thickness of a layer of a molecular gas

$$\tau = -\ln D \tag{1.59}$$

the expression for the total emissivity of a gas can be written as

$$\varepsilon = \frac{\pi}{\sigma T^4} \cdot \int_0^\infty B(\omega) \cdot [1 - \exp(-\tau)]\, d\omega \tag{1.60}$$

When the spectral transmission of a molecular gas at all its optical thicknesses is expressed by the Mayer-Goody equation (1.27), we obtain for the spectrum with the nonoverlapping bands the expression

$$\tau = \frac{CR}{\sqrt{1 + CR/\beta}} \tag{1.61}$$

Application of the multiplicity rule for spectral segments with overlapping bands using models of line groups is not justified. It was shown using the case of isolated lines [Eq. (1.19)] that in the strong-line approximation, summation of optical parameters should be performed under the square root sign. In conjunction with this, the effective optical thickness of a gas for overlapping bands is expressed as

$$\tau = \frac{\sum_{m=1}^{M} C_m R}{\sqrt{1 + \left(\sum_{m=1}^{M} C_m R\right)^2 \Big/ \sum_{m=1}^{M} C_m \beta_m R}} \tag{1.62}$$

Then in the weak-line approximation

$$\tau = \sum_{m=1}^{M} C_m R \tag{1.63}$$

and in the strong-line approximation

$$\tau = \sqrt{\sum_{m=1}^{M} C_m \beta_m' R} \tag{1.64}$$

A more detailed verification of the familiar expression (1.62) will be given in analyzing the radiation of nonuniform gases.

The total emissivity was calculated from Eqs. (1.60)–(1.62) using band parameters determined from data listed in Tables 1.4 and 1.5.

The total emissivity of gases is an important integral parameter of radiation. We shall consider it for each of the principal radiating gases separately.

Water vapor occupies an important place in combustion processes, and the study of its emissivity has been attracting attention for quite a while. Studies of water vapor performed up to 1945 and systematized by Hottel [7] in the form of a nomogram still remain significant. In addition, nomograms were also compiled by the staff of the Dzherzhinskiy Heat Engineering Institute [46, 47] and by Nevskiy [3, 48].

The aforementioned nomograms yield satisfactory results in calculating the emissivity of water vapor over the range of optical thicknesses and temperatures explored in experiments. However, modern high-temperature engineering requires the use of data on optical properties at values of these parameters that are far from the limits of experimental values that served as a basis for compiling the nomogram. In addition, these nomograms are not suitable for computer calculations. A great deal of work on approximating nomogram data by expressions suitable for use with computers was performed by Detkov et al. [49, 50].

To verify the feasibility of extrapolating data given by Hottel [7], for higher values of the gas layer thickness and its temperature Penner [1, 51] and Edwards [45] made an attempt to develop models of radiation of water vapor, based on the Mayer–Goody statistical model. However, this approach to the solution of this problem was unsuccessful, due to inavailability of experimental data for a wide range of parameters.

Detailed spectral and integral studies of the radiation of water vapor for the needs of missile technology aided in working out a new nomogram, which made a more substantial use of pressure [52]. Subsequently the range of experimentally investigated temperatures and optical thicknesses was widened significantly, and the results of the study were correlated by means of the Mayer–Goody statistical model, represented in the form of a nomogram for water vapor temperatures from 600 to 2500 K and optical thicknesses of the vapor layer from 0.1 to 1000 cm·atm [53]. In these nomograms the emissivity of water vapor at low temperatures agrees satisfactorily with the data of Hottel [7]; however, at higher temperatures these were found to differ significantly, which was confirmed by analytic calculations, since in general Hottel's nomograms are based on graphical extrapolations. On the basis of the aforementioned studies, Leckner [41] compiled a nomogram of the total emissivity of water vapor for a temperature of 2473 K and optical thicknesses of the layer to 1000 cm·atm. At certain values of parameters all the aforementioned nomograms yield different values of the total emissivity of water vapor (Table 1.7). The results of calculations performed in this study from Eqs. (1.60)–(1.62) with parameters presented in Tables 1.4 and 1.5 at $T = 600$ K are intermediate between the data of Boynton

**TABLE 1.7** Emissivity of Water Vapor at $p = 1$ atm According to the Results of Various Investigators

| Temperature, K | Layer thickness, cm | Present study | Reference [7] | | Reference [53] | | Reference [41] | |
|---|---|---|---|---|---|---|---|---|
| | | $\varepsilon_*$ | $\varepsilon$ | $\left(\frac{\varepsilon}{\varepsilon_*}-1\right)\cdot 100$ | $\varepsilon$ | $\left(\frac{\varepsilon}{\varepsilon_*}-1\right)\cdot 100$ | $\varepsilon$ | $\left(\frac{\varepsilon}{\varepsilon_*}-1\right)\cdot 100$ |
| 600 | 1.0 | 0.076 | 0.075 | −1 | 0.055 | −28 | 0.085 | +12 |
| | 2.0 | 0.115 | 0.115 | 0 | 0.080 | −31 | 0.125 | +9 |
| | 5.0 | 0.181 | 0.190 | +5 | 0.119 | −34 | 0.199 | +10 |
| | 10.0 | 0.238 | 0.258 | +8 | 0.172 | −28 | 0.272 | +14 |
| | 20.0 | 0.300 | 0.371 | +23 | 0.226 | −25 | 0.359 | +20 |
| | 50.0 | 0.385 | 0.476 | +26 | 0.325 | −16 | 0.489 | +27 |
| | 100.0 | 0.452 | 0.554 | +22 | 0.385 | −15 | 0.583 | +29 |
| 1500 | 1.0 | 0.028 | 0.020 | −29 | 0.021 | −25 | 0.024 | −14 |
| | 2.0 | 0.047 | 0.041 | −13 | 0.037 | −21 | 0.042 | −10 |
| | 5.0 | 0.087 | 0.076 | −13 | 0.072 | −17 | 0.082 | −6 |
| | 10.0 | 0.130 | 0.128 | −2 | 0.121 | −7 | 0.128 | −2 |
| | 20.0 | 0.185 | 0.197 | +6 | 0.175 | −5 | 0.181 | −2 |
| | 50.0 | 0.273 | 0.294 | +7 | 0.270 | −1 | 0.301 | +10 |
| | 100.0 | 0.348 | 0.370 | +6 | 0.360 | +3 | 0.402 | +16 |
| 2500 | 1.0 | 0.010 | 0.005 | −50 | 0.006 | −38 | 0.007 | −30 |
| | 2.0 | 0.017 | 0.013 | −24 | 0.013 | −24 | 0.014 | −18 |
| | 5.0 | 0.035 | 0.029 | −17 | 0.030 | −14 | 0.031 | −11 |
| | 10.0 | 0.056 | 0.055 | −2 | 0.051 | −9 | 0.054 | −4 |
| | 20.0 | 0.086 | 0.088 | +2 | 0.086 | 0 | 0.090 | +5 |
| | 50.0 | 0.137 | 0.154 | +12 | 0.170 | +24 | 0.165 | +20 |
| | 100.0 | 0.186 | 0.211 | +13 | 0.250 | +34 | 0.249 | +34 |

and Ludwig [53] and Leckner [41], although the nomograms in both studies were compiled on the basis of the same results of spectral measurements. These results are underestimated as compared with Hottel's data at high optical thicknesses of the water vapor layer and overestimated at small optical thicknesses. On the average the difference between the analytic values of the total emissivity and data of various nomograms is as much as ±25%. This is due both to experimental errors and to nonconformance of the models to the actual situation.

The experimental results obtained at the Kazan' Chemical Technology Institute also deviate from analytic results [54–56]. As compared with results of our calculations they are 21% higher on the average and exhibit a significant scatter (Fig. 1.7). The same applies to the sparse experimental data of Chukanova and Nevskiy [57]. The reason for the difference between analytic and experimental data with rising pressure may be dimerization of water vapor, which was not incorporated in the present calculations.

The total emissivity of carbon dioxide is frequently calculated from the nomogram from Hottel's work [7]. Approximation of these data by equations suitable for computer use is presented in papers by Schack [8] and Detkov with his co-workers [49, 58]. The data on total emissivity carbon dioxide were satisfactorily confirmed in analytic and experimental studies [59, 60]. However, all these data are for a highly limited range of optical thicknesses and temperatures.

On the basis of spectral data Leckner compiled a nomogram for calculating the total emissivity of carbon dioxide, suitable for the temperature range from 500 to 2300 K and optical thicknesses to 1000 cm.atm [41, 42].

Leckner [42] and Panfilovich et al. [61] analyzed in detail the possible errors of experimental data underlying the Hottel diagram. They found that

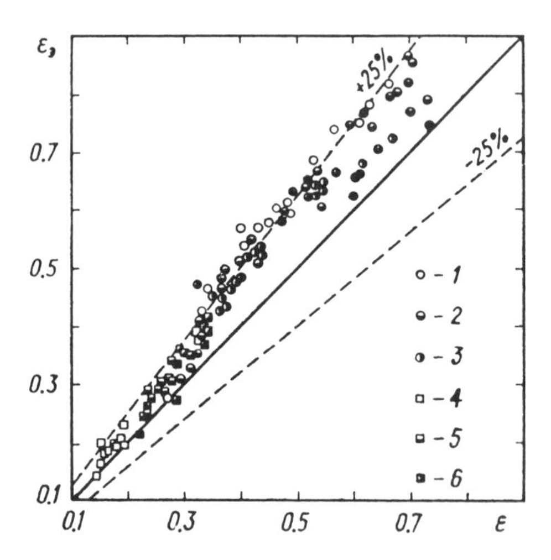


**FIG. 1.7** Comparison of experimental data on the total emissivity $\varepsilon_{meas}$ of water vapor, obtained by Panfilovich and his co-workers [54–56] (circles $p = 1 - 21$ atm, $T = 535 - 1020$ K) and [63] (squares, $R = 32.6$ cm), with the calculated values of ε. Values of $R$ are: (1) 16 cm; (2) 26.5 cm; (3) 40 cm, Values of $pR$ are: (4) 10 cm atm; (5) 20 cm·atm; (6) 30 cm·atm.

one of the reasons for the errors is neglect of absorption in the air layer between the volume being measured and the equipment. However, according to estimates of the above studies, the aforementioned nomograms should be regarded as sufficiently reliable over the temperature range under study.

In the approximation of total line overlap, nomograms were compiled for the total emissivity of carbon dioxide at temperatures to 4000 K and optical thicknesses to 5000 cm atm [5].

Data of calculations for carbon dioxide performed using Eqs. (1.60)–(1.62) (Table 1.8) are in satisfactory agreement with the nomograms of Hottel [7] and Leckner [41]. At higher temperatures the analytic results are intermediate between the results of Leckner [41] and Panfilovich et al. [61]. The agreement between the results obtained by the present authors and the experimental data obtained in the Kazan' Chemical Technology Institute [62] should also be regarded as satisfactory. The directly measured total emissivity of carbon dioxide [62] is on the average 15% lower than the analytic results, the scatter of data here being relatively small, ±6% (Fig. 1.8). This means that the results are in agreement within the limits of experimental error. The sparse experimental data of Chukanova and Nevskiy [63] exhibit a significant scatter and are also on the low side as compared with the analytic results (Fig. 1.8).

The results of the first experimental studies of the total emissivity of carbon monoxide, CO, were presented by Hottel [7] in the form of nomograms. However, in experiments underlying these nomograms the carbon monoxide contained some quantity of carbon dioxide, which is responsible for an overestimate of the emissivity of the monoxide. A new nomogram for CO was compiled by Detkov et al. on the basis of careful spectral measurement of absorption [64], and is currently regarded as the most accurate. A detailed analysis of the optical properties of carbon monoxide was performed by Detkov and his co-workers [64, 65].

The study of the emission of nitric oxide, NO, and also of other diatomic gases, is related to the presence of a fundamental band and a number of harmonics in the IR spectrum.

Nomograms of the total emissivity of NO were constructed from experimental measurements of the spectral absorptivity of the fundamental band at 300–1200 K and pressures from 1 to 4 atm [66]. The integral absorption for the band was calculated by numerical integration of the spectral absorption coefficient using the regular Elsasser model. The results were interpreted within the framework of the wide-band Edwards model. The values of the correlation constants are listed in Table 1.4.

In addition to the above gases, the total emissivity of combustion products at high temperatures may be affected by emission from the OH radical. For this gas the spectral parameters of the Edwards wide-band model were not determined. Experimental data on absorption are also not

**TABLE 1.8** Emissivity of Carbon Dioxide at $p = 1$ atm According to Results of Various Investigators

| Temperature, K | Layer thickness, cm | Present study | Reference [7] | | Reference [41] | | Reference [61] | |
|---|---|---|---|---|---|---|---|---|
| | | $\varepsilon_*$ | $\varepsilon$ | $\left(\frac{\varepsilon}{\varepsilon_*}-1\right)\cdot 100$ | $\varepsilon$ | $\left(\frac{\varepsilon}{\varepsilon_*}-1\right)\cdot 100$ | $\varepsilon$ | $\left(\frac{\varepsilon}{\varepsilon_*}-1\right)\cdot 100$ |
| 600 | 1.0 | 0.060 | 0.049 | −18 | 0.052 | −13 | 0.052 | −13 |
| | 2.0 | 0.076 | 0.062 | −18 | 0.066 | −13 | 0.065 | −14 |
| | 5.0 | 0.098 | 0.086 | −12 | 0.086 | −12 | 0.080 | −18 |
| | 10.0 | 0.117 | 0.099 | −16 | 0.102 | −13 | 0.105 | −10 |
| | 20.0 | 0.136 | 0.120 | −12 | 0.120 | −12 | 0.120 | −12 |
| | 50.0 | 0.166 | 0.150 | −10 | 0.150 | −10 | 0.144 | −13 |
| | 100.0 | 0.192 | 0.160 | −17 | 0.170 | −12 | 0.167 | −13 |
| 1500 | 1.0 | 0.044 | 0.037 | −16 | 0.040 | −5 | 0.038 | −16 |
| | 2.0 | 0.057 | 0.048 | −16 | 0.054 | −4 | 0.052 | −9 |
| | 5.0 | 0.078 | 0.068 | −13 | 0.074 | −5 | 0.068 | −13 |
| | 10.0 | 0.097 | 0.089 | 8 | 0.092 | −5 | 0.092 | −5 |
| | 20.0 | 0.118 | 0.110 | −7 | 0.110 | −7 | 0.115 | −3 |
| | 50.0 | 0.148 | 0.140 | −6 | 0.150 | +1 | 0.145 | −2 |
| | 100.0 | 0.173 | 0.180 | +4 | 0.190 | +10 | 0.170 | −2 |
| 2500 | 1.0 | 0.018 | 0.013 | −28 | 0.017 | −6 | 0.013 | −28 |
| | 2.0 | 0.024 | 0.018 | −25 | 0.024 | 0 | 0.017 | −29 |
| | 5.0 | 0.034 | 0.029 | −15 | 0.038 | +12 | 0.027 | −21 |
| | 10.0 | 0.044 | 0.040 | −9 | 0.051 | +16 | 0.035 | −20 |
| | 20.0 | 0.057 | 0.058 | +2 | 0.066 | +16 | 0.045 | −21 |
| | 50.0 | 0.075 | 0.074 | −1 | 0.091 | +21 | 0.063 | −16 |
| | 100.0 | 0.089 | 0.095 | −7 | 0.115 | +29 | 0.078 | −12 |

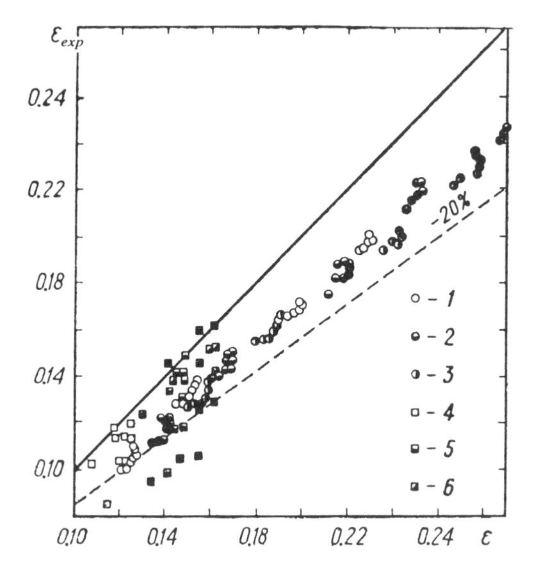


**FIG.1.8** Comparison of experimental data on the total emissivity $\varepsilon_{exp}$ of carbon dioxide obtained by Akhunov et al. [62] (circles, $p$ from 1 to 21.42 atm, $T$ between 666 and 1077 K) and Chukanova and Nevskiy [57] (squares), with the analytic values of $\varepsilon$. Values of $R$: (1) 10 cm; (2) 16 cm; (3) 26.5cm. Values of $pR$: (4) 10 cm·atm; (5) 20 cm·atm; (6) 39 cm·atm.

available. The results of calculation of the spectral parameters of OH are listed in Ludwig et al. [10].

The data on total emissivity of other polyatomic gases needed for practical calculations are given by Hottel [7], Ludwig et al. [10], Chan and Tien [39], Balakrishnan and Edwards [40], and Detkov et al.[64]

## 1.5. RADIATION OF NONUNIFORM GASES

The previously examined relationships for determining the emission and absorption in vibrational–rotational bands of molecular gases are valid only for a homogeneous layer of gas. However, in practice one always has to deal with nonisothermal or nonhomogeneous layers of radiating and absorbing gases. In calculations of radiation of nonhomogeneous gases one currently uses basically the ideas of the Curtis–Godson approximation, according to which the transmittance of a nonuniform gas is replaced by the transmittance of an equivalent layer of a uniform gas:

$$D = \exp\left(-\frac{\langle CR\rangle}{\sqrt{1+\left\langle \frac{CR}{\beta}\right\rangle}}\right) \tag{1.65}$$

According to the Curtis–Godson approximation, the structural band parameters for a nonuniform gas should be averaged using the formulas [1, 10]:

$$\langle CR \rangle = \int_0^R C dr \tag{1.66}$$

$$\langle \beta CR \rangle = \int_0^R \beta C dr \tag{1.67}$$

Using Eqs. (1.66) and (1.67) we obtain the following expression for the mean parameter of spectral line intensity $U$:

$$2\langle U \rangle \equiv \left\langle \frac{CR}{\beta} \right\rangle = \frac{\langle CR \rangle^2}{\langle \beta CR \rangle} = \frac{\left( \int_0^R C dr \right)^2}{\int_0^R \beta C dr} \tag{1.68}$$

In this averaging of structural parameters of the bands the transmittance of a nonuniform and a uniform gas in the limiting cases of strong and weak spectral lines is found to be the same. When the structural parameters are replaced by their expressions averaged using Eqs. (1.66) and (1.68), we obtain in the Mayer–Goody statistical model [Eq. (1.65)] in the weak spectral line approximation ($U << 1$) the expression

$$D = \exp\left( -\int_0^R C dr \right) \tag{1.69}$$

which in the case of uniform gases becomes the familiar Beer law (1.31). For strong spectral lines ($U >> 1$), an analogous replacement yields the expression

$$D = \exp\left( -\sqrt{\int_0^R \beta C dr} \right) \tag{1.70}$$

which for homogeneous conditions becomes an expression known as the square root law.

The conformance of the Curtis–Godson approximation to actual conditions has been extensively discussed [9, 10, 68–77]. Many investigators arrived at the conclusion that it yields satisfactory results not only in limiting cases, but also at intermediate values of parameter $U$. However, certain studies note serious deviations between the above approximation and the actual situation. For this reason three-parameter averaging methods were developed to replace the Curtis–Godson approximation, which is a two-parameter model.

Unlike the two-parameter approximation, the three-parameter one permits making allowance for the difference between the actual line shape and the Lorentz profile and for the real band envelope. The most general formulas for such an averaging within a wide band have the form [75]

$$\langle SR\rangle = \int_0^R S\,dr, \quad \langle S\gamma^m R\rangle = \int_0^R S\gamma^m\,dr, \quad \langle S\gamma^m \beta^n\rangle = \int_0^R S\gamma^m \beta^n\,dr \tag{1.71}$$

Here the exponent $n$ is controlled by the line shape whereas the exponent $m$ is a function of the shape of the band envelope. It can be easily noted that under certain conditions Eq. (1.71) becomes identical to (1.66) and (1.67), i.e., the Curtis–Godson approximation is a particular case of the Detkov three-parameter approximation. One's attention is attracted by the fact that according to Detkov [75] the effect of $m$ is an order of magnitude higher than that of $n$. This means that allowance for the real shape of the band envelope is more important than allowance for the actual line shape.

The question of suitability of the Curtis–Godson approximation for nonuniform gas mixtures was not examined previously, for which reason it will be considered here in more detail.

We shall now analyze the transmission of radiation by a nonuniform layer of a gas mixutre for a line group with Lorentz shape (1.4). We shall restrict ourselves to transmission in the center of the interval of wave numbers $Nd^*$, which is located in the vicinity of wave number $\omega_0$; here the equivalent distance $d^*$ between the lines for a gas mixture containing $M$ constituents will be defined by the expression

$$\frac{1}{d^*} = \sum_{m=1}^{M} \frac{1}{d_m} \tag{1.72}$$

Let the lines over the range of wave numbers under study be arranged randomly with probability of the presence of lines in the interval between $\omega_i$ and $\omega_i + d\omega_i$ equal to $d\omega_i/d^*$. Then the probability of the presence of lines in any of intervals from $\omega_i$ to $\omega_i + d\omega_i$, *where* $i = 1, 2, 3, \ldots, N$, will be expressed by the product

$$\prod_{i=1}^{N} \frac{d\omega_i}{d^*}$$

If the absorption coefficient at wave number ω obtained from $i$ spectral lines of all the gas components is equal to

$$\varkappa = \sum_{m=1}^{M} \frac{s_{m,i} S_{m,i}}{\gamma_{m,i}} \tag{1.73}$$

then the transmissivity of the gas layer in the direction of beam $R$ at wave number ω is

$$D_i = \exp\left(-\int_0^R \sum_{m=1}^{M} \frac{s_{m,i} S_{m,i}}{\gamma_{m,i}} dr\right) \tag{1.74}$$

Let us assume that the intensity of the spectral lines can be written as

$$S_{m,i} = S_i \cdot \varphi_{m,i}(T, p, p_1, p_2, \ldots, p_M) \tag{1.75}$$

with the distribution of quantity $S_i$ being random, independent of the nonhomogeneity of the gas along the beam, and $\varphi_{m,i}$ making allowance for its dependence on the local parameters of the gas. We shall define quantity $P(\bar{S}, S)dS$ as the probability that the intensity of the $i$th spectral line lies between $S$ and $S + dS$. Then the probability of obtaining a set of $N$ lines, the $i$th of which has an intensity $S_i$ over the range of wave numbers $d\omega_1, d\omega_2$, . . . , $d\omega_N$ is given by the product

$$\prod_{i=1}^{N} \frac{d\omega_i}{d^*} \cdot P(\bar{S}_i, S_i)\, dS_i$$

Here the mean transmissivity at wave number ω for all the possible locations of lines within different intervals of wave numbers is

$$\bar{D} = \frac{\prod_{i=1}^{N} \int_{-\frac{1}{2}Nd^*}^{+\frac{1}{2}Nd^*} \int_0^{\infty} \frac{d\omega_i}{d^*} \cdot P(\bar{S}_i, S_i) \cdot \exp\left(-S_i \int_0^R \sum_{m=1}^{M} \frac{\varphi_{m,i} s_{m,i}}{\gamma_{m,i}} dr\right) dS_i}{\prod_{i=1}^{N} \int_{-\frac{1}{2}Nd^*}^{+\frac{1}{2}Nd^*} \frac{d\omega_i}{d^*}} \tag{1.76}$$

where the $N$ integrals in the numerator and denominator are identical. Then

Eq. (1.76) can be represented as

$$\bar{D}=\left[\frac{1}{N}\cdot\int_{-Nd^*/2}^{+Nd^*/2}\int_0^{\infty}\frac{d\omega}{d^*}\cdot P(\bar{S},\ S)\cdot\exp\left(-S\times\right.\right.$$

$$\left.\left.\times\int_0^R\sum_{m=1}^{M}\frac{\varphi_m s_m}{\gamma_m}\,dr\right)dS\right]^N \tag{1.77}$$

After mathematical manipulations according to Mayer and Goody [1, 6], we obtain for the case of $N\to\infty$ the following approximate expression for the mean transmissivity of a group of spectral lines, situated near wave number $\omega_0$:

$$\bar{D}\simeq\exp\left\{-\frac{1}{d^*}\int_{-\infty}^{+\infty}\int_0^{\infty}P(\bar{S},\ S)\times\right.$$

$$\left.\times\left[1-\exp\left(-S\cdot\int_0^R\sum_{m=1}^{M}\frac{\varphi_m s_m}{\gamma_m}\,dr\right)\right]d\omega\,dS\right\} \tag{1.78}$$

For the exponential intensity distribution within the group of spectral lines under study,

$$P(\bar{S},\ S)=\frac{1}{\bar{S}}\exp\left(-\frac{S}{\bar{S}}\right) \tag{1.79}$$

Integration of Eq. (1.78) with respect to $S$ yields

$$\bar{D}=\exp\left[-\frac{1}{d^*}\int_{-\infty}^{+\infty}\frac{d\omega_0}{1+\left(\int_0^R\sum_{m=1}^{M}\frac{s_m\bar{S}_m}{\gamma_m}\,dr\right)^{-1}}\right] \tag{1.80}$$

in which allowance is made for the fact that Eq. (1.75) is suitable also for the mean values of intensity, i.e.,

$$\bar{S}_m=\bar{S}\cdot\varphi_m(T,\ p,\ p_1,\ p_2,\ \ldots,\ p_M) \tag{1.81}$$

Equation (1.80) was obtained on the assumption that the mean distance $d^*$ between the lines is a constant quantity.

The particular case of Eq. (1.80) for a nonuniform gas was obtained by the present author [77].

In this study, Eq. (1.80) is the basic expression for the study of radiation of nonuniform gases. It can be readily noted that it makes possible making allowance for interaction between molecules of different type.

Equation (1.80) cannot be integrated analytically for arbitrary distributions $\bar{S}(r)$ and $s(r)$. Useful information on the transmissivity of nonuniform gases is obtained from the following particular cases, when Eq. (1.80) is integrated analytically with respect to frequency. Let us consider the case of a uniform gas, when $M=1$.

1. $S$ = *constant,* $\gamma$ = constant

$$\bar{D}=\exp\left[-\frac{\bar{S}R}{d\cdot\sqrt{1+\bar{S}R/\pi\gamma}}\right] \tag{1.82}$$

If we convert in Eq. (1.82) to structural parameters $C$ and $\beta$ for bands, we obtain the Mayer–Goody equation (1.27) for uniform gases.

2. $\gamma$ = constant

$$\bar{D}=\exp\left[-\frac{\int_0^R \bar{S}\,dr}{d\sqrt{1+\left(\int_0^R \bar{S}\,dr\right)/\pi\gamma}}\right] \tag{1.83}$$

It can be easily noted that Eq. (1.83) becomes identical to Eqs. (1.65)–(1.68) of the Curtis–Godson approximation when $d$ = constant. This means that $\beta$ = constant the Curtis–Godson approximation is exact for any temperature dependence of intensity $\bar{S}$.

3. $\gamma << 1$. Then, neglecting the square of the half-width of the lines as compared with $\gamma$ in Eq. (1.4), we obtain an expression analogous to the Curtis–Godson approximation for strong lines (1.70):

$$\bar{D}=\exp\left[-\frac{1}{d}\sqrt{\int_0^R \pi\gamma\,\bar{S}\,dr}\right] \tag{1.84}$$

Exact agreement between Eqs. (1.84) and (1.70) can be obtained only when $d$ = constant.

4. $\gamma >> 1$. In this case, neglecting unity in the denominator and the integrand in Eq. (1.80) and changing the order of integration, we obtain

$$\bar{D}=\exp\left(-\frac{1}{d}\int_0^R \bar{S}\,dr\right) \tag{1.85}$$

It is found also that in the case of weak spectral lines Eq. (1.69), which follows from the Curtis–Godson approximation, can be obtained at $d$ = constant.

5. The radiation in a band is due to radiation near the center of the spectral lines. Then expanding $(\int_0^R s\bar{S}dr)^{-1}$ in a Taylor series in argument $z \not\equiv (\omega - \omega_0)^2$ in the vicinity of $z = 0$ and retaining the linear term, we obtain [77]:

$$\left(\int_0^R s\bar{S}\,dr\right)^{-1} = \left[\int_0^R \frac{\gamma\bar{S}dr}{\pi(z+\gamma^2)}\right]^{-1} \simeq \left(\int_0^R \frac{\bar{S}dr}{\pi\gamma}\right)^{-1} + \frac{\left(\int_0^R \frac{\bar{S}dr}{\pi\gamma^3}\right) z}{\left(\int_0^R \frac{\bar{S}dr}{\pi\gamma}\right)^3} \tag{1.86}$$

Integrating Eq. (1.75) together with (1.81) and performing a simple transformation, we can obtain an expression for the transmittance:

$$\bar{D} = \exp\left[\frac{\left(\int_0^R \bar{S}/\pi\gamma\,dr\right)^{3/2}}{d\left(\int_0^R \bar{S}/\pi^3\gamma^3\,dr\right)^{1/2}\cdot\left(1+\int_0^R \bar{S}/\pi\gamma\,dr\right)^{1/2}}\right] \tag{1.87}$$

In the expression obtained in this manner, averaging of spectral parameters over the thickness of the layer differs significantly from the averaging in the Curtis–Godson approximation. In the limiting case of a homogeneous gas, Eq. (1.87) becomes the Mayer–Goody equation (1.27); however, it does not satisfy the limiting case of strong lines (1.70). A numerical check of this approximation at $d$ = constant showed that it is suitable only at small temperature differences in the gas layer. This again confirms the signficant contribution of the wings of spectral lines in the course of radiation, the effect of which was taken into account only very approximately in deriving Eq. (1.87).

The above particular cases of determining the transmittance of a nonuniform gas layer in substance confirm the Curtis–Godson approximation only in the case of constant distance $d$ between spectral lines within their group. Such a condition is apparently valid only for relatively small intervals of temperatures of molecular gases. In such a case the Curtis–Godson approximation can be represented by a relationship satisfying all the previously examined particular cases:

$$\bar{D} = \exp\left[\frac{\int_0^R \frac{\bar{S}}{d}\,dr}{\sqrt{1+\left(\int_0^R \frac{\bar{S}}{d}\,dr\right)^2 \Big/ \int_0^R \frac{\pi\gamma}{d}\,\frac{\bar{S}}{d}\,dr}}\right] \tag{1.88}$$

Certain particular solutions of Eq. (1.80) for a gas mixture can be also obtained by the above method.

1. $\gamma_m >> 1$. Then, neglecting unity in the denominator of the integrand in Eq. (1.80) and changing the order of integration, we obtain

$$\bar{D} = \exp\left(-\frac{1}{d^*}\int_0^R \sum_{m=1}^M \bar{S}_m\, dr\right) \tag{1.89}$$

2. $\gamma_m << 1$. Then, neglecting the square of the half-width of the lines as compared with half-width $\gamma$ in Eq. (1.4), we obtain

$$\bar{D} = \exp\left(-\frac{1}{d^*}\sqrt{\int_0^R \sum_{m=1}^M \pi\gamma_m \bar{S}_m\, dr}\right) \tag{1.90}$$

3. The half-width of lines for all the gas components is the same and constant. Then ratio $s/\gamma$ in Eq. (1.80) can be eliminated from the integrand and one can find that

$$\bar{D} = \exp\left(-\frac{\int_0^R \sum_{m=1}^M \bar{S}_m\, dr}{d^* \cdot \sqrt{1+\frac{\int_0^R \sum_{m=1}^M \bar{S}_m\, dr}{\pi\gamma}}}\right) = \exp\left[-\frac{\int_0^R \sum_{m=1}^M \bar{S}_m\, dr}{d^* \cdot \sqrt{1+\frac{\left(\int_0^R \sum_{m=1}^M \bar{S}_m\, dr\right)^2}{\int_0^R \sum_{m=1}^M \pi\gamma \bar{S}_m\, dr}}}\right] \tag{1.91}$$

Note that $d^*$ in Eqs. (1.89)–(1.91) must be determined from Eq. (1.72). It can be easily seen that at $d^*$ = constant and isothermal conditions, Eqs. (1.89) and (1.90) correspond to the previously assumed expressions for optical thicknesses of the gas mixture in limiting cases of weak and strong lines [Eqs. (1.63) and (1.64)].

In the most general case we can recommend the expression

$$\bar{D}=\exp\left[-\frac{\int_0^R \sum_{m=1}^M \frac{\bar{S}_m}{d^*}\,dr}{\sqrt{1+\dfrac{\left(\int_0^R \sum_{m=1}^M \frac{\bar{S}_m}{d^*}\,dr\right)^2}{\int_0^R \sum_{m=1}^M \frac{\pi\gamma_m}{d^*}\,\frac{\bar{S}_m}{d^*}\,dr}}}\right] \tag{1.92}$$

which satisfies all the particular cases examined above and also corresponds to the previously assumed expression for optical thickness [Eq. (1.62)] under isothermal conditions. This means that we have proven that the multiplicity rule cannot be used in calculating the absorptivity of a mixture of gases over narrow spectral ranges, when the Mayer–Goddy statistical model is used for description of optical properties.

Some specific idea about the degree of the effect of temperature conditions on the transmittance of gases may be obtained from numerical calculations (Figs. 1.9–1.13). Such calculations were performed for the case when $M=1$, and the structural parameters of the bands are the following power-law functions of the temperature:

$$C=C_0\cdot\Theta^k \tag{1.93}$$

and

$$\beta=\beta_0\cdot\Theta^p \tag{1.94}$$

The calculations were performed for relative optical thickness $\tau/C_0H$ for a gas layer of height $H$, with a different temperature distribution, on assumption of absence of limiting surfaces,

$$\frac{\tau}{C_0\,H}=\frac{1}{\pi}\int_{-\infty}^{+\infty}\frac{dv}{2U_0+\left[\int_0^1 \Theta^{k+p}/(v^2+\Theta^{2p})\,d\eta_*\right]^{-1}} \tag{1.95}$$

where

$$v=\frac{\pi}{\beta_0}\cdot\frac{\omega-\omega_0}{d}$$

$$\Theta=\frac{T}{T_0},\quad \eta_*=\frac{y}{H}\ \text{and}\ \tau=-\ln D \tag{1.96}$$

In calculating linear, two-step, and trapezoidal temperature distributions, these were integrated analytically with respect to $\eta_*$.

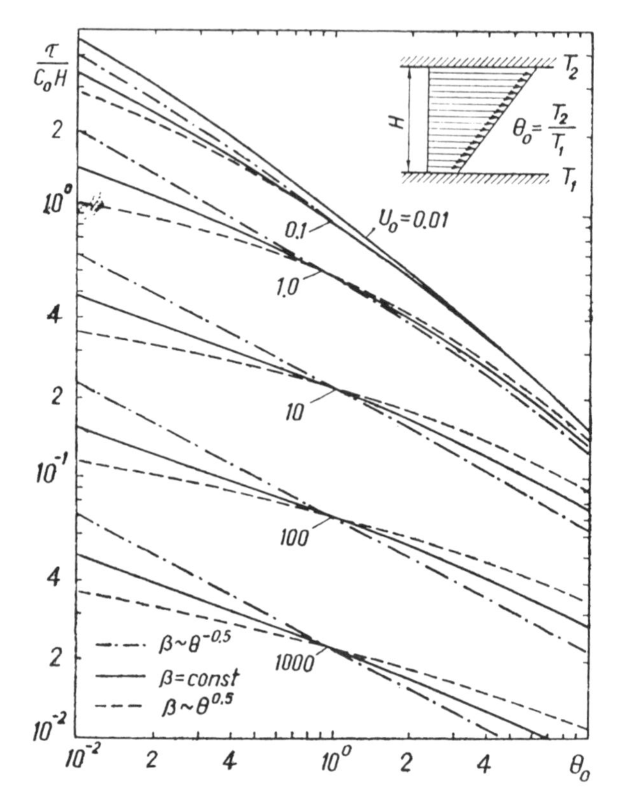


**FIG. 1.9** Relative optical layer thickness at linear temperature distribution as a function of line strength parameter $U_0$ and relative temperature $\Theta_0$ for different behaviours of the second parameter of band structure $\beta(C = C_0 \cdot \Theta^{-1.5})$.

According to Eq. (1.95) the relative optical layer thickness $\tau/C_0H$ is basically a function of the parameter of spectral line strength $U_0$ within the band and temperature variation.

Calculations show (Figs. 1.9 and 1.10) that in the case of a linear temperature distribution the temperature dependence of the relative optical thickness of the layer is in inverse proportion to the spectral line strength parameter $U_0$. For weak lines ($U_0 << 1$) this dependence is due only to the temperature dependence of line intensity parameter $C$ and is independent of the value of parameter $\beta$ (Fig. 1.9).

Analogous conclusions can also be drawn from calculations for a two-stepped temperature distribution (Figs. 1.11 and 1.12). In this case there appears an additional parameter — the ratio of thicknesses of layers with different temperatures.

In temperature distributions characteristic of cross-sections in channels with boundary layers (Fig. 1.13, curve 2), the optical thickness of the layer

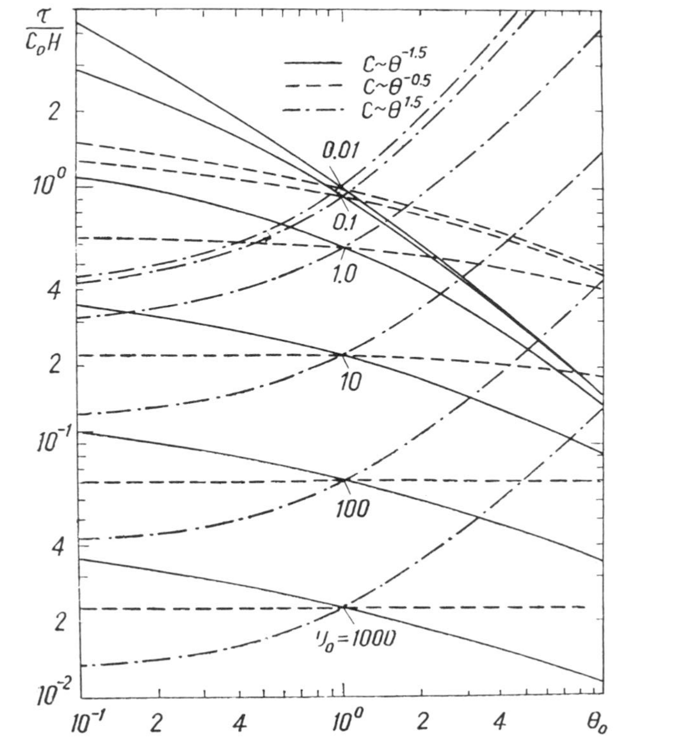


**FIG. 1.10** Relative optical thickness of layer as a function of line strength parameter $U_0$ and relative temperature $\Theta_0$ for different temperature dependences of the first band structure $C$ as a function of temperature $\beta = \beta_0 \Theta^{-0.5}$). The temperature profile is linear, as in Fig. 1.9.

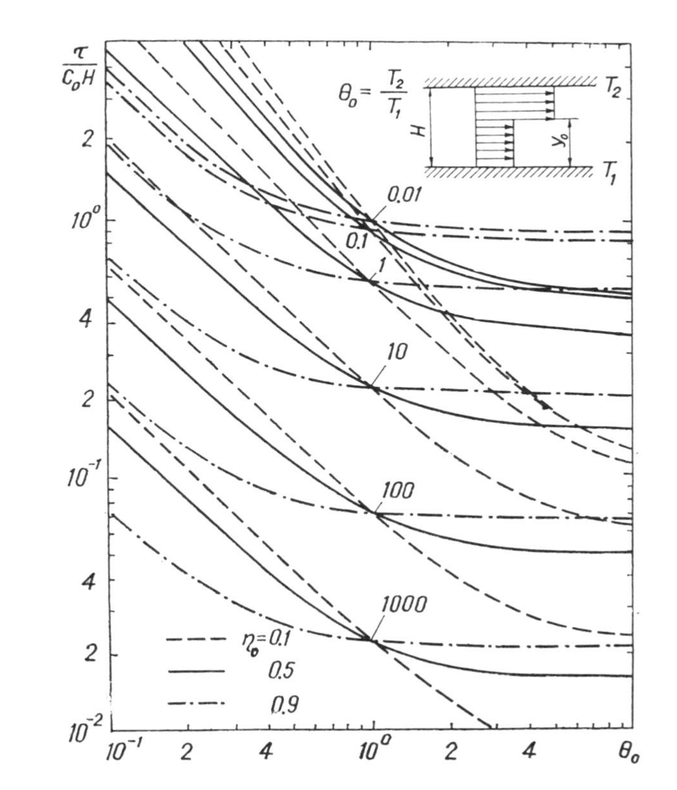


**FIG. 1.11** Relative optical thickness of layer as a function of line strength parameter $U_0$, relative temperature $\Theta_0$ and relative layer thickness $\eta_0 = y_0/H$ for a stepwise temperature distribution $(C = C_0 \cdot \Theta^{-1.5},\ \beta = \beta_0 \cdot \Theta^{-0.5})$

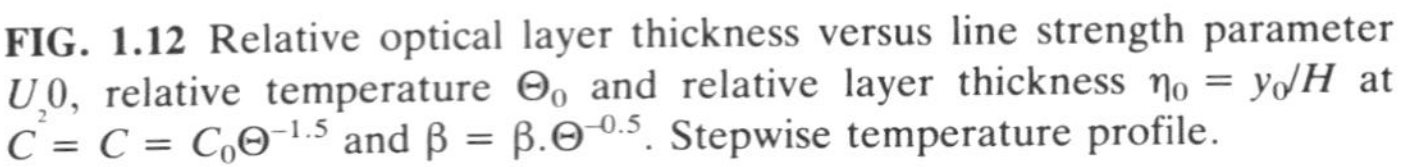

**FIG. 1.12** Relative optical layer thickness versus line strength parameter $U_20$, relative temperature $\Theta_0$ and relative layer thickness $\eta_0 = y_0/H$ at $C = C = C_0\Theta^{-1.5}$ and $\beta = \beta.\Theta^{-0.5}$. Stepwise temperature profile.

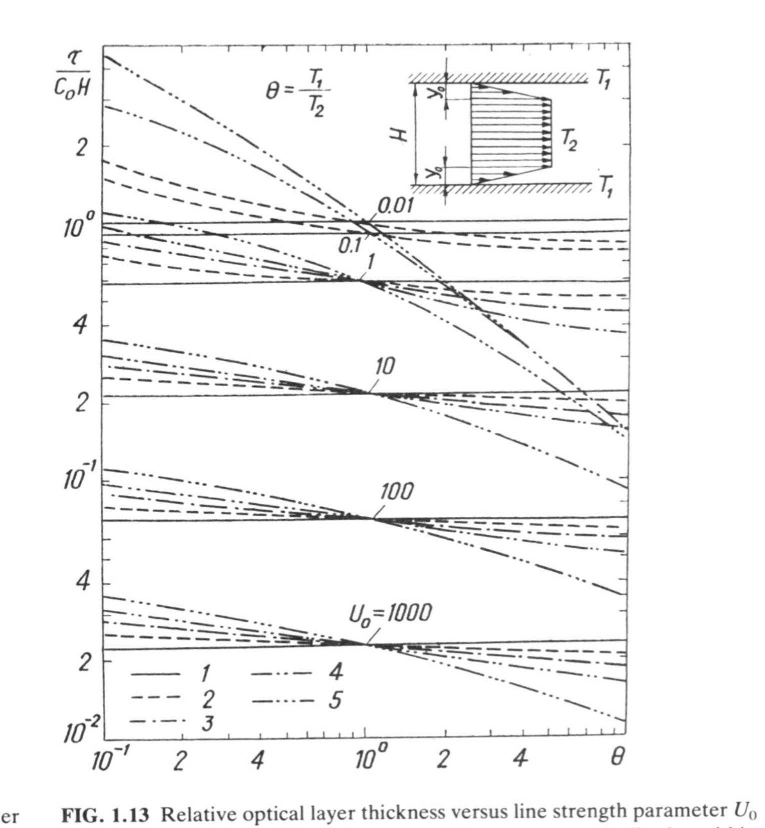


**FIG. 1.13** Relative optical layer thickness versus line strength parameter $U_0$ and relative temperature $\Theta_0$ for a trapezoidal temperature distribution within the layer. $y_0/H$ = (1) 0; (2) 0 0.10; (3) 0.2; (4) 0.3; (5) 0.5.

depends relatively little on temperature. Thus, the presence of a boundary layer with linear temperature distribution and relative thickness amounting to 10% of the channel height increases by only several per cent the relative optical thickness of a layer at high values of $U$. The effect of temperature increases somewhat with reduction in $U$. However, even an insignificant change in the relative optical thickness as a function of temperature may significantly change the magnitude of the spectral transmittance, due to the exponential relationship [Eq. (1.96)] between $D$ and $\tau$.

Since the optical thickness of the layer depends to a significant degree on the line strength, the effect of temperature is not the same even within a single band. Given this, nonisothermicity of a layer should be accounted for as early as possible in calculating the spectrum of radiation parameters.

The results obtained from Eq. (1.95) and by the Curtis–Godson method at $d$ = constant are in satisfactory agreement with one another. For the majority of cases represented in Figs. 1.9–1.13 the difference between the results does not exceed 1% even at $U_0$ close to 1, but may increase somewhat (to 2–3%) in two-step temperature distributions at $\eta_0$ close to 0 or 1. This means that the Curtis–Godson approximation reproduces Eq. (1.80) rather satisfactorily.

The agreement between the Curtis–Godson approximation and the analytic expression (1.80) does not yet mean that it corresponds to actual conditions. An experimental check of the Curtis–Godson approximation was performed by Edwards and his co-workers [28], Chukanova and Nevskiy [57, 63], Cess and Wang [70], Chan and Tien [71], and Lin and Greif [72]. Due to certain experimental errors and relative moderate effects of nonisothermicity under the experimental conditions, no significant deviations from the Curtis–Godson approximation were detected. In cases when this approximation yields results that do not agree with the experimental results, the reason for this disagreement should be sought in nonconformance of the analytic conditions to the real optical properties of the gas, or one should use more general approximations given by Detkov [75] or Khodyko and his co-workers [76].

## 1.6. RESULTS AND RECOMMENDATIONS

The following conclusions can be drawn from systematic workup of published literature and also from analysis of individual effects influencing the optical properties of molecular gases:

1. At atmospheric pressure and gas temperatures not exceeding 3000 K the effects due to Voigt spectral line broadening are insignificant and can be neglected.

2. The most general relationships for spectral transmittance for a group of spectral lines is given by the Mayer–Goody statistical model [Eq. (1.27)].

3. The use of a multistep band envelope allows integration of the ex-

pressions of total absorption of bands or the total emissivity by the method of rectangles with accuracy sufficient for practical purposes. This involves correlation of parameters in expressions for spectral band parameters.

4. The equations suggested for calculating the absorption or total emissivity, given the significant difference between the experimental results and nomogram data, are in satisfactory agreement with data of other investigators.

5. It is shown that in calculating radiation in nonisothermal gases the Curtis–Godson approximation is valid for more than the limiting cases of weak and strong spectral lines, provided that the distance $d$ between the lines in their groups is not a function of the optical path. In these cases the error of the Curtis–Godson approximation does not exceed 1–3% as compared with exact numerical calculations over a wide range of parameters.

6. The value of the spectral line strength parameter depends to a large degree on the temperature distribution within the gas layer, for which reason nonisothermicity must be considered in calculations of spectral absorption or radiation.

In this study we have analyzed in detail the optical properties of gases that are components of combustion products of organic fuels. The spectral parameters of these gases needed for calculating radiation employing a multistep band envelope were described using equations analogous to those used in the Edwards wide-band model. Allowance for asymmetry of spectral parameters in bands can be easily performed within the framework of the multistep wide-band model, using data from the book by Ludwig et al. [10], tabulated as a function of the wave number and temperature.

The following are recommended in calculating absorption or total emission of molecular gases.

1. Equations (1.54) and (1.60), respectively, which can be used in calculating the total emissivity of gases.

2. Equation (1.62), which can be used for determining the effective optical thickness $\tau$ in the presence of a multiband spectrum.

3. The values of spectral band parameters $C$ and $\beta$ as a function of temperature, pressure, and wave number, which can be calculated from data in Tables 1.4 and 1.5.

4. Equation (1.31), which can be used when it is necessary to estimate the effect of Voigt broadening.

5. The two-parameter Curtis–Godson approximation, which can be used for making allowance for the nonisothermicity of the gas. The nonisothermicity of gas mixtures can be corrected for by Eq. 1.92).

CHAPTER

# TWO

# TRANSFER EQUATIONS FOR COMBINED HEAT TRANSFER

The term combined heat transfer denotes transfer of heat simultaneously by conduction, convection, and radiation, the latter never being zero.

The mathematical description of energy transport in a radiating medium is based on Boltzmann equations for a mixture of particles of different types, the derivation of which incorporates allowance for the relativistic nature of photons [14]. In spite of the fact that the Boltzmann equations are mathematically complex and are used only infrequently for solving practical problems, they can serve for deriving all the principal gas dynamic equations for both radiating and nonradiating gases [14]. In addition, these equations yield useful information on the relationship between transport coefficients, i.e., viscosity, thermal conductivity, absorption, etc. coefficients, with macroscopic quantities such as energy, mass, and momentum fluxes. In this chapter are presented the principal expressions for a radiating gas in the approximation of the boundary-layer theory [78–80], including transformations needed for their numerical calculation.

## 2.1. PRINCIPAL EQUATIONS

The interaction between the flow and conductive–convective heat transfer and the radiation intensity field of sufficiently dense media can be investigated by treating the gas as a continuous medium. In this case the principal independent field variables are pressure $p$, density $\varrho$, temperature $T$, and the three velocity components, $u$, $v$, and $w$. In this study we are concerned with two-dimensional fluxes ($w = 0$) of media with uniform mass composition. For a radiating medium one adds to these variables the spectral radiation intensity $I(\omega)$. This means that a description of combined heat transfer requires availability of six equations. The boundary layer theory uses the following equations for steady-state nondissipative flows of gas along a nonpermeable surface in the absence of mass forces:

1. The equation of state, which relates the pressure, density, and temperature of the gas. In many cases a satisfactory approximation for gases at sufficiently high temperatures is the equation of state of the ideal gas,

$$p = \rho RT \tag{2.1}$$

2. The continuity equation, expressed by the law of conservation of mass of the medium,

$$\frac{\partial (\rho u)}{\partial x} + \frac{1}{y^k} \frac{\partial (y^k \rho v)}{\partial y} = 0 \tag{2.2}$$

Equation (2.2) was written for planar ($k = 0$) and axisymmetric ($k = 1$) systems, in which the radius of curvature is independent of the $x$ coordinate. Equation (2.2) does not contain the term expressing the change in mass due to emission of energy, since in ordinary problems of the flow of a radiating medium it is negligible.

3. The equation of motion, expressed by the law of momentum conservation. If the radiant stresses are neglected, then the equation of motion for the longitudinal velocity component becomes

$$\rho \left( u \frac{\partial u}{\partial x} + v \frac{\partial u}{\partial y} \right) = - \frac{dp}{dx} + \frac{1}{y^k} \frac{\partial (y^k \tau)}{\partial y} \tag{2.3}$$

In the general case of turbulent flow, the shear stress is given by the expression

$$\tau = (\mu + \rho \varepsilon_\tau) \frac{\partial u}{\partial y} \tag{2.4}$$

where the dynamic viscosity of the gas $\mu$ is a known function of pressure and temperature.

To be able to determine the eddy (turbulent) viscosity coefficient $\varepsilon_\tau$ one must use additional equations [81, 83] or establish the relationship between it and other gas dynamic parameters on the basis of the Prandtl empirical theory of turbulence [78, 79, 82].

According to the Prandtl hypothesis, the eddy viscosity coefficient is expressed in terms of the mixing length:

$$\varepsilon_\tau = l_T^2 \left| \frac{\partial u}{\partial y} \right| \tag{2.5}$$

Expressions for mixing length $l_T(x, y)$ specific cases of flow will be considered separately.

The main shortcoming of the Prandtl hypothesis is insufficient generality of function $l_T(x, y)$; however, this is frequently compensated for by the fact that two constants suffice for this turbulence model. Thus, for the case of a turbulent boundary layer on a flat plate.

$$l_T = \begin{cases} N_T \varkappa_T y & \text{at} \quad N_T \varkappa_T y \leqslant K_T \delta_x \\ K_T \delta_x & \text{at} \quad N_T \varkappa_T y > K_T \delta_x \end{cases} \tag{2.6}$$

where $\varkappa_T = 0.41$ is the von Karman constant.

The second constant contained in Eq. (2.6) is the coefficient of proportionality of the mixing length in the outer part of the boundary layer. It is usually assumed that $K_T = 0.08$–$0.09$. The damping factor $N_T$ makes allowance for the reduction in the mixing length as the wall is approached.

4. The equation of motion for the transverse velocity component, which, according to the boundary layer theory, becomes the condition

$$\frac{\partial p}{\partial y} = 0 \tag{2.7}$$

which basically yields no information on the pressure distribution in the axial direction. Hence in the case of external flows one must make another assumption, which relates the longitudinal velocity outside the boundary layer to the pressure

$$U_f = F(p) \tag{2.8}$$

and in the case of internal flow, the constant flow rate condition

$$G = 2\pi^k \int_0^Y \rho u y^k \, dy = \text{const}. \tag{2.9}$$

Usually Eq. (2.8) is replaced by the Bernoulli equation for potential flow:

$$\frac{dp}{dx} = -\rho_f U_f \frac{dU_f}{dx} \tag{2.10}$$

Integral equation (2.9) makes possible nonunique solutions of problems of channel flow. For this reason analysis of such flows requires the use of additional assumptions, for example of symmetry of the velocity field of the flow.

5. The energy equation, expressed by the law of energy conservation in the absence of internal heat sources:

$$\rho c_p \left( u \frac{\partial T}{\partial x} + v \frac{\partial T}{\partial y} \right) = -\frac{1}{y^k} \frac{\partial (y^k q)}{\partial y} - \frac{1}{y^k} \frac{\partial (y^k q_p)}{\partial y} \tag{2.11}$$

In the general case of turbulent flow the heat flux is given by the equation

$$q = -(\lambda + \rho c_p \varepsilon_q) \frac{\partial T}{\partial y} \tag{2.12}$$

where specific heat $c_p$ and thermal conductivity $\lambda$ are known functions of pressure and temperature.

Certain advances were achieved in making allowance for the difference between turbulent transfer of momentum and energy on the basis of solution of the equations of balance of fluctuating velocity and temperature components [83]. Within the framework of the semiempirical Prandtl theory, allowance for this difference can be performed using the turbulent Prandtl number,

$$\mathrm{Pr}_T = \frac{\varepsilon_\tau}{\varepsilon_q} \tag{2.13}$$

Analysis performed by Cebeci [84] showed that allowance for the value of $\mathrm{Pr}_T$, which is variable with respect to the $y$ coordinate, has little effect on the results of calculation of eddy heat transfer.

6. The equation of spectral intensity of radiation under conditions of local thermodynamic equilibrium expressing the radiant energy balance in unit solid angle about beam $r$:

$$\frac{\partial I}{\partial r} = \varkappa (n_\omega^2 B - I) \tag{2.14}$$

Equation (2.14) is valid for a medium with negligible scattering in which the processes occur at a rate significantly smaller than the speed of light.

At boundary condition $I = I_0(r_0)$, Eq. (2.14) can be used for obtaining a solution in the following form:

$$I(r) = I_0(r_0) \exp(-\tau_{r_0}^r) + \int_{r_0}^{r} n_\omega^2 B(r_1) \exp(-\tau_{r_1}^r) \varkappa \, dr_1 \tag{2.15}$$

where the spectral optical thickness $\tau$ is defined in terms of the spectral absorption coefficient of the medium $\varkappa$ using the formulas

$$\tau_{r_0}^r \equiv \int_{r_0}^{r} \varkappa \, dr_1 \quad \text{and} \quad \tau_{r_1}^r \equiv \int_{r_1}^{r} \varkappa \, dr_2 \tag{2.16}$$

Equation (2.16) can be used only for a medium with known spectral absorption coefficient. In the case of determination of the optical thickness for molecular gases, when using the model representation for the spectrum, one should employ the pertinent equations of optical thickness, presented in Chapter 1.

In Eq. (2.14) the term making allowance for black-body radiation is expressed by the Planck function

$$B = \frac{2c_0^2 \omega^3 h}{\exp(c_0 \omega h/kT) - 1} \qquad (2.17)$$

In gas dynamic problems, as a rule, one uses the integral equation (2.15) rather than the differential equation (2.14).

The radiant heat flux $q_r$ is defined by integrating the spectral radiation intensity $I$ over the wave number $\omega$, azimuthal angle $\varphi$, and angle $\theta$ between the opposite direction of the normal to the surface element and the incident beam:

$$q_r = \int_0^\infty \int_0^{2\pi} \int_0^\pi I(\theta, \varphi) \sin\theta \cos\theta \, d\theta \, d\varphi \, d\omega \qquad (2.18)$$

Given the integral nature of Eq. (2.18), energy equation (2.11) is an integrodifferential equation.

The above equations fully describe the flow and heat transfer in a radiating medium, with allowance for the above assumptions.

The special problems of radiation gas dynamics, with allowance for coherence of radiation, luminescence, and other processes, are formulated in the book by Adzerikho [85].

In addition to the method for describing radiative heat transfer presented in this book, extensive use is also made of methods such as the diffusion, tensor, or algebraic approximations, and also various forms of integral equations of radiant heat transfer. A detailed presentation of these methods can be found in studies by Adrianov [86], Özisik [87], Siegel and Howell [4], Viskanta [88], and others. Note that a large contribution to the development of classical methods of calculation of radiant heat transfer has been made by Soviet scientists such as Polyak [89], Surinov [90], and others.

## 2.2. BOUNDARY CONDITIONS

Each particular problem of transport can be solved only at certain initial and boundary conditions.

The term initial conditions, in the case of steady-state processes, is used to describe the distributions of velocity, state variables of the gas, and the spectral radiation intensity in a given location along the $x$-coordinate axis.

Flow and heat transfer for a continous medium are controlled not only by initial but also by boundary conditions. In the most general form and with allowance for the entire set of physical and chemical phenomena arising on a surface (sublimation, chemical reactions, etc.), the boundary conditions are examined by Motulevich [91, 92].

As a rule it is assumed that in external flow all the free-steam parameters of the flow are known, i.e.,

$$u = U_f \tag{2.19}$$

$$T = T_f \tag{2.20}$$

The boundary conditions on the wall of a flow-washed body are controlled by the formulation of the problem. The only previously known condition that is valid in any case of continuous flow over a body is the condition

$$u = 0 \quad \text{at} \quad y = 0 \tag{2.21}$$

The values of the remaining parameters at the wall are in general determined from the solution of the entire problem.

For an impermeable wall,

$$v = 0 \quad \text{at} \quad y = 0 \tag{2.22}$$

Depending on the nature of the heat transfer between the medium and the surface, the boundary temperature conditions may be of the first, second, or third kind.

For very dense media, in which the mean free-length of a photon is very small, the boundary conditions for the spectral radiation intensity are replaced by a boundary condition for temperature. In all remaining cases it is necessary to make allowance for the interaction between radiation and the interface between two media. In macroscopic study of radiant energy transfer one does not consider the behavior of each beam at the interface between two media separately, but the behavior of a statistically large number of beams. In conjunction with this, all the surfaces are subdivided into two groups: optically smooth and optically rough. Optically smooth surfaces obey Fresnel laws. In practice the majority of cases are not optically smooth. Their optical properties are described by phenomenological coefficients, which describe the directional absorptive, reflective, and emitting power of the surface. For many cases it is possible to use the diffusion approximation, when the radiant energy reflected or emitted by them is distributed uniformly in all directions. This approximation is used in all the problems examined in this study.

In describing optical properties of surfaces, the coefficients of spectral hemispherical absorptivity $a_\omega$, spectral hemispherical emissivity $e_\omega$, spectral hemispherical reflectivity $r_\omega$, and spectral hemispherical transmissivity $t_\omega$

are interrelated by the expression

$$a_\omega + r_\omega + t_\omega = 1 \qquad (2.23)$$

Under conditions of thermodynamic equilibrium between the surface and the medium,

$$a_\omega = e_\omega \qquad (2.24)$$

For a nontransparent surface, $t_\omega = 0$.

For the surface of an ideal black body the reflectivities and transmissivities are equal to zero; consequently,

$$a_\omega = e_\omega = 1 \qquad (2.25)$$

The spectral radiation intensity of the surface of an ideal black body is given by the formula

$$I(T_w) = n_\omega^2 B(T_w) \qquad (2.26)$$

For a nonblack but nontransparent surface, the spectral intensity at the wall is

$$I(T_w) = n_\omega^2 e_\omega B(T_w) = n_\omega^2 (1 - r_\omega) B(T_w) \qquad (2.27)$$

For the majority of engineering problems in gaseous media, the refractive index $n_\omega$ is very close to unity and can be neglected.

Recommendations on determining the optical properties of solids can be found in the studies by Sheyndlin [93] and Khrustalev [94].

## 2.3. BOUNDARY-LAYER EQUATIONS IN A WEAKLY RADIATING MEDIUM

Study of relationships governing transfer of energy in high-temperature flows, in which the natural radiation is either small or entirely absent, plays an important role in gaining knowledge of physical processes in combined heat transfer. According to Pai [80], at radiant heat fluxes not exceeding 10% of the convective flux, the total energy transfer can be regarded as an additive sum of the radiative and convective transfer without allowance for interaction between them. Here the mathematical description of the problem simplifies significantly, since the integrodifferential equation of energy transfer becomes an ordinary partial differential equation, and radiant heat transfer can be calculated separately using the temperature field obtained from solving the purely convective problem.

Calculations of heat transfer in nonradiating gases forms the subject of many studies, which present expressions for heat transfer as a function of a number of controlling variables [78, 79, 82, 95–100]. Lately in addition to analysis of the dependence of heat transfer on such familiar dimensionless numbers as the Reynolds, Prandtl, and Grashof numbers, temperature factor, etc., a great deal of attention has been paid to the effect of the free-stream turbulence structure [101–107]. This is particularly important for analyzing relationships governing heat transfer in high-temperature flows, since the latter, as a rule, have a high turbulence level. Although as of now very little is known about the turbulence structure of high-temperature flows, allowance for it makes possible explaining a number of governing relationships for flow of high-temperature media in inlet regions of channels [105–107]. These relationships are used also in analyzing combined heat transfer.

Consideration of the effect of the turbulence of the external medium is based on comparison of experimental results with some mathematical model of heat transfer constructed on the basis of numerical calculations at different values of turbulence parameters.

Let us consider the technique used in this book for calculating heat transfer under different flow conditions.

Heat transfer in the inlet length of a channel is examined in the approximation of gradientless flow over a planar or axisymmetric surface. In this formulation the problem of the laminar boundary layer is reduced to solving the following set of equations:

$$\left.\begin{aligned}
&\frac{\partial(\rho u)}{\partial x}+\frac{1}{y^k}\,\frac{\partial(y^k\rho v)}{\partial y}=0\\
&\rho u\,\frac{\partial u}{\partial x}+\rho v\,\frac{\partial u}{\partial y}=\frac{1}{y^k}\cdot\frac{\partial}{\partial y}\left(y^k\mu\,\frac{\partial u}{\partial y}\right)\\
&\rho c_p u\,\frac{\partial T}{\partial x}+\rho c_p v\,\frac{\partial T}{\partial y}=\frac{1}{y^k}\cdot\frac{\partial}{\partial y}\left(y^k\lambda\,\frac{\partial T}{\partial y}\right)
\end{aligned}\right\}\tag{2.28}$$

with the pertinent boundary conditions.

Integrating in Eqs. (2.28) with respect to coordinate $y$ makes it possible to transform them into easily solved integral equations of momentum and energy [79]. Such a solution allows determining the dependence of the thickness of the velocity and thermal boundary layers on the axial coordinate $x$, and also yields useful information on the dependence of the effect of the Prandtl number on the flow conditions [108]. However, the results obtained in solving integral equations of the boundary layer for flows with variable physical properties differ from those that follow from solving the differential equations.

The system of differential equations for a planar laminar boundary layer in the case of variable physical properties is usually nondimensionalized, using some stream function that satisfies the conditions of flow continuity

[78, 79]. Thus, for constant velocity flows at the outer edge of the velocity boundary layer

$$U_\infty = U_f \tag{2.29}$$

one introduces a stream function in the form [109]

$$f_l = \psi(x) \cdot \varphi(\eta_l) \tag{2.30}$$

where

$$\varphi = \frac{u}{U_\infty}, \quad \psi(x) = \sqrt{2U_\infty \nu_w x} \tag{2.31}$$

Such an approach to solving the problem allows reducing the set (2.28) of partial differential equations to a set of ordinary differential equations.

Introducing the nondimensional temperature

$$\Theta_l \equiv \frac{T - T_w}{T_f - T_w} \tag{2.32}$$

and a new coordinate, containing the Dorodnitsyn variable in the form of the integral of density $\varrho$ with respect to coordinate $y$,

$$\eta_l = \frac{U_\infty}{\rho_w \psi(x)} \int_0^y \rho \, dy \tag{2.33}$$

we can reduce Eqs. (2.28) to the self-similar form:

$$\left.\begin{aligned} & g_\rho g_\mu \varphi'''_{\eta_l} + [(g_\rho g_\mu)'_{\eta_l} + \varphi] \cdot \varphi''_{\eta_l} = 0, \\ & g_\rho g_{c_p}^{-1} g_\lambda \Theta''_{\eta_l} + [g_{c_p}^{-1} (g_\rho g_\lambda)'_{\eta_l} + \mathrm{Pr}_w \varphi] \cdot \Theta'_{\eta_l} = 0 \end{aligned}\right\} \tag{2.34}$$

where

$$g_\rho = \frac{\rho}{\rho_w} \tag{2.35}$$

$$g_{c_p} = \frac{c_p}{c_{p,\,w}} \tag{2.36}$$

$$g_\mu = \frac{\mu}{\mu_w} \tag{2.37}$$

$$g_\lambda = \frac{\lambda}{\lambda_w} \tag{2.38}$$

Equations (2.34) are somewhat nonlinear, but this does not complicate their solution. The present author and his co-workers [109–111] investigated numerically the solution of Eqs. (2.34) for boundary conditions of the first kind for a power-law temperature dependence of the physical properties. Some results of solutions of Eqs. (2.34) for laminar, nonradiating, high-temperature flows are given in Chapter 4.

It is much more difficult to solve the set of differential equations for a turbulent boundary layer. There exists a large number of methods of calculation of the turbulent boundary layer, a detailed analysis of which is presented by Žukauskas and Šlančiauskas [82] and by Popov [112]. As a rule, these methods make various assumptions of both a physical and a mathematical nature. For example, use is made of an integral momentum equation for determining the dependence of the boundary-layer thickness $\delta_x$ on the longitudinal coordinate $x$, which cannot be used in solving the problem in the case of variable physical properties. Hence the results obtained from calculations frequently cannot be extended to the range of parameters for which no experiments were performed. No less important as a reason for lack of agreement between results of calculation of the turbulent boundary layer may be the significant nonlinearity of differential equations, requiring high accuracy of the numerical methods used. This lack of agreement in results is noted not only in calculations, but also in experiments, which in part can be attributed to the fact that they were performed at different turbulent properties of the flow. In some experiments it was found that the boundary-layer structure depends significantly on the free-stream turbulence [101–104]. This made it necessary to extend the Prandtl hypothesis on the mixing length [105].

To allow for the free–stream turbulence one must make certain assumptions concerning the relationship between its parameters and the parameters of turbulence within the boundary layer. When using the Prandtl hypothesis it suffices to specify the relationship between the mixing length and the free-stream turbulence parameters.

It was assumed in studies by the present author and his co-workers [105–107] that quantity $K_T$, which is a part of Eq. (2.6) for the mixing length, is some function of the free-stream turbulence; however, due to lack of information on this dependence, at the first stage of studies $K_T$ was treated as a parameter controlling the conditions of turbulence outside the boundary layer.

The parameter used for characterizing the conditions of energy transfer in the turbulent boundary layer was the turbulent Prandtl number $\mathrm{Pr}_T$, which was regarded as constant over the entire thickness of the boundary layer.

Let us consider a planar boundary layer without a pressure gradient on the basis of Eqs. (2.1)–(2.6) and (2.11)–(2.13) for a transparent medium at the following boundary conditions (Fig. 2.1):

$$u=0, \quad v=0, \quad T=T_w \quad \text{at} \quad y=0 \tag{2.39}$$

$$u=U_f, \quad v=0, \quad T=T_f \quad \text{at} \quad y \geqslant \delta_x \tag{2.40}$$

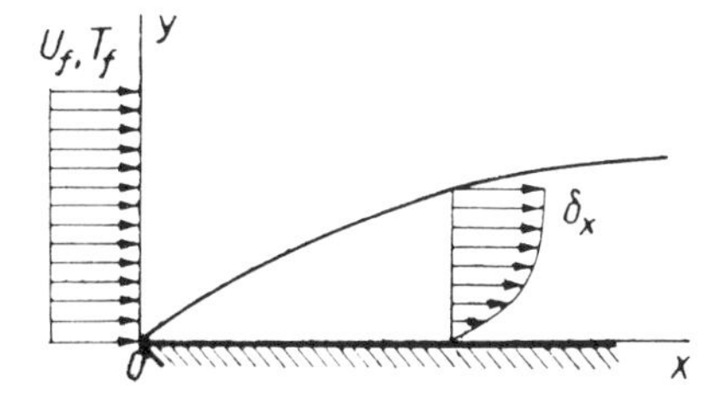


**FIG. 2.1** Coordinate system for the flow over a planar body.

The above equations were transformed by a method suggested by Clauser [113] for the case of constant physical properties. If in laminar flow the stream function depends on $x$ in the manner given by Eq. (2.30), then it follows from the paper by Clauser that for turbulent flow one can successfully use the specification of constant friction velocity ($u_* =$ constant), which allows introducing a stream function for turbulent flow in the form

$$f'_{\eta_T} = \frac{U_f - u}{u_*} \tag{2.41}$$

and new coordinates, containing the Dorodnitsyn variable,

$$x = \xi \text{ and } \eta_T = \frac{1}{D_x} \int_0^y \rho \, dy \tag{2.42}$$

where

$$D_x \equiv \int_0^{\delta_x} \rho \, dy \tag{2.43}$$

Here the derivatives of the new coordinates have the form

$$\left.\begin{aligned} \frac{\partial \eta_T}{\partial x} &= -\eta_T \frac{D'_x}{D_x}, \quad \frac{\partial \xi}{\partial x} = 1 \\ \frac{\partial \eta_T}{\partial y} &= \frac{\rho}{D_x}, \quad \frac{\partial \xi}{\partial y} = 0 \end{aligned}\right\} \tag{2.44}$$

Using the familiar expressions for conversion from old to new coordinates, we can write continuity equation (2.2) as

$$\rho v = D'_x u_* (f - \eta_T f'_{\eta_T}) \tag{2.45}$$

Allowance was made in deriving this expression for the fact that

$$f(0) = 0 \tag{2.46}$$

Given that

$$\tau_w = \rho_w u_*^2 \tag{2.47}$$

on the basis of Eqs. (2.3), (2.4), and (2.8) we obtain the expression

$$\frac{\partial \bar{\tau}}{\partial \eta_T} = \frac{D'_x}{\rho_w}\left(\eta_T f''_{\eta_T}\frac{U_f}{u_*} - f f''_{\eta_T}\right) \tag{2.48}$$

where

$$\bar{\tau} \equiv \frac{\tau}{\tau_w} \tag{2.49}$$

The derivatives of the stream function with respect to $x$ were neglected in deriving Eq. (2.48), since the universality of the velocity deficit in the turbulent boundary layer was verified experimentally [113].

Integrating Eq. (2.48) once with respect to coordinate $\eta_T$, using the boundary condition

$$\bar{\tau} = 0 \quad \text{at} \quad \eta_T = 1 \tag{2.50}$$

we obtain

$$\bar{\tau} = \frac{D'_x}{\rho_w}\cdot\left\{\frac{U_f}{u_*}\left[\eta_T f'_{\eta_T} - f + f(1)\right] - f f'_{\eta_T} + \int_1^{\eta_T} f'^2_{\eta_T}\, d\eta_T\right\} \tag{2.51}$$

The use of the additional boundary condition

$$\bar{\tau} = 1 \quad \text{at} \quad \eta_T = 0 \tag{2.52}$$

makes it possible to obtain from Eq. (2.51) an expression governing the growth of the velocity boundary layer:

$$D'_x = \frac{\rho_w}{\dfrac{U_f}{u_*} f(1) - \displaystyle\int_0^1 f'^2\, d\eta_T} \tag{2.53}$$

Using Eq. (2.53) it is possible to express quantity $\bar{\tau}$, which is independent of the conditions of boundary-layer growth. Equations (2.51) and (2.53) yield

$$\bar{\tau} = \frac{\dfrac{U_f}{u_*}\cdot\left[\eta_T f'_{\eta_T} - f + f(1)\right] - f f'_{\eta_T} + \displaystyle\int_1^{\eta_T} f'^2_{\eta_T}\, d\eta_T}{\dfrac{U_f}{u_*}\cdot f(1) - \displaystyle\int_0^1 f'^2_{\eta_T}\, d\eta_T} \tag{2.54}$$

Coordinate $\eta_T$ used in this study made it possible to obtain Eqs. (2.53) and (2.54) analogous to the Clauser equations for a planar turbulent boundary layer in the case of constant physical properties [113].

Equation (2.54) is valid for any description of turbulent transport. If we limit ourselves to the Prandtl hypothesis on the mixing length, then according to Eqs. (2.4) and (2.5), we obtain the following expression for the relative shear stress:

$$\bar{\tau} = -\frac{g_\rho g_\mu}{\mathrm{Re}_{\delta,\,\omega}} \cdot \frac{\rho_w \delta_x}{D_x} \cdot \frac{U_f}{u_*} \cdot f''_{\eta_\mathrm{T}} - g_\rho^3 \cdot \left(\frac{\rho_w \delta_x}{D_x}\right)^2 \cdot \left(\frac{l_\mathrm{T}}{\delta_x}\right)^2 \cdot |f''_{\eta_\mathrm{T}}| \cdot f''_{\eta_\mathrm{T}} \tag{2.55}$$

The two-layer method of solution of the resulting equations used by Clauser has a number of serious shortcomings. In particular, it is impossible to make allowance for the variability of physical properties in the logarithmic zone of the boundary layer.

For numerical solution of the equation of motion we reduce Eq. (2.55) to the form

$$f''_{\eta_\mathrm{T}} = \frac{-2\bar{\tau} \cdot \mathrm{Re}_{\delta,\,w} \cdot \left(\frac{D_x}{\rho_w \delta_x}\right) \cdot \frac{u_*}{U_f}}{g_\rho g_\mu + \sqrt{g_\rho^2 g_\mu^2 + 4 g_\rho^3 \cdot \bar{\tau} \cdot \mathrm{Re}^2_{\delta,\,w} \cdot \left(\frac{l_\mathrm{T}}{\delta_x}\right)^2 \cdot \left(\frac{u_*}{U_f}\right)^2}} \tag{2.56}$$

where quantity $\bar{\tau}$ is related to stream function $f'_{\eta T}$ by Eq. (2.54).

Using the boundary conditions

$$f'_{\eta_\mathrm{T}} = \frac{U_f}{u_*} \quad \text{at} \quad \eta_\mathrm{T} = 0 \tag{2.57}$$

$$f'_{\eta_\mathrm{T}} = 0 \quad \text{at} \quad \eta_\mathrm{T} = 1 \tag{2.58}$$

and Eq. (2.56), we obtain

$$f'_{\eta_\mathrm{T}} = \frac{U_f}{u_*} + \int_0^{\eta_\mathrm{T}} f''_{\eta_\mathrm{T}} \, d\eta_\mathrm{T} \tag{2.59}$$

Such an expression of the equation of motion is quite convenient for constructing an iterative computational scheme.

Analogously, introducing the nondimensional temperature

$$\Theta_\mathrm{T} = \frac{T - T_f}{T_*} \tag{2.60}$$

we can transform Eqs. (2.11) and (2.12) thus:

$$\Theta_\mathrm{T} = \mathrm{Pr}_w \, \mathrm{Re}_{\delta,\,w} \cdot \frac{u_*}{U_f} \cdot \frac{D_x}{\rho_w \delta_x} \int_{\eta_1}^{\infty} \frac{\exp(F_1)}{g_\rho g_{\lambda,\,\mathrm{T}}} \, d\eta_{\mathrm{T},\,1} \tag{2.61}$$

where

$$F_1 \equiv \mathrm{Pr}_w \, \mathrm{Re}_{\delta,\,w} \cdot \frac{u_*}{U_f} \cdot \frac{D_x \cdot D'_x}{\rho_w^2 \delta_x} \int_0^{\eta_{\mathrm{T},\,1}} \frac{g_{c_p}}{g_\rho g_{\lambda,\,\mathrm{T}}} \left(f - \frac{U_f}{u_*} \eta_{\mathrm{T},\,1}\right) d\eta_\mathrm{T}, \tag{2.62}$$

$$T_* \equiv \frac{q_w}{\rho_w c_{p,\,w} u_*} \tag{2.63}$$

$$g_{\lambda,\,\mathrm{T}} \equiv \frac{\lambda}{\lambda_w} \cdot \left(1 + \frac{\mathrm{Pr}}{\mathrm{Pr}_\mathrm{T}} \cdot \frac{\varepsilon_\tau}{\nu}\right) \tag{2.64}$$

Equation (2.61) satisfies the boundary conditions:

$$\Theta_{T,w} = \frac{T_w - T_f}{T_*} \quad \text{at} \quad \eta_T = 0 \tag{2.65}$$

and

$$\Theta_T = 0 \quad \text{at} \quad \eta_T = \infty \tag{2.66}$$

In the above coordinates the relative eddy viscosity is

$$\frac{\varepsilon_\tau}{\nu} = g_\rho \, \mathrm{Re}_\delta \, \frac{u_*}{U_f} \cdot \left(\frac{l_T}{\delta_x}\right)^2 \cdot \frac{\rho_w \delta_x}{D_x} \cdot |f''_{\eta_T}| \tag{2.67}$$

In Eq. (2.67) the mixing length was determined from Eq. (2.6) on the assumption that damping factor $N_T$ is a function of the local values of physical properties.

Numerical calculations of the frictional drag and heat transfer at a given thickness $\delta_x$ of the velocity boundary layer were performed by the method of successive approximations. The zero approximation assumed a linear dependence of $\bar{\tau}$ on $\eta_T$. The numerical values of the integrals were determined in calculations by the method of parabolas or trapezoidal rule with uniform subdivision of the argument with respect to the variable

$$z = \ln\left(1 + \eta_T \cdot \mathrm{Re}_{\delta,w} \cdot \frac{u_*}{U_f}\right) \Big/ \ln\left(1 + \mathrm{Re}_{\delta,w} \cdot \frac{u_*}{U_f}\right) \tag{2.68}$$

The values of $f''_{\eta T}$ and $f'_{\eta T}$ were calculated from Eqs. (2.56) and (2.59) at a given temperature and physical properties field. Boundary conditions (2.57) were satisfied by iterations with respect to parameter $U_f/u_*$. Once the profiles of $f'_{\eta T}$ and $f$ were determined, it was possible to solve the energy equation and to determine the new values of the field of relative shear stress from Eq. (2.54). When the specified accuracy with respect to equation of motion (2.59) at $\eta_T = \eta_{T,\max}$ and from energy equation (2.61) at $\eta_T = 0$ was achieved, the iterations were discontinued.

The relationship between transverse coordinate $y$, integral boundary layer thickness $D_x$, and the coordinate $\eta$ is given by the expressions

$$y = \delta_x \cdot \frac{\int_0^{\eta_T} \frac{d\eta_1}{\rho}}{\int_0^1 \frac{d\eta_1}{\rho}} \tag{2.69}$$

and

$$D_x = \frac{\delta_x}{\int_0^1 \frac{d\eta_T}{\rho}} \tag{2.70}$$

The calculations in determining $\delta_x = f(x)$ were performed for several cross-sections of the boundary layer, with subsequent numerical integration of Eq. (2.53).

The solution of Eq. (2.53) is written as

$$D_x = \rho_w \cdot \int_0^x \frac{dx}{\frac{U_f}{u_*} \cdot f(1) - \int_0^1 f'^2 \, d\eta_T} \tag{2.71}$$

which was determined numerically for a number of precalculated values of the integrand at several arbitrarily selected values of $\delta_x$.

The attainment of acceptable accuracy required from nine to 12 steps along the $x$ coordinate with coordinate $\eta_T$ subdivided into 150–200 parts.

The results of calculations of eddy heat transfer obtained using the above technique for a wide range of flow variables are described in Chapter 4.

The boundary-layer conditions become more complicated in the presence of radiation, so that other methods must be used for solving them.

## 2.4. EQUATIONS OF COMBINED HEAT TRANSFER OF A MEDIUM IN CHANNELS

The steady-state flow of a selectively radiating and absorbing medium with variable physical properties in a planar or axisymmetric channel is described by Eqs. (2.1)–(2.6), (2.9), and (2.11)–(2.14).

Let us consider the flow in a channel at the following initial and boundary conditions of the first kind (Fig. 2.2):

at $x = 0$ and $-Y \leqslant y \leqslant Y$

$$T = T_0(y), \quad u = U_f(y) \tag{2.72}$$

at $x > 0$ and $y = \pm Y$

$$T = T_w, \quad u = 0, \quad v = 0 \tag{2.73}$$

at $x > 0$ and $y = 0$

$$\frac{\partial T}{\partial y} = 0, \quad \frac{\partial u}{\partial y} = 0, \quad v = 0 \tag{2.74}$$

Boundary conditions (2.72)–(2.74) correspond to flows that are symmetrical with respect to axis $y = 0$. It is assumed that the planar channel is of unit width.

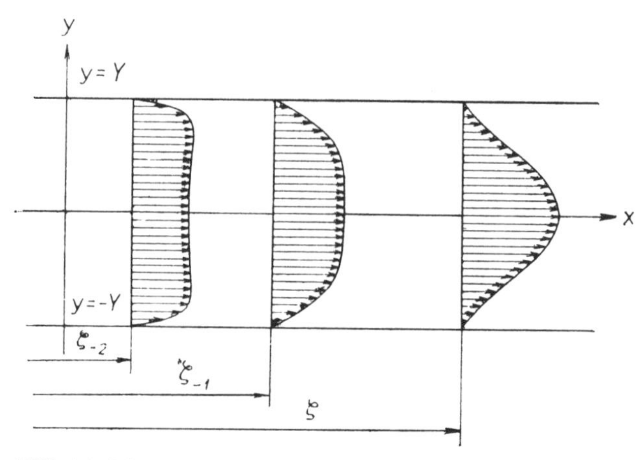


**FIG. 2.2** Schematic for calculating flow in channels.

Due to divergence of the radiation, a similar solution of the energy equation is impossible even when the flow in the channel is laminar. In a radiating medium it is difficult to predict the dependence of stream function $f$ or of the friction velocity $u_*$ on the $x$ coordinate. It follows from the rather few experiments performed at relatively moderate radiant heat fluxes that the radiation has little effect on the velocity field, which does not apply at high flux densities. In conjunction with this it is not advisable to seek solutions using the quite complex expressions for the nondimensional coordinate or stream functions, as is done in the case of a nonradiating boundary layer. Even the introduction of the Dorodnitsyn variable complicates the expressions to such an extent that it cannot be justified.

To investigate the flow of a selectively radiating and absorbing medium in channels, we introduce the nondimensional coordinates

$$\eta = \frac{y}{Y}, \quad \zeta = \frac{x}{H} = \frac{x}{2Y} \tag{2.75}$$

Using these coordinates we have from continuity equation (2.2)

$$\rho v = -\frac{1}{2\eta^k} \cdot \frac{\partial}{\partial \zeta} \int_0^{\eta} \rho u \eta_1^k \, d\eta_1 \tag{2.76}$$

and momentum equation (2.3) becomes

$$\frac{1}{\eta^k} \cdot \frac{\partial (\eta^k \tau)}{\partial \eta} = \frac{1}{2} \frac{dp}{d\zeta} + \frac{\rho u}{2} \cdot \frac{\partial u}{\partial \zeta} - \frac{1}{2\eta^k} \cdot \frac{\partial u}{\partial \eta} \cdot \frac{\partial}{\partial \zeta} \int_0^{\eta} \rho u \eta_1^k \, d\eta_1 \tag{2.77}$$

Equation (2.77) is valid for both laminar and turbulent flows.

Numerical solution of Eq. (2.77), which is a partial differential equation, is very difficult. For studies of channel flow it is convenient to replace derivatives with respect to coordinate $\zeta$ by finite differences. Then in each section the derivatives in the direction of flow, contained in the right-hand side of Eq. (2.77), can be replaced by difference approximations of Lagrange for forward interpolation [114, 115]. Then the derivatives in the direction of motion are written as

$$\frac{dp}{d\zeta}=a_0 p+a_{-1}p_{-1}+a_{-2}u_{-2} \tag{2.78}$$

$$\frac{\partial u}{\partial \zeta}=a_0 u+a_{-1}u_{-1}+a_{-2}p_{-2} \tag{2.79}$$

$$\frac{\partial(\rho u)}{\partial \zeta}=a_0 \rho u+a_{-1}(\rho u)_{-1}+a_{-2}(\rho u)_{-2} \tag{2.80}$$

Subscripts −1 and −2 in Eqs. (2.78)–(2.80) designate two locations (Fig. 2.2) preceding the coordinate of independent variable $\zeta$, whereas quantities $a_0$, $a_{-1}$, and $a_{-2}$ are functions of the selected distributions of the values of $\zeta$. In the case of a three-point approximation,

$$a_0=\frac{1}{\zeta-\zeta_{-1}}+\frac{1}{\zeta-\zeta_{-2}} \tag{2.81}$$

$$a_{-1}=\frac{\zeta_{-2}-\zeta}{(\zeta-\zeta_{-1})\cdot(\zeta_{-1}-\zeta_{-2})} \tag{2.82}$$

$$a_{-2}=\frac{\zeta-\zeta_{-1}}{(\zeta-\zeta_{-2})\cdot(\zeta_{-1}-\zeta_{-2})} \tag{2.83}$$

and for a two-point approximation,

$$a_0=\frac{1}{\zeta-\zeta_{-1}} \tag{2.84}$$

$$a_{-1}=-\frac{1}{\zeta-\zeta_{-1}} \tag{2.85}$$

$$a_{-2}=0. \tag{2.86}$$

The flow field in the zero cross-section is specified by initial conditions. The flow field in the subsequent cross-section can be calculated from the two-point Lagrange approximation. As $\zeta$ is moved in the positive direction, the flow field can be calculated in the remaining cross-sections of the channel.

Using Eqs. (2.4) and (2.81)–(2.86), Eq. (2.77) is reduced to the form

$$\frac{1}{\eta^k}\cdot\frac{\partial}{\partial\eta}\left[\eta^k\left(\frac{\mu}{\mu_w}+\frac{\rho\varepsilon_\tau}{\mu_w}\right)\cdot\frac{\partial u}{\partial\eta}\right]=\Phi_U(\eta,\ \rho,\ u,\ p) \tag{2.87}$$

where

$$\Phi_U(\eta,\ \rho,\ u,\ p) \equiv \frac{Y}{2\mu_w} \cdot \Bigg\{ a_0 p + a_{-1} p_{-1} + a_{-2} p_{-2} +$$

$$+ \rho u\,(a_0 u + a_{-1} u_{-1} + a_{-2} u_{-2}) - \frac{1}{\eta^k} \cdot \frac{\partial u}{\partial \eta} \times$$

$$\times \left[ a_0 \int\limits_0^\eta \rho u \eta^k\, d\eta + a_{-1} \int\limits_0^\eta (\rho u)_{-1} \eta^k\, d\eta + a_{-2} \int\limits_0^\eta (\rho u)_{-2} \eta^k\, d\eta \right] \Bigg\} \tag{2.88}$$

The solution of Eq. (2.87) requires using the remaining equations for the flow field and also the temperature dependence of physical properties. The only unknown quantities in the right-hand side of Eq. (2.87) are $p$ and $u$, the latter making up a part of various mathematical operators. Quantities $p_{-1}, p_{-2}, u_{-1}$, and $u_{-2}$ and also $(\rho u_{-1})$ and $(\rho u_{-2})$ are known from initial conditions of the problem or from preceding solutions. Since the right-hand side of Eq. (2.87) is nonlinear, it is best to solve it by the method of iterations. For this purpose we introduce the auxiliary function $S_u$, defined as

$$\frac{\partial S_u}{\partial \eta} = \left( \frac{\mu}{\mu_w} + \frac{\rho \varepsilon_\tau}{\mu_w} \right) \cdot \frac{\partial u}{\partial \eta} \tag{2.89}$$

It is assumed that the dependence of the eddy viscosity coefficient $\varepsilon_\tau$ on other parameters of the field is known. Using Eq. (2.88), Eq. (2.89) is reduced to the form

$$\frac{1}{\eta^k} \cdot \frac{\partial}{\partial \eta} \left( \eta^k \frac{\partial S_u}{\partial \eta} \right) = \Phi_U(\eta,\ \rho,\ u,\ p) \tag{2.90}$$

It is assumed in iterative solution of the problem that the zero approximations for quantities $p$, $\varrho(\eta)$ and $u(\eta)$ for the section under study are known. Then Eq. (2.90) can be treated as a linear differential equation of the second kind,

$$\frac{1}{\eta^k} \cdot \frac{\partial}{\partial \eta} \left( \eta^k \frac{\partial S_u}{\partial \eta} \right) = \Phi_U(\eta,\ \rho^0,\ u^0,\ p^0) \tag{2.91}$$

where the zero in the superscript means that the right-hand side of the equation is defined as a function of coordinate $\eta$ at any step of iteration from values of the fields of the preceding approximation. The general solutions of homogeneous equations of this kind are known [115]. The technique of numerical calculation will be considered in more detail in analyzing specific problems.

Energy equation (2.11) can, using Eqs. (2.12), (2.75), and (2.76), be rewritten in the form

$$\frac{1}{\eta^k} \cdot \frac{\partial}{\partial \eta} \left[ \eta^k \left( \frac{\lambda}{\lambda_w} + \frac{\rho c_p \varepsilon_q}{\lambda_w} \right) \frac{\partial \Theta}{\partial \eta} \right] = \Phi_\Theta(\eta,\ \Theta,\ u,\ \rho,\ c_p,\ \lambda,\ \bar{q}_r) \tag{2.92}$$

where

$$\Phi_\Theta(\eta, \Theta, u, \rho, c_p, \lambda, \bar{q}_r) = \frac{\rho c_p}{\lambda_w} \cdot \frac{uY}{2} \cdot \frac{\partial \Theta}{\partial \zeta} + \frac{1}{\eta^k} \frac{\partial(\eta^k \bar{q}_r)}{\partial \eta} -$$

$$-\frac{c_p Y}{2\eta^k \lambda_w} \cdot \frac{\partial \Theta}{\partial \eta} \cdot \frac{\partial}{\partial \zeta} \int_0^\eta \rho u \eta^k \, d\eta \tag{2.93}$$

$$\Theta = \frac{T}{T_w} \tag{2.94}$$

$$\bar{q}_r = \frac{q_r \cdot Y}{\lambda_w T_w} \tag{2.95}$$

Using the forward Lagrange equation for the arbitrary nondimensional temperature $\theta$ along the longitudinal coordinate $\zeta$,

$$\frac{\partial \Theta}{\partial \zeta} = a_0 \Theta + a_{-1} \Theta_{-1} + a_{-2} \Theta_{-2} \tag{2.96}$$

and Eq. (2.80), the right-hand side of Eq. (2.92) can be written as

$$\Phi_\Theta(\eta, \Theta, u, \rho, c_p, \lambda, \bar{q}_r) = \frac{\rho c_p}{\lambda_w} \frac{uY}{2} \cdot (a_0 \Theta + a_{-1} \Theta_{-1} + a_{-2} \Theta_{-2}) +$$

$$+ \frac{1}{\eta^k} \cdot \frac{\partial(\eta^k \bar{q}_r)}{\partial \eta} - \frac{Y}{2\eta^k} \cdot \frac{c_p}{\lambda_w} \cdot \frac{\partial \Theta}{\partial \eta} \times$$

$$\times \int_0^\eta [a_0 \rho u + a_{-1} (\rho u)_{-1} + a_{-2} (\rho u)_{-2}] \eta_1^k \, d\eta_1 \tag{2.97}$$

Equation (2.97) was written for any section $\zeta$ of the channel. In solving Eq. (2.90) one must have the values of the temperature and velocity fields of the flow in the preceding sections, $\zeta_{-1}$ and $\zeta_{-2}$.

For numerical solution of Eq. (2.92), as in the case of the solution of the equation of motion, we shall use the iterative scheme of calculation with linearizing by means of the auxiliary function

$$\frac{\partial S_\Theta}{\partial \eta} = \left( \frac{\lambda}{\lambda_w} + \frac{\rho c_p \varepsilon_q}{\lambda_w} \right) \cdot \frac{\partial \Theta}{\partial \eta} \tag{2.98}$$

Assuming that the right-hand side of Eq. (2.97) is defined as a function of $\eta$ from values of the temperature and velocity fields from the preceding approximation, Eq. (2.92) can be written as

$$\frac{1}{\eta^k} \cdot \frac{\partial}{\partial \eta} \left( \eta^k \frac{\partial S_\Theta}{\partial \eta} \right) = \Phi_\Theta(\eta, \Theta^0, u^0, \rho^0, c_p^0, \lambda^0, \bar{q}_r^0) \tag{2.99}$$

This means that the solution of Eqs. (2.91) and (2.99) can be reduced to an iterative solution of identical nonhomogeneous second-order differential equations.

In the case of turbulent channel flows, equations of motion (2.91) and energy (2.99) must be supplemented by expressions for the eddy viscosity and eddy thermal conductivity. In the simplest approximation these relationships can be expressed by the algebraic equations (2.5) and (2.13), which follow from the Prandtl hypothesis.

Successive determination of the velocity and temperature fields in individual sections along the $\zeta$ coordinate for a radiating medium requires a great amount of time. It is hence of great interest to investigate less work-consuming problems of combined heat transfer. The solution of the problem becomes significantly simplified in the case of stabilized flow in channels that begins at a significant distance from the inlet to the channel. For stabilized flow, the equations of motion and energy can be written as

$$\frac{1}{y^k} \cdot \frac{\partial}{\partial y} \left[ (\mu + \rho \varepsilon_\tau) \, y^k \, \frac{\partial u}{\partial y} \right] = \frac{dp}{dx} \tag{2.100}$$

$$\frac{1}{y^k} \cdot \frac{\partial}{\partial y} \left[ y^k \left( \lambda + \frac{\rho c_p \varepsilon_\tau}{\mathrm{Pr}_\mathrm{T}} \right) \frac{\partial T}{\partial y} \right] = \frac{1}{y^k} \, \frac{\partial (y^k q_r)}{\partial y} + \rho c_p u \, \frac{\partial T}{\partial x} \tag{2.101}$$

with the previously examined boundary conditions.

Following Viskanta [88], we can replace the derivative of temperature with respect to $x$ by the expression

$$\frac{\partial T}{\partial x} = - \frac{q_w Y^k}{\int\limits_0^Y \rho u c_p y^k \, dy} \cdot \frac{T_w - T}{T_w - T_m} \tag{2.102}$$

where the bulk temperature of the flow is

$$T_m = \frac{\int\limits_0^Y \rho u c_p T y^k \, dy}{\int\limits_0^Y \rho u c_p y^k \, dy} \tag{2.103}$$

Such a transformation allows us to obtain expressions for stabilized flow, which is not explicitly dependent on the longitudinal coordinate.

Using analogous transformations, the expressions (2.100) and (2.101) for stabilized flow can be written as

$$\frac{1}{\eta^k} \, \frac{\partial}{\partial \eta} \left( \eta^k \, \frac{\partial S_u}{\partial \eta} \right) = 2 \, \frac{\partial u}{\partial \eta} \bigg|_{\eta = 1} \tag{2.104}$$

$$\frac{1}{\eta^k} \, \frac{\partial}{\partial \eta} \left( \eta^k \, \frac{\partial S_\Theta}{\partial \eta} \right) = \Phi_\mathrm{T} (\eta, \, \Theta, \, \Theta_m, \, u, \, \rho, \, c_p, \, \lambda, \, \bar{q}_r) \tag{2.105}$$

where quantities $S_u$ and $S_\Theta$ are defined by Eqs. (2.89) and (2.98), and

$$\Phi_{\text{T}} = -\frac{Y q_w}{\lambda_w T_w} \cdot \frac{\rho c_p u}{\int_0^1 \rho c_p u \eta^k \, d\eta} \cdot \frac{1-\Theta}{1-\Theta_m} + \frac{1}{\eta^k} \frac{\partial (\eta^k \bar{q}_r)}{\partial \eta} \tag{2.106}$$

The above equations of stabilized flow were written for the most general case. They simplify significantly in problems of laminar flow of a medium with constant physical properties. In this case analytic solutions of the problem are available for a nonradiating medium [98, 99].

Numerical methods of iterative calculation of the aforementioned second-order nonhomogeneous equations will be examined in Chapter 6 in solving specific problems. The presence in functions $\phi_U$ and $\phi_\Theta$, in addition to velocity $u$ and temperature $\Theta$, of terms with derivatives of these quantities with respect to coordinate $\eta$ does not complicate the numerical calculations, since the general solution of the problem contains an integral of function $\phi_U$ or $\phi_\Theta$ and it is possible to obtain analytic expressions for integrals of terms containing derivatives with respect to coordinate $\eta$. Such a procedure can also be performed for the term containing the divergence of the radiative heat flux. This means that when using this iterative scheme it suffices to determine only the field of values of radiative heat flux when the temperature field is known from the preceding approximation.

## 2.5. EQUATION OF RADIATIVE HEAT FLUX IN FUNDAMENTAL FORMULATION OF THE PROBLEM

Calculations of radiative heat flux at specified temperature and concentration fields (fundamental formulation of problem) are important in gaining insight into relationships governing transport of radiative energy. Such solutions are usually based on Eqs. (2.14)–(2.18), using the values of the applicable optical properties of the medium and of the bounding surfaces. Calculation of the radiative heat flux of a nonisothermal medium within spaces of complex geometric shape using Eq. (2.18) involves great mathematical difficulties in evaluating multiple integrals. A large number of investigators were and are concerned with finding sufficiently accurate and simple methods of calculation; these are surveyed by Siegel and Howell [4], Adrianov [86], Özisik [87], Viskanta [88], Polyak [89], and Surinov [90].

The radiation of gray media are most frequently calculated by zonal methods [86–90, 116], whereas selective radiation is calculated by finite-difference schemes, the Monte Carlo method, and also methods based on stepwise functions and orthogonal polynomials. A survey of studies on the use of the Monte Carlo method in problems of radiative energy transfer is given by Howell [117]. In spite of its simplicity, the Monte Carlo method is not economical with respect to the use of machine time. The feasibility of using orthogonal polynomials for numerical calculation of the radiative flux is examined by Shmiglevskiy [118]. Certain difficulties in the use of this method appear when approximating the spectral absorption coefficient by orthogonal polynomials.

Finite-difference computational schemes are used for determining the radiative heat flux in media with very complex spectral optical properties [119, 120].

As of now spectral radiation can be calculated exactly only in the case of relatively simple geometries of the problem. Extensive use is made of methods of calculation for an infinite plane-parallel layer [4, 80, 87, 88, 121] or for a cylindrical volume of infinite length [119, 120, 122–124] in the one-dimensional approximation.

In this study the finite-difference computational scheme is used both for media with continuum radiation and for those with a band-type spectrum. When the optical properties of the medium are highly selective, the total radiative heat flux is determined by dividing the spectrum into $(N-1)$ intervals of arbitrary length. Each of these intervals is additionally subdivided into $[L(n)-1]$ equal parts. The use of the method of rectangles for integrating over the spectrum allows us to obtain the expression

$$q_r = \int_0^\infty q_{r,\omega}\, d\omega = \sum_{n=1}^{N-1} \frac{\omega_n - \omega_{n-1}}{L(n)-1} \sum_{l=1}^{L(n)-1} q_{r,\omega} \tag{2.107}$$

In the one-dimensional approximation of radiation in a planar layer between infinite parallel walls (Fig. 2.3), one neglects the effect of the longitudinal temperature gradient as compared with the temperature gradients in the $y$ direction. This allows replacing integration over azimuthal angle $\varphi$ in

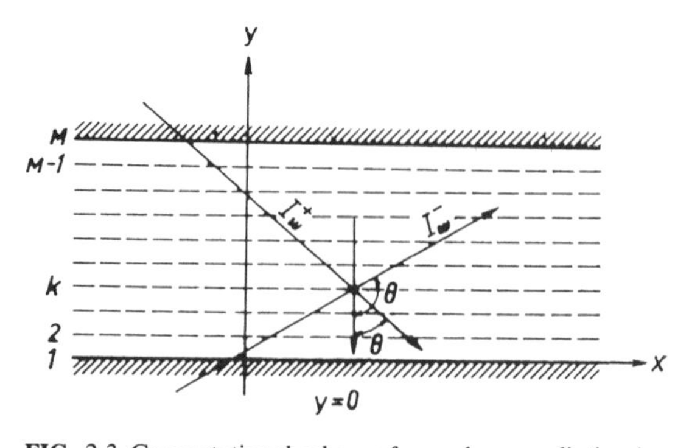


FIG. 2.3 Computational scheme for a planar radiating layer between infinite plates.

Eq. (2.18) by calculation of the exponential integral function,

$$E_n(\tau) = \int_0^1 z^{n-2} \exp(-\tau/z)\, dz \tag{2.108}$$

Then the spectral radiative flux along the $y$ coordinate will be expressed as [80]

$$q_{r,\omega}(y) = q^w_{r,\omega}(y) + 2\pi \int_y^H B(\tau_1) E_2(\tau_y^{y_1})\, d\tau_1 -$$

$$-2\pi \int_0^y B(\tau_1) E_2(\tau_{y_1}^y)\, d\tau_1 \tag{2.109}$$

where the spectral optical thickness for continuum and line radiation is given by the expression

$$\tau_{y_1}^y = \int_{y_1}^y \varkappa\, dy_2 \cong \sum_{i=j}^{k-1} \varkappa_i \Delta y_i \tag{2.110}$$

i.e., upon integration along the beam by the method of rectangles with incorporation of the spectral absorption coefficient $\varkappa_i$. The dependence of media with band spectrum on the spectral optical thickness was examined in Chapter 1.

Quantity in $f^w_{r,\omega}(y)$ in Eq. (2.109) corresponds to the radiant flux at a point $k$ due to radiation of the enclosures. To determine this quantity it is convenient to subdivide the spectral radiation intensity into two components: that related to upward directed beams, $I^-$, for which angle $\theta$ takes values from $\pi/2$ to $\pi$, and that associated with downward directed beams, $I^+$, for which angle $\theta$ ranges from 0 to $\pi/2$ (Fig. 2.3). Then the flux in the case of perfectly black enclosures ($a_\omega = 1$) due to their radiation is determined from the equation

$$\mathbf{q}^w_{r,\omega}(y) = 2\pi [I_H^+ \cdot E_3(\tau_y^H) - I_0^- \cdot E_3(\tau_0^y)] \tag{2.111}$$

In the more general case of diffusely reflecting and radiating walls, with allowance for attenuation and reflection of various beams, the magnitude of this flux is given by the expression [80]

$$q^w_{r,\omega}(y) = \frac{2\pi}{1 - 4r_{\omega,H} \cdot r_{\omega,0} \cdot E_3^2(\tau_0^H)} \Big\{ e_{\omega,H} \cdot I_H^+ \cdot E_3(\tau_y^H) - e_{\omega,0} \cdot I_0^- \cdot E_3(\tau_0^y) +$$

$$+ e_{\omega,0} \cdot I_0^- \cdot E_3(\tau_0^H) \cdot 2r_{\omega,H} \cdot E_3(\tau_y^H) - e_{\omega,H} \cdot I_H^+ \cdot E_3(\tau_0^H)\, 2r_{\omega,0}\, E_3(\tau_0^y) +$$

$$+ 2r_{\omega,H} \cdot E_3(\tau_0^H - \tau_0^y) \cdot \int_0^{\tau_0^H} B(\tau_1) \cdot E_2(\tau_0^H - \tau_1)\, d\tau_1 -$$

$$-2r_{\omega,0}\cdot E_3(\tau_0^y)\cdot\int_0^{\tau_0^H} B(\tau_1)\cdot E_2(\tau_1)\,d\tau_1+$$

$$+4r_{\omega,0}\cdot r_{\omega,H}\cdot E_3(\tau_0^H)\cdot E_3(\tau_y^H)\int_0^{\tau_0^H} B(\tau_1)\,E_2(\tau_1)\,d\tau_1-$$

$$\left.-4r_{\omega,0}\cdot r_{\omega,H}\cdot E_3(\tau_0^H)\cdot E_3(\tau_0^y)\cdot\int_0^{\tau_0^H} B(\tau_1)\,E_2(\tau_0^H-\tau_1)\,d\tau_1\right\}\tag{2.112}$$

Equation (2.112) does not incorporate the effect of reflected beams on the optical thickness of the layer. Allowance for this phenomenon resulted in more complex equations [125].

In numerical evaluation of integrals in Eqs. (2.109) and (2.112), the layer thickness was subdivided into $(M-1)$ parts of different length in such a manner that within each interval the intensity of radiation $B(\tau_i)$ of an ideal black body can be regarded as constant. Then Eq. (2.109) becomes

$$q_{r,\omega}(y)=q^w_{r,\omega}(y)+2\pi\sum_{i=k}^{M-1}\int_{\tau_i}^{\tau_{i+1}} B(\tau_1)\cdot E_2(\tau_k^i)\,d\tau_1-$$

$$-2\pi\sum_{i=1}^{k-1}\int_{\tau_i}^{\tau_{i+1}} B(\tau_1)\cdot E_2(\tau_i^k)\,d\tau_1\tag{2.113}$$

Since the value of $B(\tau_1)$ changes little within one layer, analytic integration of exponential function $E_2(\tau)$ over the thickness of the layer in each of the segments in Eq. (2.113) yeilds the expression

$$q_{r,\omega}(k)=q^w_{r,\omega}(k)+2\pi\sum_{i=k}^{M-1} B_i\cdot[E_3(\tau_k^i)-E_3(\tau_k^{i+1})]-$$

$$-2\pi\sum_{i=1}^{k-1} B_i\cdot[E_3(\tau_{i+1}^k)-E_3(\tau_i^k)]\tag{2.114}$$

where the Planck function was defined as the mean value at the boundaries of the layer:

$$B_i=(B_i+B_{i+1})/2\tag{2.115}$$

An analogous transformation of integrals over the thickness of the layer can be performed also for Eq. (2.112). Then the flux due to enclosure radiation is

$$q^{w}_{r,\omega}(k) = \frac{2\pi}{1-4r_{\omega,H}\cdot r_{\omega,0}\cdot E_3^2(\tau_1^M)}\cdot\Big\{ e_{\omega,H}\cdot I_H^+\cdot E_3(\tau_k^M) -$$
$$- e_{\omega,0}\cdot I_0^-\cdot E_3(\tau_1^k) + e_{\omega,0}\cdot I_0^-\cdot E_3(\tau_1^M)\cdot 2r_{\omega,H}\cdot E_3(\tau_k^M) -$$
$$- e_{\omega,H}\cdot I_H^+\cdot E_3(\tau_1^M)\cdot 2r_{\omega,0}\cdot E_3(\tau_1^k) +$$
$$+ 2r_{\omega,H}\cdot E_3(\tau_k^M)\cdot \sum_{i=1}^{M-1} B_i\cdot[E_3(\tau_{i+1}^M) - E_3(\tau_i^M)] -$$
$$- 2r_{\omega,0}\cdot E_3(\tau_1^k)\cdot \sum_{i=1}^{M-1} B_i\cdot[E_3(\tau_1^i) - E_3(\tau_1^{i+1})] +$$
$$+ 4r_{\omega,0}\cdot r_{\omega,H}\cdot E_3(\tau_1^M)\, E_3(\tau_k^M)\cdot \sum_{i=1}^{M-1} B_i\cdot[E_3(\tau_1^i) - E_3(\tau_1^{i+1})] -$$
$$- 4r_{\omega,0}\cdot r_{\omega,H}\cdot E_3(\tau_1^M)\cdot E_3(\tau_1^k)\cdot \sum_{i=1}^{M-1} B_i\cdot[E_3(\tau_{i+1}^M) - E_3(\tau_i^M)]\Big\} \quad (2.116)$$

For numerical calculation of radiative heat flux from Eqs. (2.114) and (2.116) one must have convenient expressions for determining the exponential integral functions. Exponential integral function $E_3(\tau)$ was expressed in terms of $E_1(\tau)$ by familiar expressions. Quantity $E_1(\tau)$ in its turn was determined from approximations due to Abramowitch and Stegun [126], with an error $[\varepsilon(\tau)] < 2\times10^{-7}$ over the range from $\tau$ to 0 and with an error of $[\varepsilon(\tau)] < 5\times10^{-5}$ at the other values of the argument.

In the case of an infinite cylindrical volume filled by a radiating and absorbing medium, integration over angle $\varphi$ should be replaced by integration over angle $\gamma$ (Fig. 2.4). A method for such an approximation was

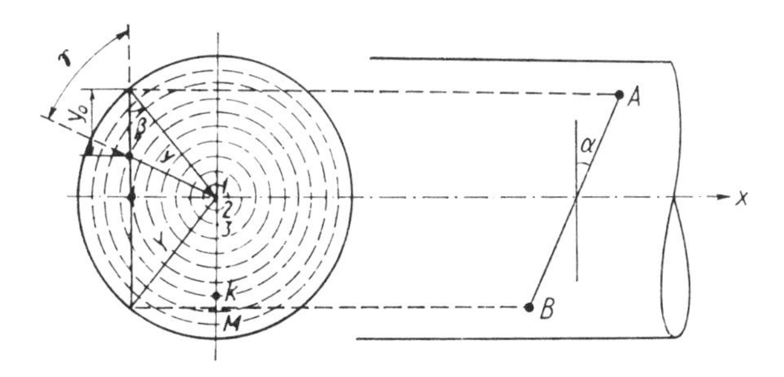


**FIG. 2.4** Computational scheme for a cylindrical volume of an infinitely long radiating medium according to Kesten [122].

suggested by Kesten [122], and was subsequently used by Gianaris [119], Lee [120], Habib and Greif [123], and Edwards and Wassel [124]. Unlike a planar layer, the radiative heat flux for a cylindrical volume has a more complex expression:

$$q_{r,\omega}(y) = 4 \int_{\gamma=0}^{\gamma=\pi/2} \cos\gamma \cdot (q_1 + q_2 + q_3 + q_4 + q_5)\, d\gamma \tag{2.117}$$

where

$$q_1 = I_Y \cdot D_3 \left(\tau^{Y}_{y\sin\gamma} + \tau^{y}_{y\sin\gamma}\right) \tag{2.118}$$

$$q_2 = -I_Y \cdot D_3 (\tau^{Y}_{y}), \tag{2.119}$$

$$q_3 = -\int_{y}^{Y} D_2 (\tau^{y_1}_{y})\, B(y_1)\, d\tau^{y_1}_{y} \tag{2.120}$$

$$q_4 = \int_{y\sin\gamma}^{Y} D_2 \left(\tau^{y}_{y\sin\gamma} + \tau^{y_1}_{y\sin\gamma}\right) B(y_1)\, d\tau^{y_1}_{y} \tag{2.121}$$

$$q_5 = \int_{y\sin\gamma}^{y} D_2 (\tau^{y}_{y_1})\, B(y_1)\, d\tau^{y_1}_{y} \tag{2.122}$$

Equations (2.117)–(2.122) employ the following symbolic expressions for the optical thickness and its derivative:

$$\tau^{y_j}_{y_i} = \int_{y_i}^{y_j} d\tau^{y_1}_{y_i} = \int_{y_i}^{y_j} \frac{\varkappa\, dy_1}{\sqrt{1 - \left(\frac{y}{y_1}\right)^2 \cdot \sin^2\gamma}} = \int_{y}^{y_j} \varkappa\, d\sqrt{y_1^2 - y^2 \sin^2\gamma} \tag{2.123}$$

$$\frac{d\tau^{y_1}_{y}}{dy_1} = \frac{\varkappa}{\sqrt{1 - \left(\frac{y}{y_1}\right)^2 \cdot \sin^2\gamma}} = \varkappa \frac{d\sqrt{y_1^2 - y^2 \cdot \sin^2\gamma}}{dy_1} \tag{2.124}$$

where $y_i$ and $y_j$ correspond to limits of integration over the radius of the cylindrical volume.

Equations (2.117)–(2.122) were obtained with allowance for symmetry of the infinitely long cylindrical volume and contain integration over the radius and angle $\gamma$ (Fig. 2.4). In the one–dimensional approximation, as in the case of a planar layer, the temperature gradient along the $x$ coordinate can be neglected. In the presence of cylindrical symmetry use was made of the integral function

$$D_n(\tau) = \int_0^{\pi/2} (\cos^{n-1}\alpha) \cdot \exp(-\tau/\cos\alpha)\, d\alpha \tag{2.125}$$

which in numerical calculations was first approximated by polynomials that yield a sufficiently high accuracy (to the fifth decimal point) [127]. Detkov et al. [128] present approximations of this function in the form of fractional polynomials.

Subsequently, as in the case of a planar layer, one subdivides the radius into $(M-1)$ parts of arbitrary length. Since $B(y)$ changes little within each annular section, analytic integration of function $D_2(\tau)$ over the layer thickness in each of the annular sections makes it possible to represent the radiative heat flux components (2.118)–(2.122) in a cylindrical space in the form:

$$q_1 = I_Y \cdot D_3\left(\tau^Y_{y\sin\gamma} + \tau^y_{y\sin\gamma}\right) \tag{2.126}$$

$$q_2 = -I_Y \cdot D_3\left(\tau^Y_y\right) \tag{2.127}$$

$$q_3 = -\sum_{j=k}^{M-1} B_j \cdot \left[D_3\left(\tau^{y_j}_y\right) - D_3\left(\tau^{y_{j+1}}_y\right)\right] \tag{2.128}$$

$$q_4 = B_{j_0} \cdot \left[D_3\left(\tau^y_{y\sin\gamma}\right) - D_3\left(\tau^y_{y\sin\gamma} + \tau^{y_{j_0}}_{y\sin\gamma}\right)\right] +$$

$$+ \sum_{j=j_0}^{M-1} B_j \cdot \left[D_3\left(\tau^y_{y\sin\gamma} + \tau^{y_j}_{y\sin\gamma}\right) - D_3\left(\tau^y_{y\sin\gamma} + \tau^{y_{j+1}}_{y\sin\gamma}\right)\right] \tag{2.129}$$

$$q_5 = B_{j_0} \cdot \left[D_3\left(\tau^y_{y_{j_0}}\right) - D_3\left(\tau^y_{y\sin\gamma}\right)\right] +$$

$$+ \sum_{j=j_0}^{k-1} B_j \cdot \left[D_3\left(\tau^y_{y_{j+1}}\right) - D_3\left(\tau^y_{y_j}\right)\right] \tag{2.130}$$

Equation (2.117) is integrated over angle $\gamma$ using the five-point Gauss scheme [115], whereas optical thickness $\tau$ is integrated over the radius by the method of rectangles. The integrals of optical thicknesses, the limits of which are a function of angle $\gamma$, are rounded off to the nearest value of radius $y_{j0}$ satisfying the condition

$$y_{j_0-1} < y\sin\gamma \leqslant y_{j_0} \tag{2.131}$$

on the assumption that in any case $j_0 \geqslant 2$. The optical thickness over the range from $y\sin\gamma$ to $y_{j0}$ is obtained by analytic integration of function $D_2(\tau)$. In this form, Eqs. (2.126)–(2.130) can already be used for calculating the spectral radiative heat flux in a medium for which the spectral absorption coefficient $\varkappa$ is known, although Eq. (2.123) for the optical thickness of the layer allows performing further transformations. Assuming that absorption coefficient $\varkappa_i$ for each of the annular sections is constant, it is possible to write the optical thickness of the layer from $y$ to $Y$ in the form

$$\tau_y^Y = \sum_{i=k}^{M-1} \varkappa_i \, \Delta y_i \tag{2.132}$$

where $$\Delta y_i = \sqrt{y_{i+1}^2 - y^2 \cdot \sin^2\gamma} - \sqrt{y_i^2 - y^2 \cdot \sin^2\gamma} \tag{2.133}$$

The expressions for other optical thicknesses contained in Eqs. (2.126)–(2.130) can be transformed analogously. Thus, in general an arbitrary thickness of layer from $y \sin\gamma$ to $y_l$ is given by

$$\tau_{y\sin\gamma}^{y_l} = \tau_{y\sin\gamma}^{y_{j_0}} + \tau_{y_{j_0}}^{y_l} = \varkappa_{j_0} \sqrt{y_{j_0}^2 - y^2 \sin^2\gamma} + \sum_{i=j_0}^{l-1} \varkappa_i \, \Delta y_i \tag{2.134}$$

The above expressions of optical thickness of the layer show that at any values of angle $\gamma$, with the exception of the case of $\gamma = 0$, when

$$\Delta y_i = y_{i+1} - y \tag{2.135}$$

the optical thickness of the layer is a function of the beam direction. This means that in calculating radiation in molecular gases one cannot use Eqs.(1.61) and (1.62) for optical thickness derived only for a beam in a single direction $y$. According to Zachor [129], the optical thickness of a layer of molecular gases can be defined as the formal vector sum of expressions of reciprocals of squares of optical thickness in the approximation of weak and strong lines:

$$\frac{1}{\tau^2} = \frac{1}{\tau_w^2} + \frac{1}{\tau_{st}^2} \tag{2.136}$$

We shall express the optical thickness of the layer $\tau_w$ in the weak-line approximation for molecular gases in the form

$$\tau_{w\ y\sin\gamma}^{y_l} = C_{j_0} \cdot \sqrt{y_{j_0}^2 - y^2 \cdot \sin^2\gamma} + \sum_{i=j_0}^{l-1} C_i \, \Delta y_i \tag{2.137}$$

and in the approximation for strong spectral lines,

$$\tau_{st \cdot y\sin\gamma}^{y_l} = \sqrt{\beta_{j_0} C_{j_0} \sqrt{y_{j_0}^2 - y^2 \sin^2\gamma} + \sum_{i=j_0}^{l-1} \beta_i C_i \, \Delta y_i} \tag{2.138}$$

Then, using Eq. (2.136), the optical thickness of a cylindrical layer of molecular gases will be determined from the expression

$$\tau^{y_l}_{y\sin\gamma} = \frac{\tau^{y_l}_{W\ y\sin\gamma}}{\sqrt{1+\left(\dfrac{\tau^{y_l}_{W\ y\sin\gamma}}{\tau^{y_l}_{st\ y\sin\gamma}}\right)^2}} \tag{2.139}$$

For greater accuracy the spectral intensity of radiation in Eqs. (2.128)–(2.130) must be determined from Eq. (2.115) as a mean for each of the annular sections. The spectral absorption coefficient $\varkappa$ and the band parameters should be defined analogously:

$$\varkappa_l = (\varkappa_{l+1} + \varkappa_i)\,/\,2 \tag{2.140}$$

$$C_i = (C_{l+1} + C_l)\,/\,2 \tag{2.141}$$

$$\beta_i\, C_i = (\beta_{l+1}\, C_{l+1} + \beta_i\, C_i)\,/\,2 \tag{2.142}$$

The above expressions for an infinite planar and cylindrical layer should be used in investigating relationships governing radiative energy transfer, although in many cases from the point of view of geometry such conditions do not conform to situations encountered in industrial heating facilities. In general the radiative flux to a given area $F$ in a closed space of arbitrary shape must be calculated from Eq. (2.18). Naturally, in such cases one must seek an approximate solution of the problem.

The simplest solution of the problem is obtained in the case when it is assumed that the integral intensity of radiation is independent of the solid angle. Then analytic integration of Eq. (2.18) yields

$$q_r = \int\limits_{2\pi} \hat{I} \cos\Theta\, d\Omega = \pi \hat{I} \tag{2.143}$$

More exact expressions for the magnitude of radiative heat flux within an arbitrarily shaped space can be obtained by making a certain, although approximate, allowance for the solid angle dependence of the radiation intensity. According to Eq. (2.15), the integral intensity of radiation at a wall at arbitrary beam direction with allowance for natural radiation of the enclosures is given by the expression

$$\hat{I}(\Theta,\ \varphi) = \frac{1}{\pi} \cdot q_r^W(0) + \int\limits_0^\infty \int\limits_0^R B(r) \cdot \exp\left(-\int\limits_0^r \varkappa\, dr_1\right) \varkappa\, dr\, d\omega \tag{2.144}$$

For an arbitrarily shaped space (Figs. 2.5) filled with an isothermal or close-to-isothermal medium, we can assume that the radiation intensity basically depends on the geometric beam length $R(\Theta, \phi)$. We expand

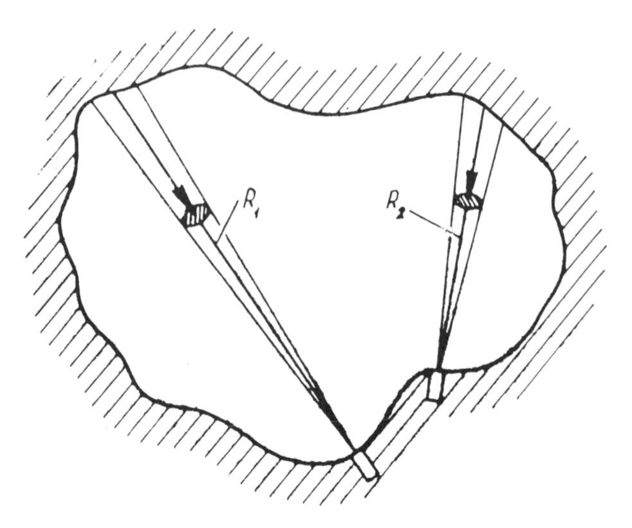


**FIG. 2.5** For calculating radiation in a closed space of arbitrary shape filled by a uniform gas.

$I(\Theta, \phi)$ in a Taylor series in the vicinity of $R=R_1$ and retain the linear term. Then

$$\hat{I}(R) \simeq \hat{I}(R_1) + \left.\frac{d\hat{I}(R)}{dR}\right|_{R=R_1} \cdot (R-R_1) \tag{2.145}$$

Replacement of differentiation by finite differences yields

$$\hat{I}(R) \simeq \hat{I}(R_1) + [\hat{I}(R_2) - \hat{I}(R_1)] \cdot \frac{R-R_1}{R_2-R_1} \tag{2.146}$$

Substituting Eq. (2.146) into (2.18) and writing

$$R_e = \frac{1}{\pi F} \int_F \int_0^{2\pi} \int_0^{\pi/2} R(\Theta, \varphi) \cdot \sin\Theta \cdot \cos\Theta \, d\Theta \, d\varphi \, dF \tag{2.147}$$

we obtain an expression for the radiant flux from a radiating and absorbing medium in an arbitrarily shaped space [131].

$$q_r = \pi \frac{\hat{I}(R_1) \cdot (R_2 - R_e) + \hat{I}(R_2) \cdot (R_e - R_1)}{R_2 - R_1} \tag{2.148}$$

This means that in order to approximately determine the total radiant heat flux within a space of arbitrary shape, it suffices to determine the intensity of integral radiation in two directions, corresponding to different beam lengths $R_1$ and $R_2$ and the effective beam length $R_e$, which, for a space of any geometric shape, can be calculated from Eq. (2.147), for example, by the Monte Carlo method [130].

The integral radiation intensity $\hat{I}$ used in Eq. (2.148) is independent of the solid angle, and consequently it can be determined from expressions for radiation in a hemispheric layer. Since

$$q_{r,\hat{h}} = \pi \hat{I} \tag{2.149}$$

Eq. (2.107) can be used for determining the radiant heat flux $q_{r,h}$ in the hemispherical layer. Here the expression for the spectral radiative heat flux simplifies significantly:

$$q_{r,h,\omega} = q^{w}_{r,h,\omega}(0) + \pi \sum_{i=0}^{M-1} B_i [\exp(-\tau_1^i) - \exp(-\tau_1^{i+1})] \tag{2.150}$$

It was assumed, in obtaining Eq. (2.150) as in the case of a planar layer, that the Planck function changes less in each $i$th layer than the exponent as a function of the optical layer thickness.

In the case of black-body enclosures,

$$q^{w}_{r,h,\omega}(0) = \pi \cdot [I_H^+ \cdot \exp(-\tau_1^M) - I_0^-] \tag{2.151}$$

For selectively emitting and reflecting surfaces, the spectral radiative heat flux due to radiation of the enclosures can be determined from Eq. (2.116) at $k = 1$, replacing the exponential function $E_3(\tau)$ in it by the expression

$$E_3(\tau) = \frac{1}{2} \exp(-\tau) \tag{2.152}$$

Equation (2.148) is suitable also for calculating a gray medium located between two arbitrarily arranged surfaces with different emission characteristics and with specified distribution of parameters. This makes it possible to obtain a technique for calculating heat-engineering facilities with allowance for the spectral properties of the medium and the enclosures. Examples of such calculations are presented in subsequent chapters.

CHAPTER

# THREE

# EXPERIMENTAL TECHNIQUE

Heat transfer, including convective energy transfer in turbulent flows and radiation, cannot at present be investigated solely analytically and must be checked experimentally. For this reason a great deal of attention is paid at our Institute to the experimental study of heat transfer from combustion products not only to walls of channels (internal problem) but also under conditions of external flow over bodies (plates, single cylinder in crossflow, tube banks, etc.). The initial experimental studies, performed in 1963 with equipment having a low thermal rating, were concerned with the study of the principal processes controlling heat transfer in high-temperature flows and were rather exploratory in nature [132–134]. In conjunction with the importance of experimental studies for solving a number of practical problems associated with design of industrial channels for MHD generators and other high-temperature equipment, an engineering basis was developed at our Institute for investigating various processes of heat transfer in high-temperature flows. The investigations were directed mainly toward obtaining experimental data on combined heat transfer from cooling of dissociated high-temperature combustion products upon significant changes in the thermophysical properties of the flow.

## 3.1. EXPERIMENTAL STANDS

Most of the experimental studies were performed on two open-loop stands that employed the products of combustion of natural gas as the working fluid. One of these stands, the ADA–2, rated at 1000 kJ/s, generated a high-temperature flow of combustion products of natural gas with oxygen or with an air–oxygen mixture. The stand was made up of fuel, air, oxygen, and cooling-water supply systems and also of a remote-control system and measuring instrumentation (Fig. 3.1).

The natural gas was supplied from the medium-pressure (up to 3 atm) municipal gas supply system. When necessary, the pressure could be raised by compressor to 6 atm.

The combustion air was supplied through a type RK-1500 M pressure reducer, from an air-line providing for a flow rate of up to 1500kg/h. The air could be heated to 850 K by two consecutively connected electrical heaters with a total rating to 200 kW.

Oxygen was supplied from a receiver station through a type DKR-500 pressure reducer; the oxygen flow rate amounted to 500 kg/h for 6 h of continuous operation at maximum flow rates.

The required stoichiometric flow rate of fuel and oxidant in experiments was determined by remote-controlled flow rate regulators, installed on the natural-gas, oxygen, and air lines.

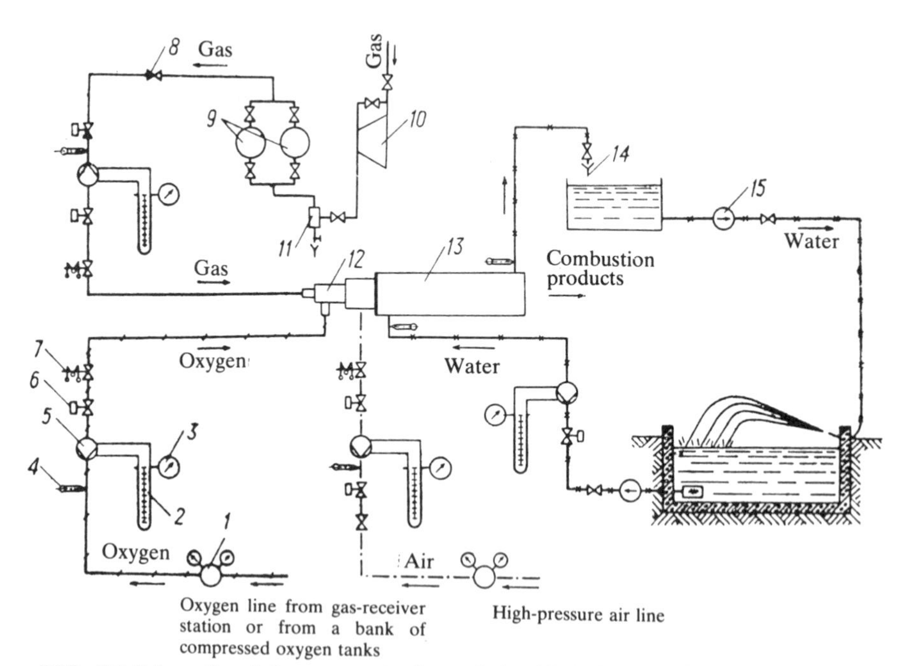


**FIG. 3.1** Schematic of the communications of the ADA–2 stand. (1) Pressure reducer; (2) differential manometer; (3) pressure gage; (4) flow-temperature measuring device; (5) flow-metering orifice; (6) remote-controlled valve; (7) solenoid cut-off valve; (8) valve; (9) receivers; (10) compressor; (11) oil trap; (12) burner device; (13) combustion chamber; (14) volumetric water flowrate meter; (15) pump.

To ensure safe and convenient performance of experiments, the control of the stand and the recording of readings of the principal monitoring instrumentation were concentrated at a remote-control panel, placed in separate premises. The premises in which the experiments were performed were equipped by an air influx and exhaust system. Provisions were made for emergency stopping of the stand by automatic shutoff of the combustible mixture should the cooling water supply cease.

The water cooling system consisted of a tank, piping, delivery and extraction pumps, and remote-controlled flow rate regulators. The cooling water flow rate was 20 $m^3$/h.

The experiments were performed with combustion chambers of various designs, developed at our Institute. All the combustion chambers and test sections were water-cooled. The heat losses were reduced by using thick-walled combustion chambers, since lining of the inner surface of the combustion chamber with heat-resistant ceramic materials made of zirconium or aluminum oxides did not ensure sufficient mechanical strength upon sudden temperature changes.

Special burning devices (Fig. 3.2) were used for combusting natural gas with oxygen–air mixtures.

Completeness of combustion was achieved by using specially designed burners, which ensured intensive mixing of the fuel and oxidant supplied to the combustion chamber in the form of intersecting jets.

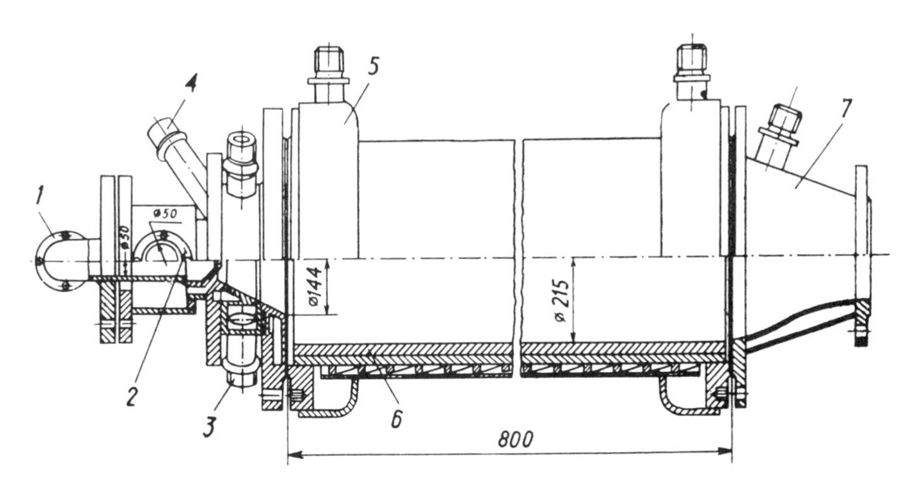


**FIG. 3.2** Gas-burning device with combustion chamber for the ADA–2. (1) Natural-gas supply connection; (2) air-supply connection; (3) oxygen-supply connection; (4) connection for supplying cooling water for the gas-burning device; (5) combustion chamber cooling water jacket; (6) thin-walled refractory-steel rings for reduction of heat losses in combustion chamber; (7) removable nozzle with transition from rectangular to circular cross-section conforming to the test-loop dimensions.

The cylindrical combustion chamber was connected to the rectangular test loops by means of a Witozynski-profiled water-cooled transition piece [135]. The degree of contraction of the transition piece was 0.0848 for the 50 × 80 mm cross-section experimental section and 0.405 for the 102 × 147 mm cross-section experimental section.

The combustion chamber together with the test loop and air heater was built into a pullout hood (Fig. 3.3). The combustion products, which leave the test section through a water-cooled box, were released to the atmosphere.

The temperature of the combustion products leaving the combustion chamber changed as a function of the temperature rise of the air and its oxygen content. The flow temperature leaving the combustion chamber ranged from 1550 to 2650 K, and the flow velocity in the test loop was up to 400 m/s.

Prior to running the main series of experiments, we determined the combustible mixture flow modes at which the combustion chamber was capable of stable operation at oxidant excess ratios from 1.05 to 1.10.

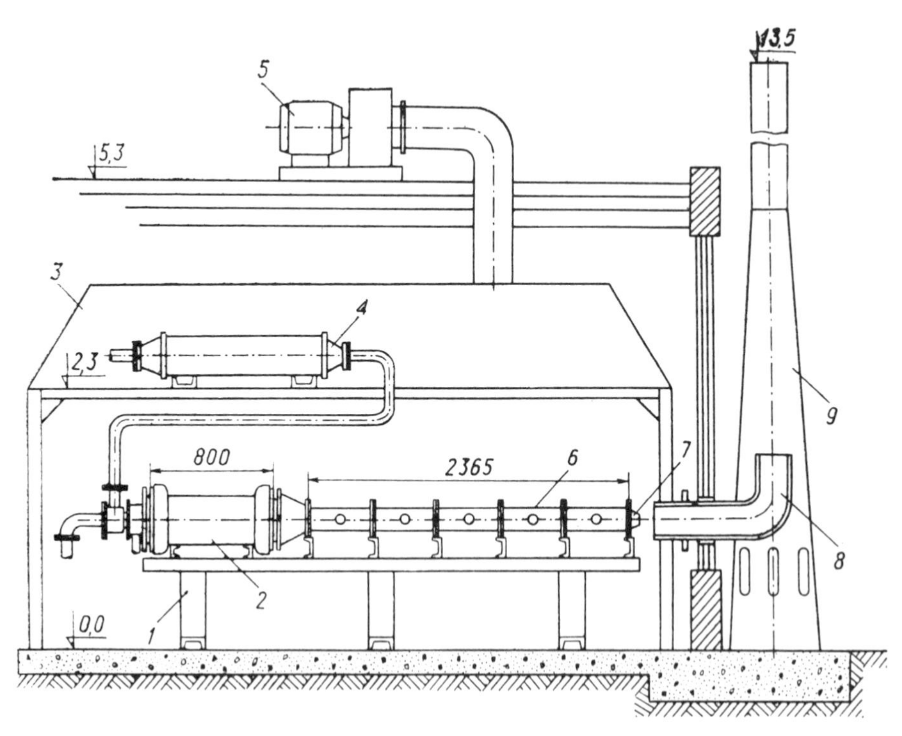


**FIG. 3.3** Arrangement of the principal component parts of the ADA-2 stand. (1) Support frame; (2) combustion device with combustion chamber; (3) pullout hood; (4) electric air preheater; (5) exhaust fan; (6) test section; (7) convergent nozzle; (8) water-cooled box for ejection of combustion products; (9) flue.

Chemical analysis performed under these conditions by chromatography of the combustion products leaving the chamber showed that the combustion was complete within the limits of experimental error.

The ADA–2 was intended for experimental study of heat transfer of high-temperature combustion products in inlet lengths of rectangular ducts of various dimensions. The flow variables were such that it was possible to obtain experimental data on heat transfer both for convection–diffusion heat transfer with highly insignificant radiant energy transfer and also under conditions when the radiant heat flux amounted to about half of the overall heat flux. The fraction of chemical energy in convection–diffusion heat transfer amounted to 58% of the total convective heat transfer.

However, due to instability of combustion in the chamber with relatively cold walls, the ADA–2 unit was limited with respect to obtaining combustion products at lower temperatures. We thus needed a unit that would allow obtaining combustion product flows. These requirements were met by the ADA-5 unit (Fig. 3.4), designed under the leadership of Filimonov and Khrustalev at the Krzhizhanovskiy Energetics Institute and delivered under the friendly cooperation agreement with our Institute for use in experimental work. The ADA-5 was designed for burning natural gas with air, diluted by inert gases or sol particles, and in addition to the natural-gas and air supply system, it is also equipped with an analogous nitrogen supply system. Nitrogen is supplied through a type DKR-500 pressure reducer, ensuring a flow rate to 500 kg/h. Provision was made for supplying nitrogen both with air and separately for injection of sols into the combustion chamber by way of a nitrogen jet. The combustion chamber of this unit was lined by fireclay and chrome-magnesium bricks and aluminium oxide rings (Fig. 3.4). Water cooling was provided only in the upper part of the combustion chamber, where the test section, consisting of circular segments maintained at constant temperature, is situated.

The combustion chamber of the ADA-5 unit is equipped with a premix gas combustion device. The gas–air mixture is supplied to the combustion chamber through a perforated device in the form of ceramic rings. The cross-section of the opening holes was selected in such a manner that the velocity of the gas–air mixture would exceed the flame velocity. Since the combustion-chamber walls were at a temperature close to the theoretical combustion temperature, completeness of combustion was ensured at an excess air factor close to unity. The temperature of the combustion products at the exit from the combustion chamber ranged from 1500 to 1900 K.

Like the ADA–2, the ADA–5 is equipped with a remote-control and data recording system.

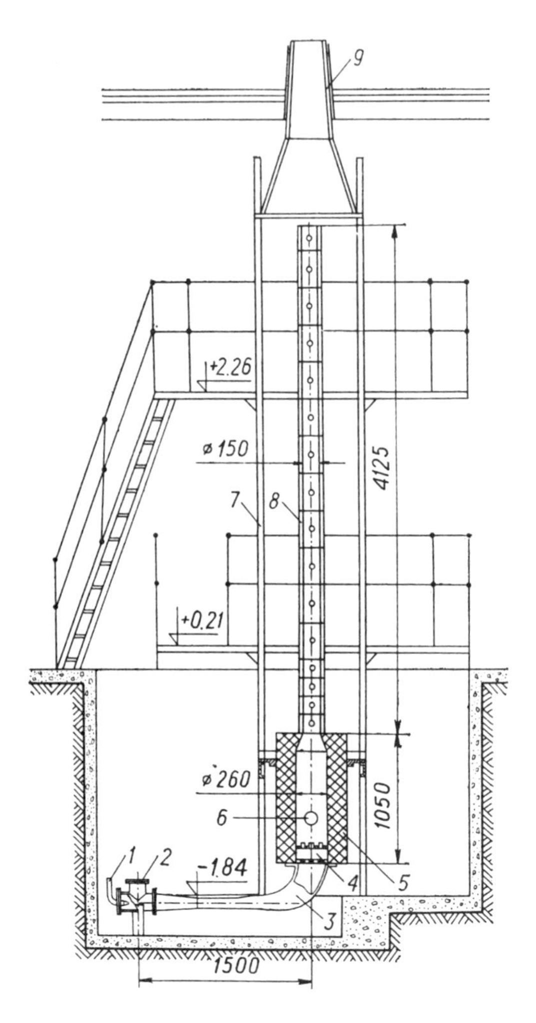


**FIG. 3.4** Arrangement of the main component parts of the ADA–5 unit. (1) Natural-gas supply connection; (2) air-supply connection; (3) mixer; (4) flame holder; (5) lined combustion chamber; (6) port for starting up the combustion chamber or for supplying an inert gas; (7) main frame with service platforms; (8) test section; (9) water-cooled pipe with cap for ejection of combustion products.

## 3.2. TEST CHANNELS

Four channels with test sections were used for investigating combined heat transfer in the flow of combustion products with different optical thickness. Two water-cooled channels, each 50×80 mm in cross-section, were used for investigating convection–diffusion heat transfer with slight radiant energy transfer. In channel number 1, made of brass, a part of one of the walls was maintained at a constant temperature separately from the remaining part of the channel, which made it possible to investigate heat transfer not only in the channel as a whole, but also at a plate in axial flow [100, 136]. Condensation of water vapor from combustion products at relatively low temperatures of the brass walls of channel number 1 interfered with performance of experiments at low velocities and temperatures. For this reason we fabricated a rectangular steel channel number 2 (Fig. 3.5) of the same cross-section, but with thicker double walls. This channel was comprised of eight milled double walls. The cooling water circulated over special channels in the double walls of the channel. The channel was subdivided over its length into 14 separately cooled segments. As in channel number 1, the cooling-water temperature was measured by copper–constantan thermocouples, placed in locations of cooling-water supply and discharge, and also in locations where water was discharged from one channel segment to another. This allowed consistent maintenance of constant temperature both along the plate and along the channel.

The temperature of the inner surface of the channel walls was measured by copper–constantan thermocouples placed at different locations along the channel.

Channel number 3, with a cross-section of 102×147 mm (Fig. 3.6), was comprised of five dismountable sections, equipped with thermocouple arrays for measuring heat losses at the different locations. The sections were provided with ports for installing radiometric transducers. This channel was

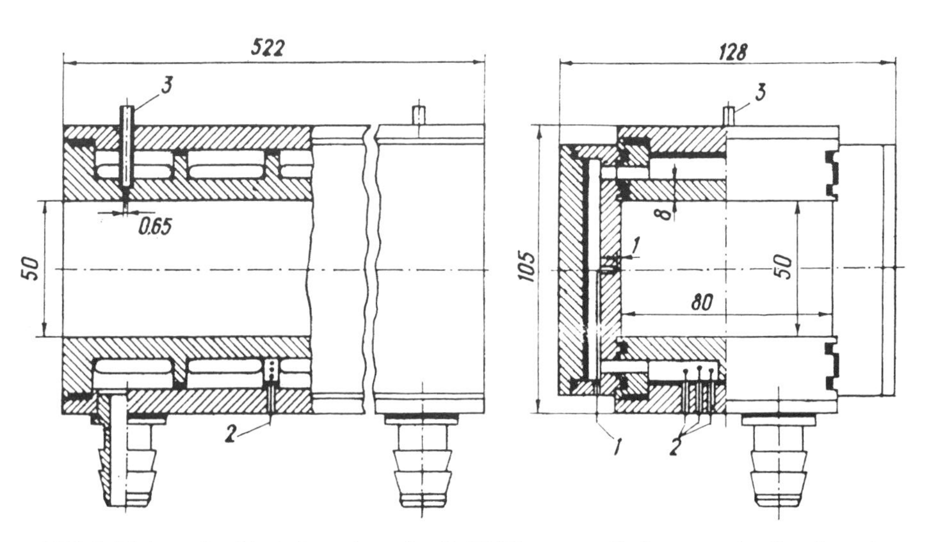


**FIG. 3.5** Schematic of test channel number 2. (1) Thermocouple for measuring the channel-wall temperature; (2) thermocouple array for measuring the cooling-water temperature upon transition from one segment to another; (3) pulse tube for pressure measurement.

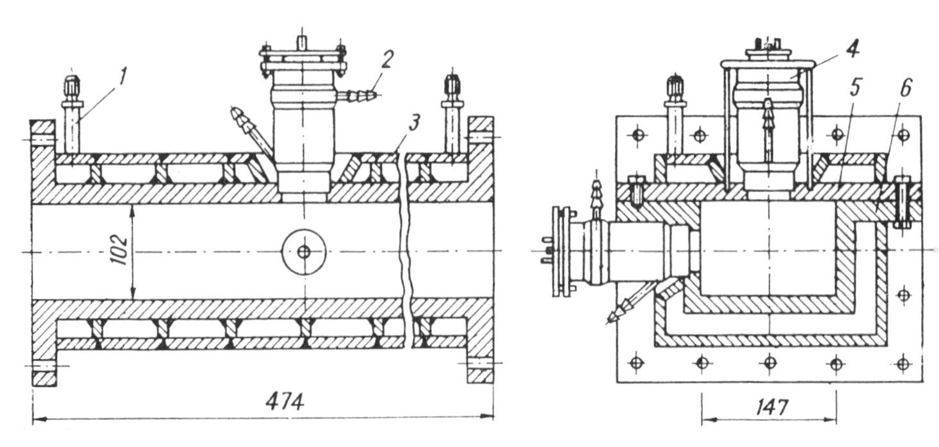


**FIG. 3.6** Schematic of sections of test channel number 3. (1) Cooling-water supply connection; (2) connection for supply of constant-temperature water for cooling of the radiometer; (3) test-channel jacket; (4) radiometers; (5) top water-cooled channel cover; (6) bottom water-cooled part of the channel.

equipped with a set of convergent exit devices, allowing raising of the pressure of the combustion products in the test section to 3 atm. This means that the optical thickness could be varied within rather wide limits.

In the rectangular channels described here it was possible to investigate heat transfer only in the inlet region of the flow.

Test section number 4, which was 4.1 m long (about 30 equivalent diameters), assembled on the ADA-5 unit of 16 constant-temperature sections of different length with inner diameter of 150 mm (Fig. 3.7), made it possible to investigate heat transfer not only of the inlet length, but also of

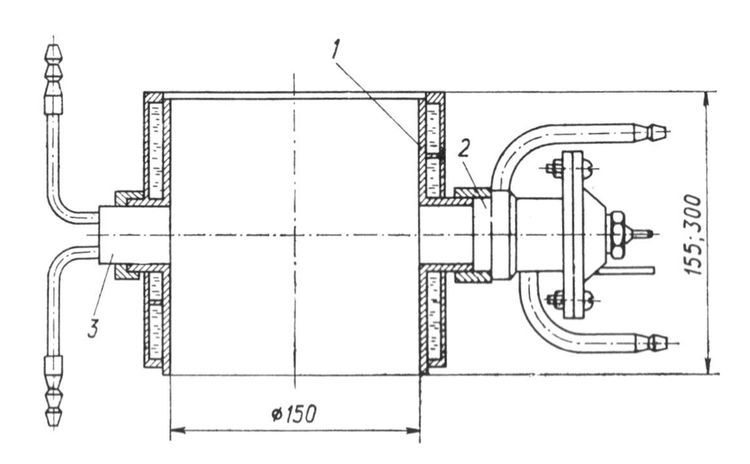


**FIG. 3.7** Schematic of sections of test channel number 4. (1) Water-cooled casing; (2) radiometer; (3) water-cooled plug.

the stabilized flow length. Each section was equipped with two thermocouples for measuring the wall temperature and also thermocouples for measuring the temperature differences in the cooling water. Each section was provided with two ports for placing radiometers or devices for measuring the velocity and temperature fields.

## 3.3 DETERMINATION OF FLOW TEMPERATURE IN THE CHANNEL

The flow temperature at the inlet to the test channel is one of the most important parameters characterizing the heat transfer. However, in investigating gas flows it is very difficult to measure. The known optical methods of measuring the temperature of gases are not always suitable for high-power test sections. It is hence necessary to use methods of measurement of flow temperatures by means of thermocouples, which are most frequently used in cases when the flow temperature is lower than the temperature of melting (or failure) of the thermocouple material. If, however, the flow temperature exceeds the thermocouple failure temperature, then one must use measurement methods based on short-term insertion of thermocouples [137].

In the first case, which is most characteristic for channel number 4, the difficulty in measuring the flow temperature consists additionally in the fact that the thermocouple is placed in a radiating and absorbing medium. Here the temperature of the thermocouple junction differs from that of the flow. This can be prevented by using two thermocouples with different junction temperature or by using shielded thermocouples (Fig. 3.8) [138]. Such thermocouples were used for measuring the flow temperature in test channel number 4.

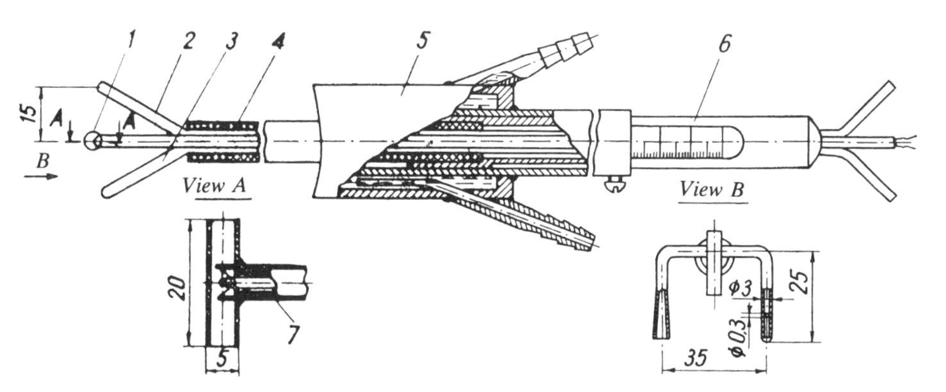


**FIG. 3.8** Schematic of combined temperature and velocity field probe. (1) Shielded thermocouple; (2) dynamic head measurement tube; (3) static head measurement tube; (4) ceramic sheath; (5) water-cooled probe casting; (6) traveling device with read-off scale; (7) ceramic thermocouple insulation.

It is even more difficult to measure the temperature of the combustion products in the ADA–2 unit when the flow temperature exceeds that at which the thermocouples fail. In this case it is best to use analytic temperature-determination methods. According to one method, one must assume that the bulk temperature of combustion products entering the test channel is equal to the combustion temperature with correction for the loss of heat in the combustion chamber. For this reason the combustion chamber of the ADA–2 unit was maintained at constant temperature by cooling water. Since the quantity that is unknown at the inlet to the test section is not only the flow temperature, but also its velocity, determination of the temperature requires solving the nonlinear equation

$$q_{org} - q_l - q_{r,b} - \frac{U^2}{2} - \int_0^{T_m} c_p \, dT = 0 \tag{3.1}$$

where $q_l$ is the heat loss in the combustion chamber at the inlet to the test section, $q_{org}$ is the heat carried into the chamber with the original combustion components, $q_{r,b}$ is the energy of the break of bonds of the organic fuel, $U_f^2/2$ is the kinetic energy of the flow entering the channel, and $\int_0^{T_m} c_p dT$ is the enthalpy of combustion products at the inlet to the channel.

Equation (3.1) was solved by computer on the assumption that the composition of the natural gas, the flow rate of the natural gas and of the oxidant, and also the temperature at the inlet to the combustion chamber are known. The accuracy in determining the temperature in this manner hinges on errors in determining the flow rates, heat losses of the combustion chamber, and, most importantly, on the assumption of completeness of combustion, which, unfortunately, cannot always be claimed to actually occur. This means that the direct measurement of temperature remains a timely problem.

## 3.4. RADIATIVE HEAT FLUX SENSORS

The total radiant flux can be measured by various radiometers [2, 86, 139, 140]. Typical designs of such radiometers employ radiation thermocouples as an indicator for sensing the radiant flux. Depending on the kind of instrument, a distinction is made between radiometers with limited viewing angle and radiometers of hemispherical radiation.

As of now there is no serial production of radiometers. Various research organizations produce and use in-house constructed thermal radiation sensors [139–141].

The radiometer equipment needed for our work was developed jointly with the Heat Measurement Division of the Engineering Thermophysics

Institute of the Ukrainian Academy of Sciences, who have accumulated a great deal of experience in constructing thermal-sensing transducers.*

The benefits of determining the radiant heat flux in configured channels by measuring the radiation intensity within narrow solid angles were listed in Chapter 2. Such measurements yield detailed information on the optical properties of the flow. In conjunction with this, the main attention in this study was paid to finding optimal designs for narrow-angle radiometers [142–144]. Of the known designs of such radiometers, we selected one with a limited viewing angle without focusing optics (Fig. 3.9). Since there were no focusing optics and glass ports in the diaphragm device, the entire radiation spectrum was received by the sensing element. Access of the radiating medium to the sensing element was prevented by a diathermal curtain, produced by blowing dry gaseous nitrogen through the diaphragm device.

Since the receiving area $F_{rec}$ of the radiation thermoelement is visible from any point $D_1D_1'$ of the input diaphragm at the same solid angle (Fig. 3.9)

$$\Omega = \frac{\pi d_1^2}{4l^2} \tag{3.2}$$

we can determine the radiant flux from a source with radiation intensity $\hat{I}$, passing through the diaphragm and reaching the sensing surface of the radiation thermoelement [145]:

$$Q_r = \Omega \hat{I} \frac{\pi d_1^2}{4} \tag{3.3}$$

On the one hand, in order to obtain information on the heat flux density in a given cross–section of the channel one should attempt to reduce the solid angle $\Omega$ by proper selection of the inlet diameter $d_1$ of the instrument's diaphragm and distance $l$ between the diaphragms. On the other hand, in order to improve the sensitivity of the instrument it is desirable to increase as

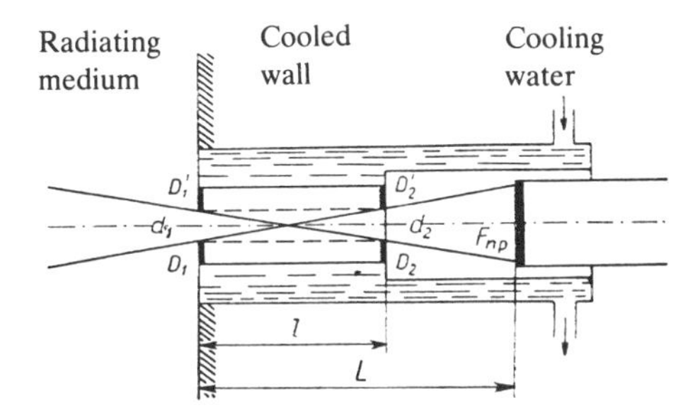


**FIG. 3.9** Operating principle of narrow-angle radiometer.

*The Engineering Thermophysics Institute of the Ukrainian Academy of Sciences was represented in the development of radiometer equipment by Professor O. Geneschenko and S. Sazhing.

far as possible the flux incident on the receiving area of the thermoelement, which can be achieved only by increasing Ω. This means that in selecting the dimensions of the input diaphragm one must satisfy two conflicting requirements.

An acceptable instrument sensitivity in this work was achieved using a narrow-angle radiometer, whose design is shown in Fig. 3.10. The blackened copper rings, placed in the inlet channel of the radiometer, prevent reflected beams from reaching the thermoelement. The radiometer casing is maintained at constant temperature by supplying cooling water from a constant-temperature bath. This radiometer design allowed using various radiation-sensing elements. In particular, this radiometer was tested out with a radiation-sensing element designed at the Leningrad Electrical Engineering Institute, and also with sensing elements designed at the Engineering Thermophysics Institute of the Ukrainian Academy of Sciences, installed in a specially water-cooled housing (Fig. 3.10). Heat flux sensors DPT-02 and DPT-05 were employed as high-sensitivity transducers [141]. One of the main shortcomings of these sensors is their long response time and inability to receive high heat fluxes (Table 3.1).

Better performance in this respect is exhibited by tapered sensing elements, made of blackened aluminum, of which the heat flux sensors were comprised. Several sensors with different cooling systems were constructed (Fig. 3.11).

The tapered radiation sensors were tested in single- and double-beam radiometers (Figs. 3.12 and 3.13). The latter allow simultaneous measurement in two beam directions. In the double-beam radiometer, unlike in the single-beam instrument, the diaphragming device channel is made by

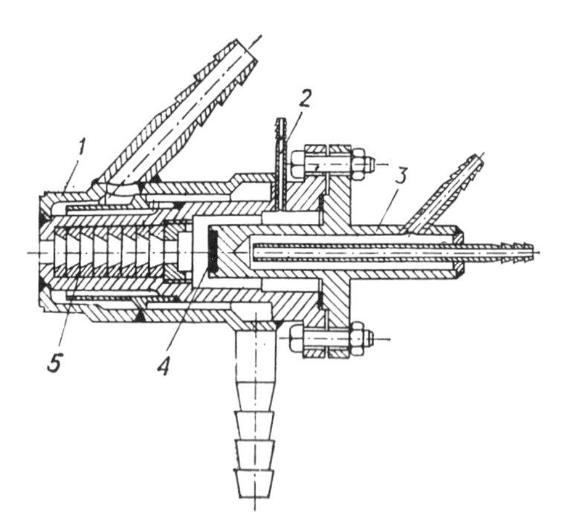


**FIG. 3.10** Schematic of narrow-angle radiometer with a DPT–02 transducer (1) Water-cooled housing of diaphragm device; (2) nitrogen-supply connection; (3) water-cooled housing of DPT–02 transducer; (4) DPT–02 transducer; (5) blackened copper rings of diaphragm device.

**TABLE 3.1** Specifications of Sensing Elements Employed in Narrow-Angle Radiometers

| Type of sensing element | Sensitivity, $MV \cdot m^2/kW$ | Integral absorptivity | Maximum permissible radiative heat flux $kW/m^2$ |
|---|---|---|---|
| Leningrad Institute of Electrical Engineering sensor | $3 \cdot 10^{-2}$ | 0.95 | 500 |
| DTP-02 | $(0.95-1.3) \cdot 10^{-2}$ | 0.95 | 500 |
| DTP-05 | $(4.0-4.7) \cdot 10^{-3}$ | 0.95 | 300 |
| Tapered heat flux sensor | $(2.9-5.0) \cdot 10^{-3}$ | 0.98 | 2000 |

cutting a standard thread on the inner surface of the steel casing of the radiometer; the thread is subsequently blackened.

Radiometers with a common system for cooling the diaphragm device and the heat flux sensors were found to be least sensitive to external perturbations.

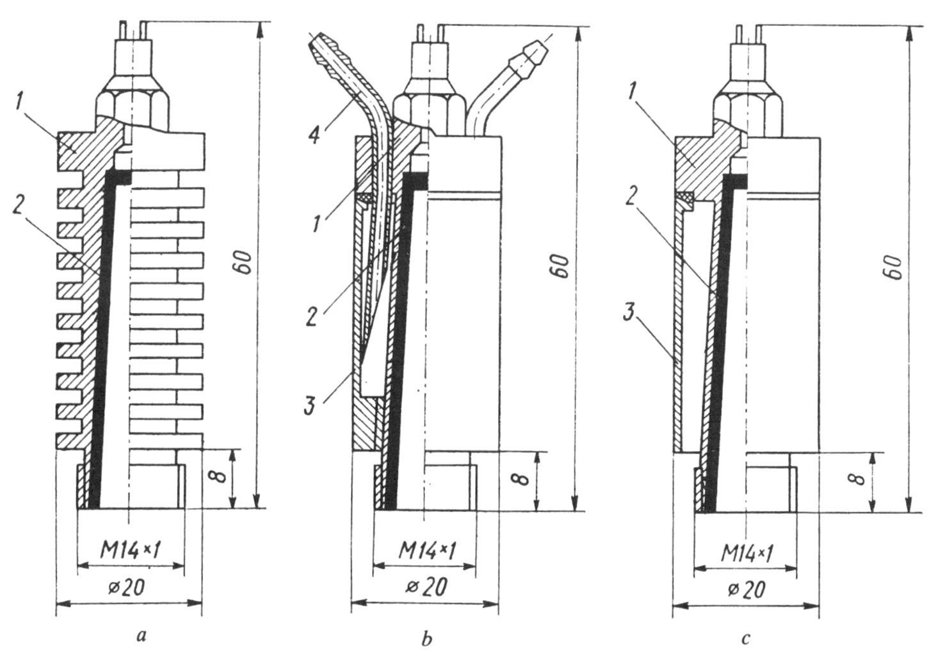


**FIG. 3.11** Schematics of radiation sensors of narrow-angle radiometers: (a) with natural cooling; (b) with individual water cooling; (c) with common cooling system. (1) Radiation sensor casing; (2) sensing element; (3) water-cooled jacket; (4) water-supply connection.

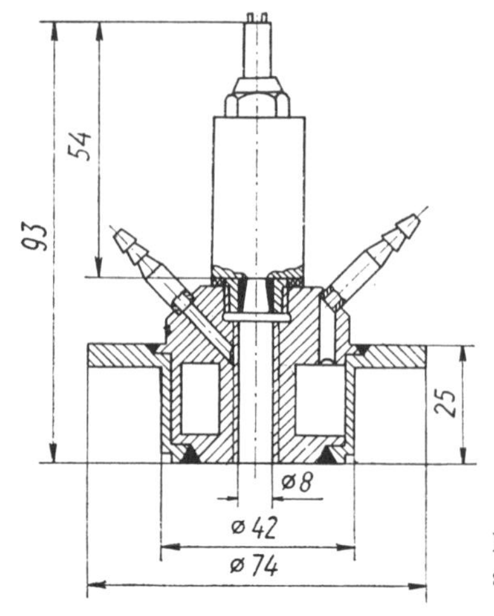


**FIG. 3.12** Schematic of diaphragm device with a single-beam sensing element.

Narrow-angle radiometers were tested out in the aforementioned test channels at heat flux densitites from 5 to 150 kW/m$^2$.

All the radiometers were calibrated before the experiments.

## 3.5. CALIBRATION OF RADIOMETERS

The narrow-angle radiometers were calibrated by the absolute method on a special stand. The thermal radiation source consisted of an ideal black body (Fig. 3.14) made of a 98–mm–diameter and 300–mm–long copper bar, in which a 60–mm–diameter hole, terminating in a cone was drilled to a depth of about 200 mm. The casing of the black body was heated by a Nichrome

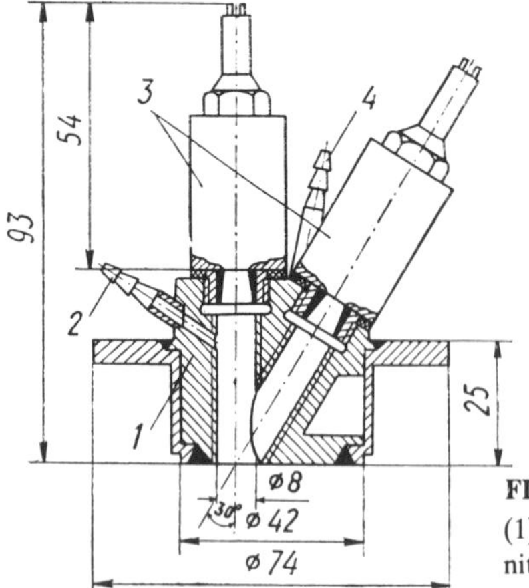


**FIG. 3.13** Schematic of double-beam radiometer. (1) Diaphragm device casing; (2) blow-down nitrogen connection; (3) radiation sensors; (4) cooling-water connection.

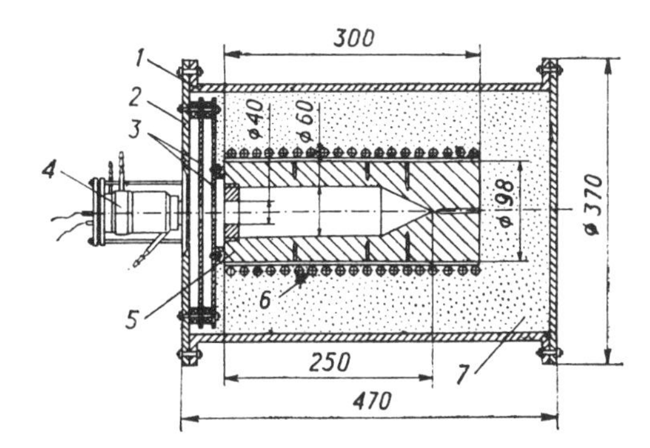


**FIG. 3.14** Schematic of ideal black body for calibrating narrow-angle radiometers. (1) Jacket, (2) bottom with device for fastening narrow-angle radiometers; (3) shields; (4) narrow-angle radiometer; (5) copper housing; (6) electric motor; (7) thermal insulation.

spiral rated at about 1.2 kW. The heating temperature was adjusted by a rheostat and monitored by six chromel–alumel thermocouples, embedded into the black-body bar. The black–body temperature was defined as the mean reading of these thermocouples, whose emf was measured by a type F–30 digital millivoltmeter.

The radiant heat flux was determined using the Stefan–Boltzmann law from the measured black-body temperature. The resulting calibration curves were used in interpreting the readings of the instrument in measuring the radiant heat flux in the test channel.

Recalibration showed that the radiometer readings are sufficiently stable.

Calibration of radiometers against an ideal black body made it possible to find the relationship between the measured signal and the radiation intensity. The radiative heat flux incident on the wall of the test channel was recalculated using the expression

$$\hat{I}=\hat{I}_m \quad (T_w)-\frac{\sigma}{\pi}\cdot T^4_{\text{rad}} \tag{3.4}$$

which makes allowance for the difference in the temperatures of the sensing element of the radiometer $T_{\text{rad}}$ and channel wall $T_w$. The latter was measured by copper–constantan thermocouples, embedded into the wall of the test channel, and the value of $T_{\text{rad}}$ was taken to be equal to that of the cooling water. Equation (3.4) was written on the assumption of unity emissivity of the channel wall.

The calibration curves for all the narrow-angle radiometers were close to linear over a rather wide range of radiative heat flux. The principal specifications of the radiometers used in this study are listed in Table 3.1.

CHAPTER

# FOUR

# CONVECTIVE HEAT TRANSFER IN HEATED GAS FLOWS

Study of combined heat transfer is impossible without gaining detailed knowledge of relationships governing heat transfer in high-temperature, weakly radiating gas flows. At high gas temperatures the relationships governing convective heat transfer are complicated by the effect of variability of physical properties of the working fluid, turbulent perturbations of flow, and also chemical changes occurring within the gas flow. A detailed study of relationships governing convective heat transfer is not the principal purpose of this study. These problems were explored elsewhere [79, 80, 82, 95–100, 146]. We shall consider only the principal features of high-temperature heat transfer, the understanding of which is needed for investigating governing relationships of combined heat transfer. Analysis of heat transfer is limited to cases of axial flow over a plate and stabilized channel flow. Such heat transfer conditions are characteristic of various elements of heat engineering equipment.

## 4.1. FEATURES OF HEAT TRANSFER IN LAMINAR FLOWS

In laminar flow, heat transfer is complicated by the effect of variability of physical properties, and also by dissociation and recombination of the medium's constituents, which are responsible for diffusion transfer of the energy of chemical properties. The effects of variability of physical properties of high-temperature flows were investigated by the present author and his co-workers [109–111]. The results of these studies are given in [100, 146], for which reason they will not be considered in detail here.

It is known that combustion products start dissociating at flow temperatures above 1900 K. In calculating heat transfer as a result of dissociation of combustion products, allowance must be made for diffusion transfer of the energy of chemical transformations. For the majority of engineering applications the dissociation–recombination of combustion products can be analyzed in the approximation of local thermodynamic equilibrium [147, 148]. Analysis of heat transfer in such a medium is best performed on the basis of the balance of energy transport to the wall,

$$q_{res} = q_c + q_{chem}, \tag{4.1}$$

according to which the resultant convection–diffusion heat flux is equal to the sum of the purely convective $q_c$ and chemical $q_{chem}$ components of heat flux. If it is remembered that convective heat transfer is driven by the temperature difference, whereas the driving force of heat transfer due to chemical transformations is the difference in energies of chemical transformations, then according to Eq. (4.1), we can obtain the following nondimensional equation:

$$Nu^*_{res,f} = Nu_f \cdot \left[1 + (\Lambda - 1) \cdot \frac{\Delta h_{chem}}{\bar{c}_p^* \cdot (T_f - T_w)}\right] \tag{4.2}$$

where $\Lambda$ is the proportionality factor between the diffusive and convective energy transports.

Equation (4.2) was written for the value of $Nu^*_{res,f}$, expressed in terms of the difference of total enthalpies. An equivalent expression for $Nu_{res,f}$, expressed in terms of the temperature difference, has the form

$$Nu_{res,f} = Nu_f \cdot \left[1 + \Lambda \frac{\Delta h_{chem}}{\bar{c}_p \cdot (T_f - T_w)}\right] \tag{4.3}$$

where $\Delta h_{chem}$ is the enthalpy difference due to chemical transformations within the flow, and $\bar{c}_p$ is the mean specific heat of the gas. The physical properties in calculating the Nusselt number were defined at the flow temperature.

Equations (4.2) and (4.3) can be used for both laminar and turbulent flows provided that $Nu_f$ and $\Lambda$ are determined in the proper manner. In the case of insignificant variation in physical properties of the gas flow $(0.2 \leqslant Pr_f \leqslant 1)$ in laminar axial flow over a plate [109],

$$Nu_f = 0.332\, Re_f^{0.5}\, Pr_f^{0.36} \tag{4.4}$$

and in turbulent flow over a plate [149] we can use the expression

$$\mathrm{Nu}_f = 0.037\ \mathrm{Re}_f^{0.8}\ \mathrm{Pr}_f^{0.43} \tag{4.5}$$

According to Fay and Riddel [150] and Probstein et al. [151], parameter $\Lambda$ is expressed as some function of the Lewis number Le. For laminar heat transfer in equilibrium-dissociated air, Fay and Riddel [150] suggested the expression

$$\Lambda = \mathrm{Le}^{0,52} \tag{4.6}$$

which was obtained on the assumption that Le is independent of the temperature.

For real gases at sufficiently high temperature differences, the physical properties of the flow change significantly. Hence the above expressions for $\mathrm{Nu}_f$ and $\Lambda$ should be supplemented by factors that make allowance for variability of physical properties of the flow.

Detailed analysis of the effect of variability of physical properties on $\mathrm{Nu}_f$ and $\Lambda$ in laminar flow of an equilibrium-dissociating binary mixture, consisting of components A and B between which is possible the chemical reaction

$$A \rightleftarrows B$$

was performed by the present author and Dagys [152], who made allowance for variability of physical properties of the reacting mixture as a function of concentration and temperature. The concentrations of flow components were determined by solving a set of algebraic equations for a thermodynamically equilibrium composition. It was assumed that the physical properties of the individual flow components are a power law function of temperature. Analysis of the effect of variability of physical properties on heat transfer was performed by numerically solving the set of boundary-layer equations (2.34).

The results of analysis for the case of laminar flow were correlated in [152] by the expression

$$\mathrm{Nu}_f = 0.332\ \mathrm{Re}_f^{0.5}\ \mathrm{Pr}_f^{0.36}\ \varphi_l \tag{4.7}$$

where $\varphi_l$ is a factor making allowance for the effect of variability of physical properties of the flow on heat transfer. This factor is frequently termed the temperature factor.

Calculation of heat transfer performed for the case of identical physical properties of both constituents showed that the expression for the temperature factor

$$\varphi_{l,A} = (g_{\rho,f,A} \cdot g_{\lambda,f,A})^{-0.28} \cdot (g_{\rho,f,A} \cdot g_{\mu,f,A})^{0.14} \cdot g_{c_p,f,A}^{-0.04} \tag{4.8}$$

remains the same as for a uniform gas [109].

In the general case of convective–diffusive heat transfer in an equilibrium-reacting binary mixture, the expression for $\varphi_l$ is supplemented by factors characterizing the variation in physical properties as a function of the completion of reaction within the flow, allowance for which can be made by the corresponding ratios of physical properties of the mixture relative to the physical properties of one of the reacting components, for example, component A:

$$\varphi_{l,mix} = \varphi_{l\ A} \cdot \left(\frac{g_{\rho,f,mix}}{g_{\rho,f,A}}\right)^{-0.27} \cdot \left(\frac{g_{c_p,f,mix}}{g_{c_p,f,A}}\right)^{0.15} \times$$

$$\times \left(\frac{g_{\lambda,f,mix}}{g_{\lambda,f,A}}\right)^{-0.41} \cdot \left(\frac{g_{\mu,f,mix}}{g_{\mu,f,A}}\right)^{0.14} \qquad (4.9)$$

Parameter $\Lambda$ of similitude of convection–diffusion energy transport is also a function of the variation in physical properties in the boundary layer:

$$\Lambda = \mathrm{Le}_w^{n_1} \cdot \left(\frac{\mathrm{Le}_f}{\mathrm{Le}_w}\right)^{n_2} \cdot g_{c_p,f,A}^{-0.52} \cdot \left(\frac{g_{c_p,f,mix}}{g_{c_p,f,A}}\right)^{-1.0} \qquad (4.10)$$

where

$$n_1 = 0.52 + 0.078 \ln \mathrm{Le}_w \qquad (4.11)$$

$$n_2 = 0.29 + 0.12 \ln \mathrm{Le}_w \qquad (4.12)$$

This means that Eq. (4.10) at constant physical properties and $\mathrm{Le}_w$ close to unity is in satisfactory agreement with Eq. (4.6).

Equations (4.9) and (4.10) are valid at $0.2 \leqslant \mathrm{Le}_w \leqslant 5.0$, $0.5 \leqslant \mathrm{Pr}_w$ 2.0, and $400 \leqslant T_f \leqslant 8000$ K upon variation in the exponents of expressions for the temperature dependence of coefficients of thermal conductivity, viscosity, specific heat, and binary diffusion from 0 to 2.

The above expressions for convective–diffusive heat transfer can also be used in the case of multicomponent reacting flow. Thus, the variation in physical properties of combustion products due to changes in the components of the mixture is insignificant [100, 153]. With allowance for this, the effect of variability of physical properties of combustion products can be incorporated by the simpler expression (4.8), and parameter $\Lambda$ can be determined from the formula

$$\Lambda = \mathrm{Le}_w^{n_1} \cdot \left(\frac{\mathrm{Le}_f}{\mathrm{Le}_w}\right)^{n_2} \cdot g_{c_p,f,mix}^{-0.52} \qquad (4.13)$$

where $n_1$ and $n_2$ are calculated from Eqs. (4.11) and (4.12), respectively.

The above expressions for laminar convective–diffusive heat transfer in combustion products are compared in this study with experimental data.

## 4.2. FEATURES OF HEAT TRANSFER IN TURBULENT FLOWS

Various experimenters found, in analyzing their data on heat transfer from a plate in axial turbulent gas flow, that this heat-transfer mode exhibits a number of specific features. Heat transfer from gas flows is affected not only by variability of the physical properties of the medium, but also by the free-stream turbulence. Experiments show that, depending on the latter, the coefficient of $Re_f$, and also the power exponent of $Re_f$ in Eq. (4.5) may vary within wide limits. In real heat engineering devices high-temperature flows are usually turbulized, so that in order to calculate heat transfer under such conditions one must have general relationships that make allowance for all the factors affecting heat transfer. At present such relationships can be obtained only analytically [105, 106].

The results of calculations of heat transfer in axial turbulent flow, using the technique presented in Section 2.3, at $Pr_T = 1$, $K_T = 0.09$, and constant physical properties are in satisfactory agreement with analytic results of Spalding, Van Driest, and Popov [155] and with experimental results obtained in the flow of air (Fig. 4.1). A more detailed comparison of analytic and experimental data is prevented by the lack in these studies of experimentally measured values of $Pr_T$ and $K_T$ which, as shown by calculations in this study, have a significant effect on heat transfer as a function of $Re_f$. The behavior of experimental results as a function of $Re_f$ can be explained only by using certain values of $K_T$ and $Pr_T$ characteristic of the conditions of these experiments.

Numerical calculations show that the variability of physical properties affects heat transfer as a function of $Re_f$ and turbulence parameters in highly different ways. For example, at $K_T = 0.09$ and $Pr_T = 1.0$ the data on the effect of physical properties as a function of $Re_f$ approach results corresponding to the asymptotic turbulence theory due to Kutateladze and Leont'yev [97] (Fig. 4.2).

The analytic results obtained on the effect of the temperature factor on heat transfer agree basically with experimental data obtained under conditions close to those used in calculations (Fig. 4.2). In certain studies, for example in the papers by Velichko [159] and Ambrazevičius et al. [161], the experimental study was performed over such a range of $Re_f$ and temperatures, over which this effect manifests itself very little, for which reason the experimental results cannot serve as the basis for obtaining nondimensional expressions for the temperature factor, suitable for a wide range of $Re_f$. Even in cases when the temperature factor was found to be a function of $Re_f$, preference was given to the workups of experimental data using a temperature factor that is independent of $Re_f$ [158].

In analyzing experimental results it is important to have a nondimensional relationship for heat transfer that incorporates all the features of

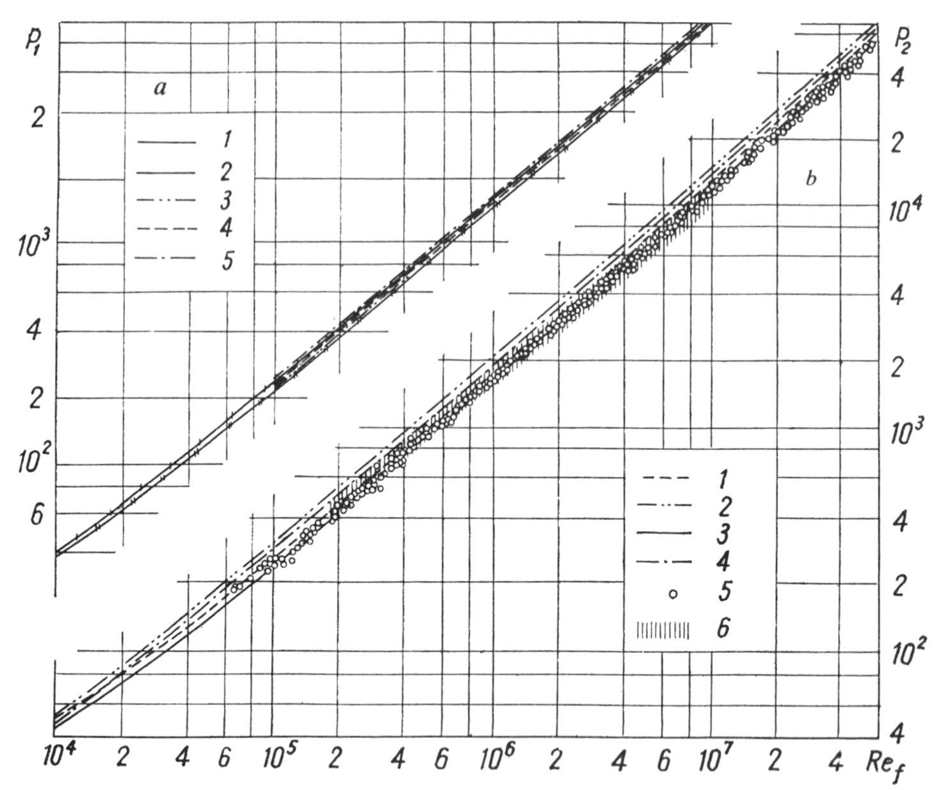


**FIG. 4.1** Comparison of the predicted heat transfer with data of others. (*a*) At constant physical properties, $P_1 = \mathrm{Nu}_0$: (1) $K_T = 0.09$ is the Van Driest damping factor [156]; (2) $K_T = 0.09$ is the Šlančiauskas damping factor [157]; (3)–(5) calculations of Popov, Spalding and Van Driest [155]. (*b*) In cooling and heating of air, $P_2 = \mathrm{Nu}/(\phi_t\phi_T)$: (1) $\mathrm{Pr} = \mathrm{Pr}_T = 1$ and $K_T = 0.07$; (2) $\mathrm{Pr} = \mathrm{Pr}_T = 1$ and $K_T = 0.12$; (3) $\mathrm{Pr} = \mathrm{Pr}_T = 0.7$ and $K_T = 0.07$; (4) $\mathrm{Pr} = \mathrm{Pr}_T = 0.07$ and $K_T = 0.12$; (5) experimental data of Vilemas, Česna and Survila [158]; (6) experimental data of Velichko [159].

turbulent flow; however, it is quite difficult to obtain such an expression.

The first attempts at obtaining such an equation for the planar turbulent layer made by the present author and his co-workers [106] were based on relatively sparse analytic data. For this reason, as shown by analysis, the expression suggested in this study for combustion products at values of parameters for which no calculations were made may result in rather significant error. This, in part, is due to an unfortunate selection of the functional relationship. A better validated expression was obtained for air flows [106].

Analysis of the dependence of heat transfer on the values of $\mathrm{Re}_f$ and $K_T$ shows that heat transfer virtually stabilizes with rise in the latter, i.e., the turbulent sublayer becomes conservative with respect to increasing scale of turbulence in the outer part of the boundary layer. Given the above, the following functional relationship was selected for the turbulent number

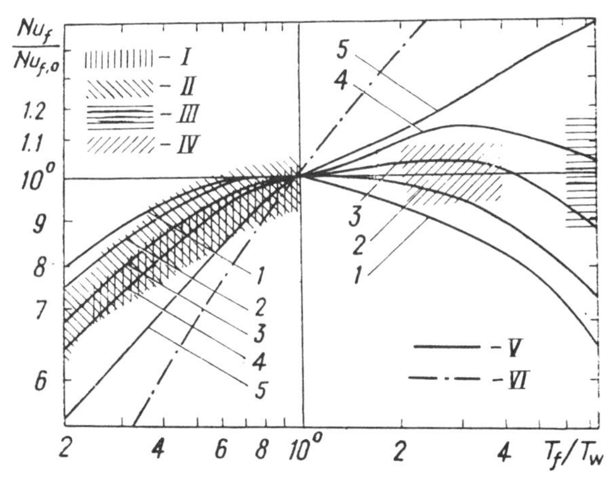


**FIG. 4.2** Comparison of the results of calculations of the effect of the temperature factor on heat transfer at different value of $Re_f$ with data of the following studies; (I) [158]; (II) [160]; (III) [161]; (IV) [159]. Analytic results: (V) [106]; (VI) [97]. Analytic curves (1–5) are presented at values of $Re_f$ of $10^5$, $10^6$, $10^7$, $10^8$, and $10^{12}$, respectively.

$Nu_{T,0}$ at constant physical properties:

$$Nu_{T,0} = \varphi_K \cdot Nu_{lim} \tag{4.14}$$

which employs the concept of the limiting Nusselt number $Nu_{lim}$, which is independent of parameter $K_T$.

Given the very complex manner in which $Nu_{lim}$ depends on $Re_f$, $Pr_f$, and $Pr_T$, it is difficult to find a simple expression for defining the relationship. Satisfactory results are obtained from an empirical expression for $Nu_{lim}$ in the form of a sum of power-law functions of $Re_f$, $Pr_f$, and $Pr_T$:

$$Nu_{lim} = Nu_1 + Nu_2 + Nu_3 + Nu_4 \tag{4.15}$$

where

$$Nu_1 = 3.1\, Re_f^{0.25}\, Pr_f^{0.052}\, Pr_T^{0.45-0.078 \ln Pr_T} \tag{4.16}$$

$$Nu_2 = 0.016\, Re_f^{0.8}\, Pr_f^{0.41}\, Pr_T^{0.072+0.013 \ln Pr_T} \tag{4.17}$$

$$Nu_3 = 0.0036\, Re_f^{0.9}\, Pr_f^{0.83}\, Pr_T^{-0.93-0.107 \ln Pr_T} \tag{4.18}$$

$$Nu_4 = 0.000099\, Re_f\, Pr_f\, Pr_T^{-1.0} \tag{4.19}$$

Calculations show that correction factor $\varphi_K$ depends little on $Re_f$ and is an asymptotic function of $K_T$:

$$\varphi_K = \frac{K_T}{0.012 + K_T} - \frac{1}{4.5 + 450\, K_T^4 \cdot \sqrt{Re_f}} \tag{4.20}$$

Equations (4.14)–(4.20) describe the analytic results at $0.04 \leqslant K_T \leqslant 0.4$ and $0.05 \leqslant \mathrm{Pr}_T \leqslant 1.0$ with a mean arithmetic error not exceeding ±3.0%. For a narrower range of variables, Eq. (4.15) can be significantly simplified.

When using Eqs. (4.14)–(4.20) the effect of variable physical properties can be represented by two factors, contained in the expression

$$\mathrm{Nu}_T = \mathrm{Nu}_{T,0} \cdot \varphi_l \cdot \varphi_T. \tag{4.21}$$

The quantity $\varphi_l$ is a temperature factor, making allowance for the variation in the transport properties of the boundary layer flow, is independent of the flow pattern, and can be determined from Eq. (4.8) or (4.9). The quantity $\varphi_T$ is characteristic only of turbulent flow and is basically a function of variations in the flow density; however, it changes it value depending on the turbulent properties of the medium. Definition of the temperature factor by Eq. (4.21) makes it possible to obtain an expression to allow for the variability of thermophysical properties that is independent of the kind of gas. Such an expression for the temperature factor in turbulent flow is given by Makarlvičius [100]. Analysis of a large number (about 2500 sets) of analytic data for various combustion products and for air yields the expression

$$\varphi_T = \varphi_1 + \varphi_2 \tag{4.22}$$

where

$$\varphi_1 = g_{\rho,f}^{\;-0.46 \cdot \ln g_{\rho,f} \cdot (0.74 + g_{\rho,f})^{2.3} \cdot \left(\frac{1.1 - K_T}{K_T}\right)^{-0.093} \cdot \mathrm{Pr}_T^{0.43 \cdot \mathrm{Pr}_T^{2.1}} \cdot \mathrm{Pr}_f^{0.63\,\mathrm{Pr}_f^{-0.28}}} \tag{4.23}$$

$$\varphi_2 = 0.0568 \cdot \ln\left(\frac{\mathrm{Nu}_{T,0}}{\mathrm{Nu}_0}\right) \cdot \left(1 - \sqrt{g_{\rho,f}}\right) \cdot \left(1 + 4.1 \cdot \mathrm{Pr}_T^{0.9}\right) \tag{4.24}$$

and $\mathrm{Nu}_0$ is given by Eq. (4.4).

For products of combustion of natural gas with air, air–oxygen mixture, and oxygen as the oxidant at an excess oxidant ratio of $0.9 \leqslant \alpha \leqslant 1.2$, Eqs. (4.22)–(4.24) at $2 \times 10^5 \leqslant \mathrm{Re}_f \leqslant 10^8$, $0.08 \leqslant K_T \leqslant 0.4$, and $0.05 \leqslant \mathrm{Pr}_T \leqslant 1$ describe analytic results with a mean arithmetic error of ±4.5%.

Calculations underlying the previously presented relationships for combustion products were performed for frozen gas, i.e., without allowance for transfer of chemical energy.

At the same time a special series of calculations was performed with allowance for transport of the energy of dissociation of the combustion-product components. The calculations were performed according to a similar technique, with, however, use of the effective values of specific heat and thermal conductivity in the approximation of local thermodynamic equilibrium. At $\mathrm{Re}_f$ from $2 \times 10^5$ to $10^8$ these calculations confirmed the experimentally verified conclusion that in turbulent flow the value of $\mathrm{Nu}_T^*$

based on the total enthalpy difference makes satisfactory allowance for transport of the chemical energy of dissociation without introduction of additional factors [136]. Hence it can be assumed that for turbulent flows the parameter of proportionality between the diffusive and convective energy transport is

$$\Lambda = 1 \tag{4.25}$$

Such a result is in agreement with conclusions from experimental data drawn by Shorin and Pechurkin [162]. This means that for dissociated combustion products

$$\mathrm{Nu}_{\mathrm{T}}^{*} = \mathrm{Nu}_{\mathrm{T},0} \cdot \varphi_{l} \cdot \varphi_{\mathrm{T}} \tag{4.26}$$

where $\mathrm{Nu}T_{,0}$ can be calculated from Eqs. (4.14)–(4.24). However such calculations may be found to be very work-consuming. Approximate values of the Nusselt number can be obtained using nomograms presented in Appendix 1 of this study.

Since at present no methods are available for experimentally determining turbulence parameters at high flow temperatures, only an indirect com-parison of the results of this theoretical study with experimental data is possible. Hence the previously obtained equations were used for solving the inverse problem, i.e., for obtaining the values of $\mathrm{Pr}_T$ and $K_T$ from data on heat transfer. Such a formulation of the problem requires validation of the results at least under such temperature conditions when direct experimental determination of $\mathrm{Pr}_T$ and $K_T$ is possible.

## 4.3. DETERMINATION OF TURBULENCE PARAMETERS OF FLOWS

Numerical analysis of the turbulent boundary-layer equations showed that the turbulence parameters of the free stream play an important role in the structure of the turbulent boundary layer. Allowance for these parameters in engineering calculations is prevented by difficulty in determining them.

At present the proportionality factor $K_T$ in the mixing length for the outer part of the boundary layer can be determined on the basis of other statistical characteristics of turbulence. Thus, Dyban and Epik [102] showed that there exists a quantitative relationship between the mixing length and the turbulence scale $L_T$, which can be relatively easily determined from experimental results. Independence of the mixing length of the transverse velocity gradient allows using its relationship to the turbulence scale also at the outer edge of the boundary layer. The turbulent Prandtl number $\mathrm{Pr}_T$ can be determined analogously on the basis of the statistical characteristics of turbulence [102]. This approach requires experimental determination of six turbulence parameters (kinetic energy of turbulence, based on three components of velocity fluctuations, energy spectrum of the transverse component, the correlation coefficient $\overline{u'v'}$, and the intermittence factor). Such measurements are highly work-consuming and can be performed only at room temperature.

Quantities $K_T$ and $\Pr_T$ can be determined by a simpler method, on the basis of the measured mean values of the velocity and temperature in the boundary layer (Figs. 4.3 and 4.4). Since these factors have a different effect on the velocity and temperature fields, there is a basis for the claim that the experimentally measured profiles of $u$ and $T$ carry some information on the values of turbulence parameters, which can be determined by the method of minimization of the square error.

Let us assume that there are available a number of measurements of velocity $u_1, u_2, \ldots, u_n$ and temperature $T_1, T_2, \ldots, T_n$ over the depth of the turbulent boundary layer at distances $y_1, y_2, \ldots, y_n$ from the wall. All the measurements were performed at distance $x$ from the start of the flow-washed body. The turbulent boundary layer calculated analytically from Eqs. (2.54)–(2.67), with allowance for variation in the physical properties of the flow at distances $y_1, y_2, \ldots, y_n$ from the wall, has the velocity $u_1^T$, $u_2^T$,

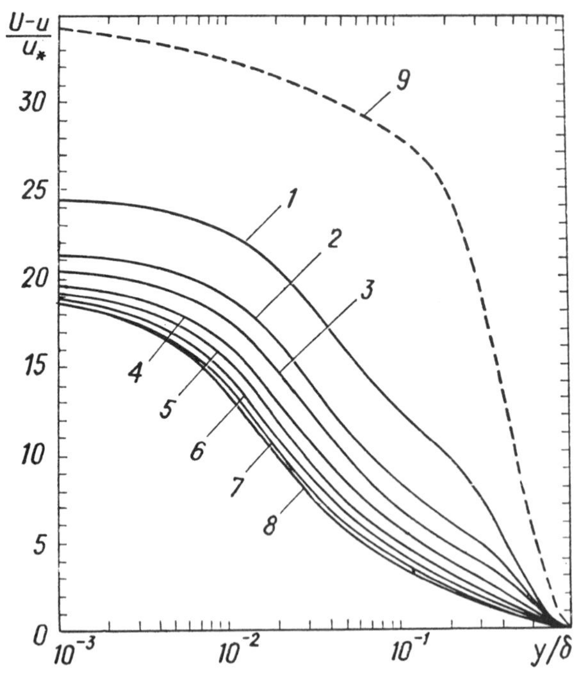


**FIG. 4.3 Excess velocity as a function of** $K_T$ at a constant physical properties of the flow. $K_T$ = 0.04; (2) 0.07; (3) 0.09; (4) 0.12 (5) 0.15; (6) 0.2; (7) 0.3; (8) 0.4; (9) 0 (calculated for a laminar boundary layer, according to Žukauskas and Žiugžda [146]).

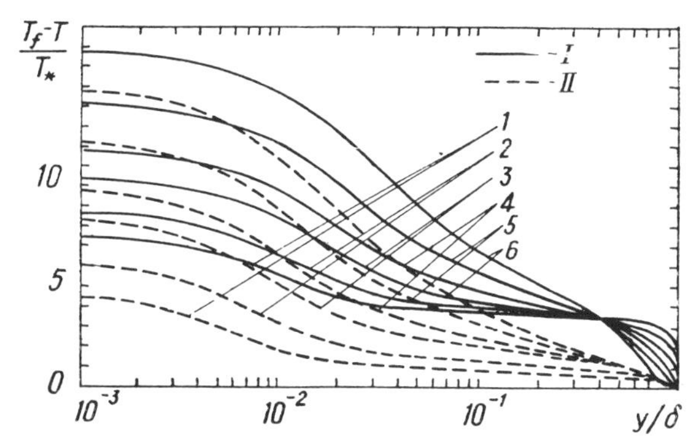


**FIG. 4.4** Excess velocity as a function of $K_T$ and $\mathrm{Pr}_T$ at constant physical properties. $\mathrm{Pr}_T$ values: (1) 0.05; (2) 0.1; (3) 0.2; (4) 0.3; (5) 0.5; (6) 0.8. (I) $K_T = 0.08$; (II) $K_T = 0.2$.

. . ., $u_n^T$ and temperature $T_1^T, T_2^T, \ldots T_n^T$. We shall assume that the values of $u_i^T$ and $T_i^T$ were obtained at some arbitrarily selected values of parameters of the turbulent boundary layer $\pi_1, \pi_2, \ldots, \pi_3$, where $\pi_j$ designates parameters such as the boundary-layer thickness $\delta_x$, von Karman constant $\varkappa_T$, parameters of expressions for the damping factor and mixing length, $\mathrm{Pr}_T$, and other quantities.

The solution of function

$$F(\pi_1, \pi_2, \ldots, \pi_k) = \sum_{i=1}^{n} \left(1 - \frac{u_i}{u_i^T}\right)^2 + \sum_{i=1}^{n} \left(1 - \frac{T_i}{T_i^T}\right)^2 = \min \tag{4.27}$$

corresponding to the minimum rms deviation of the measured [parameters of the] turbulent boundary layer from those analytically obtained, yields the best values of parameters $\pi_j$. Simultaneously with attaining the minimum according to Eq. (4.27) one determines numerically the frictional drag $c_f$ and heat flux at the wall $q_w$ for the experiment under study. At sufficiently exact experimental measurements and $n >> k$ the reliability of the given analysis will depend only on the degree of conformance of the mathematical model of the turbulent layer to the actual situation. Such an analysis can be used for checking out various hypotheses in mathematical description of turbulent transport.

The above technique was checked out for analyzing turbulent parameters in the flow of air [107, 163], which resulted in values of $\mathrm{Pr}_T$ from 0.69 to 0.81, which are very close to values determined by graphical analysis of the measured velocity and temperature profiles in weakly turbulized flow [82]. Such an analysis was performed also for velocity profiles measured by Dyban and Epik [102]* in turbulent flow of air over a plate and different

*The present analysis was performed under the program of scientific cooperation between the Engineering Thermophysics Institute of the Ukrainian Academy of Sciences and our Institute.

free-stream turbulence levels (Tu = 0.76, 1.72, 3.52, 6.85, and 9.72). The mixing length in these experiments was determined from the measured statistical parameters of turbulence. Satisfactory agreement was obtained between the measured and calculated velocity profiles (the mean arithmetic error ranged between 0.24 and 0.98%) at values of the von Karman constant and $K_T$ plotted in Fig. 4.5. The analysis was performed with a model of the turbulent boundary with the damping factor $N_T$ expressed according to Van Driest [156]. It is of interest to compare the experimentally determined value of the mixing length over the depth of the boundary layer with that calculated from the technique examined here. As seen from Fig. 4.6, at moderate free-stream turbulence the nature of variation in the mixing length is satisfactorily described by the assumed mathematical model. A significant disagreement between the results was obtained only in experiments with high free-stream turbulence. The reason for this is at present difficult to explain, since in spite of this difference in the value of the mixing length the experimental and analytic values of the velocity field are in rather satisfactory agreement (the rms error is 0.011 and the arithmetic mean error is 0.64%). This basically reemphasizes the fact that the effect of the value of the mixing length on the mean values of $u$ and $T$ decreases with rising free-stream turbulence. Unfortunately, the measurements of the mean values of profiles of $u$ and $T$ are limited to the range of relatively moderate temperatures, since it is very difficult to perform measurements at temperatures of the order of 1000 K and above.

Given the nonavailability of other methods for determining turbulence parameters at high flow temperature, one must use their estimates based on heat-transfer measurements (164).

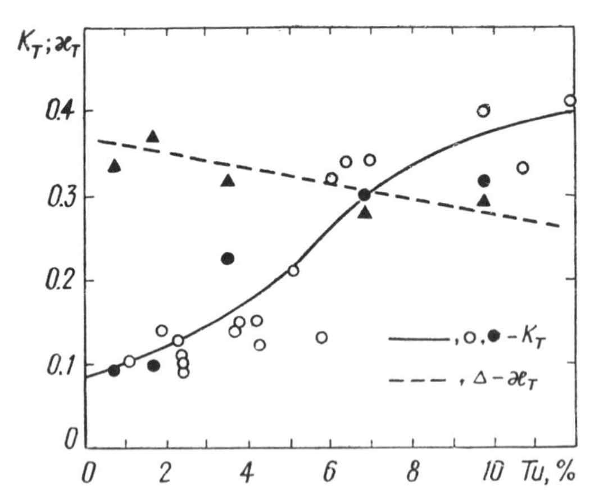


**FIG 4.5** Proportionality factor of the mixing length in the outer part of the boundary layer $K_T$ and the von Karman constant $K_T$ versus the free-stream turbulence. The darkened points represent analysis by this author on the basis of analysis of experimental data of Dyban and Epik [102], and the open symbols represent new results of Pedĭsius et al. [104].

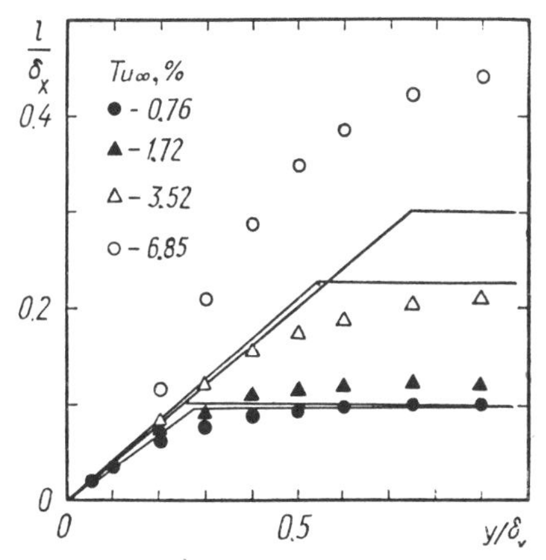


**FIG. 4.6** Comparison of analytic and experimental data on the mixing length at different values of free-stream turbulence. The solid curves represent the present analysis; the points represent calculations by Dyban and Epik.

## 4.4. EXPERIMENTAL RESULTS ON CONVECTIVE HEAT TRANSFER IN THE INLET REGION OF CHANNELS

The use of combustion products in heat-engineering studies involves the following difficulties. First, it is necessary to obtain complete combustion of the fuel in the chamber at different starting compositions and flow rate of the combustible mixture; second, heat transfer should occur without condensation of water vapor on the water-cooled channel walls. The conditions of the inlet to the test channel are closely related to the combustion of fuel in the chamber. These circumstances significantly limit the ranges of flow temperature and Reynolds number, and are in part responsible for the fact that for test channels 1 and 2 results could be obtained at $Re_f$ only from $10^4$ to $10^6$. This is a poorly explored interval, within which the development of the transition and turbulent flows is affected by the free-stream turbulence and by turbulization of the high-temperature flow by the cold wall. For the purpose of analysis the experimental data were subdivided into three groups. The first group comprised the scarce data obtained at $Re_f < 2\times10^4$, while the third group comprised the data are results obtained at $Re_f > 2\times10^5$. Data in the second group, obtained at $2\times10^4 < Re_f < 2\times10^5$, i.e. in a more or less expressed transition flow, were not analyzed. Analysis of the experimental data in the first group showed that they are satisfactorily described by analytic equations (4.3), (4.4), and (4.13) obtained for laminar flow (Fig. 4.7*a*), with the deviations not exceeding the experimental error.

At $Re_f > 2\times10^5$ one should expect turbulent flow, for which reason the data were analyzed on the basis of empirical relationships obtained in the present chapter. Analysis was performed in order to obtain estimates of

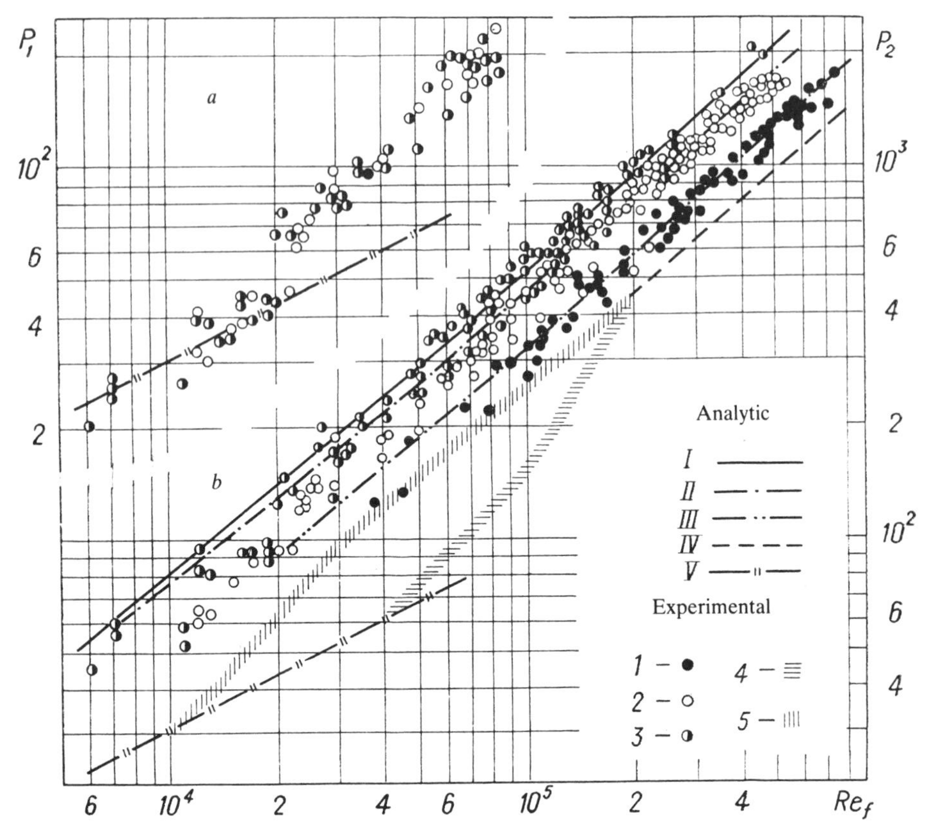


**FIG. 4.7** Local heat transfer at correlating temperature $T_f$. (*a*) $P_1 = \mathrm{Nu}^*_f/\{\phi[1+(\Lambda-1).\Delta h_{chem}/\Delta h\}$, where $\phi_l$ and $\Lambda$ were obtained from Eqs. (4.8) and (4.13). (*b*) $P_1 = \mathrm{Nu}_f(\phi_l\phi_T)$, where $\phi_T$ is from Eq. (4.22). (I)–(V) Results of calculations from Eq. (4.21) for, respectively, $\mathrm{Pr}_T = 0.13$ and $K_T = 0.19$, 0.2 and 0.14, and 0.38 and 0.12; from the study by Ambrazevičius and Žukauskas [149]; and from Eq. (4.2) for laminar flow. (1)–(3) Plate combustion products, channel combustion products, and air, respectively; the shaded region represents the experimental data of Sukomel et al. [103]. (4) Without induced turbulization. (5) air, Tu = 9%.

parameters $K_T$ and $\mathrm{Pr}_T$ corresponding to the least rms error of experimental measurements

$$\sum_{i=1}^{n}\left(\frac{\mathrm{Nu}^*_{T,i,\exp}}{\mathrm{Nu}^*_{T,i}(K_T,\ \mathrm{Pr}_T)}-1\right)^2 = \min \tag{4.28}$$

Relation (4.28) was solved by the method of steepest descent. The mathematical model for $\mathrm{Nu}^*_{T,i}$ consisted of Eqs. (4.14)–(4.24). Analysis showed that condition (4.28) for combustion products is satisfied by values of $\mathrm{Pr}_T = 0.20$ and $K_T = 0.14$ for heat transfer in a channel and $\mathrm{Pr}_T = 0.13$ and $K_T = 0.19$ for heat transfer at a plate. (Note that when the less exact equation from a previous study by the author and his co-workers [164] was used as the mathematical model, it was found that $Pr_T = 0.47$ and $K_T = 0.13$).

The experimental results from which the effect of the temperature factor was deducted are shown in Fig. 4.7. At $Re_f > 2\times10^4$ the experimental data on heat transfer start to deviate gradually from the analytic curve for laminar heat transfer. The transition flow region continues to approximately $Re_f \leqslant 6\times10^4$ (Fig. 4.7*b*). It was found that a significant difference exists between results obtained in transition flow and data from the study by Velichko [159], performed at moderate temperature differences and low turbulence (Fig. 4.7*b*). Our results are close to those obtained by Sukomel et al. [103] at high free-stream turbulence. Sukomel et al. also found that the transition flow zone shifted in the direction of lower Reynolds numbers. The use of Eqs. (4.14)–(4.24) for describing heat transfer in the transition region can result in an 80% overestimate of results.

Data on heat transfer in turbulent flow without allowance for variability of physical properties lie above results obtained by Ambrazevĭcius and Žukauskas [149], which can also be attributed to the high turbulence of combustion products.

Circumstances did not permit determining the turbulent parameters of high-temperature combustion products; however, some measurements of turbulence were performed when the channel was blown down with air. Measurements using DISA hot-element anemometer instrumentation in test channel 2 showed that Tu = 9–10%. Heat transfer in this channel was also measured with hot air blown through the combustion chamber. Data on heat transfer in this case are on the average 12–15% lower than results on heat transfer for the case of combustion products, but are still satisfactorily described by the same expressions at $Pr_T = 0.38$ and $K_T = 0.12$. This means that the values of $Pr_T$ for air at 800 K are higher than for combustion products. The higher values of $K_T$ are apparently due to conditions at the inlet to the channel, which is indirectly confirmed by turbulence measurements. Our data on heat transfer in the flow of air are on average 15% higher than the results of Ambrazevičius and Žukauskas [149]. It is also interesting to note that transition flow for air is, at the same values of $Re_f$, much less expressed than for combustion products. This is because cooling of the high-temperature flow of combustion products near cold surfaces of the test channel is accompanied by a reduction in the viscosity of gases, which induces additional flow turbulization. For this reason transition flow for combustion products starts at lower values of $Re_f$.

Our analysis indicates that in calculations of heat transfer, allowance must be made for the turbulence parameters of high-temperature flows, which have characteristically high values of $K_T$. When the free-stream turbulence is known, the value of $K_T$ can be determined from data plotted in Fig. 4.5. It is more difficult to determine $Pr_T$. At present there are grounds for assuming that $Pr_T$ decreases with rising temperature and turbulence level.

The results of analysis of values of $Pr_T$ obtained in the present studies for narrow ranges of flow temperature confirm that $Pr_T$ decreases with rising temperature (Fig. 4.8*a*). However, the values stratify significantly with changes in the flow turbulence level. The lowest values of $Pr_T$ are characteristic of the experimental results of Ambrazevičius et al. [161] obtained at free-stream turbulence. A similar value of $Pr_T$ was obtained for test channel 3; it was also found that convective heat transfer in the initial length of the channel was systematically higher than for channels 1 and 2. Since this difference exceeded the measuring error, the reasons for it were sought in the specifics of the test channel or flow structure at the inlet to it. Measurements of the turbulence level of air flow in different locations along the channel using DISA instrumentation showed that the turbulent flow structure differed significantly from the flow structure in test channels 1 and 2, for which on the average Tu = 9–10% (Fig. 4.9). According to Fig. 4.9, the flow turbulence at the inlet to test channel 3 was 30–40%, and dropped off exponentially with distance downstream. Measurements show that the turbulent flow structure in a channel is nonhomogeneous and that large vortices predominate. In our measurements the air was supplied through the gas-burning device into the combustion chamber, which was responsible for this high-flow turbulence. There is a basis for assuming that this high turbulence prevails also in combustion products. Since the experimental conditions did not change, it remains to assume that the value of $Pr_T$ is a function of Tu and by virtue of the same fact changes along the channel.

It is interesting to note that an analogous conclusion on the dependence of $Pr_T$ on the free-stream turbulence was obtained in new studies in our Institute by Pedišius, Kažimekas and Šlančiauskas in detailed analysis of the

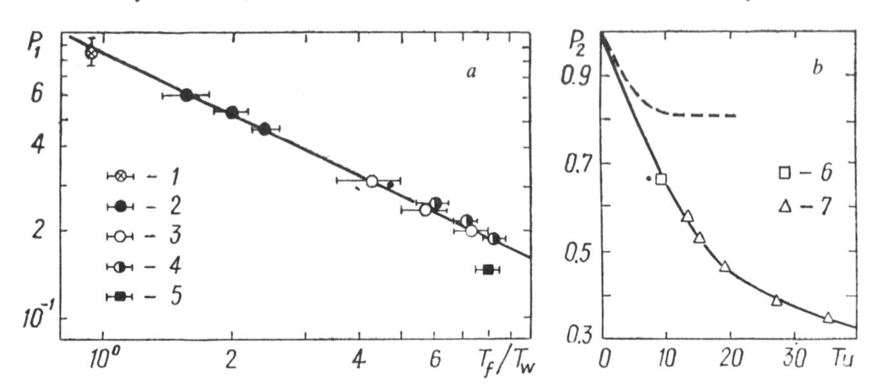


**FIG. 4.8** Estimate of $Pr_T$ at different temperature conditions (*a*) and free-stream turbulence (*b*). (1) Flow of air [82]; (2) flow of air, channel 2; (3) flow of combustion products, channel 2; (4) flow of combustion product over plate, channel 1; (5) flow of air [161]; (6) flow of combustion products, channels 1 and 2; (7) flow of combustion products, channel 3. The dashed curve represents new data of Pedišius, Kažimekas and Šlančiauskas for heating of air; $P_1 = Pr_T/[1 - 0.68th(0.056\ Tu)]$; $P_2 = Pr_T/[0.86.(T_f/T_w)^{-0.71}]$

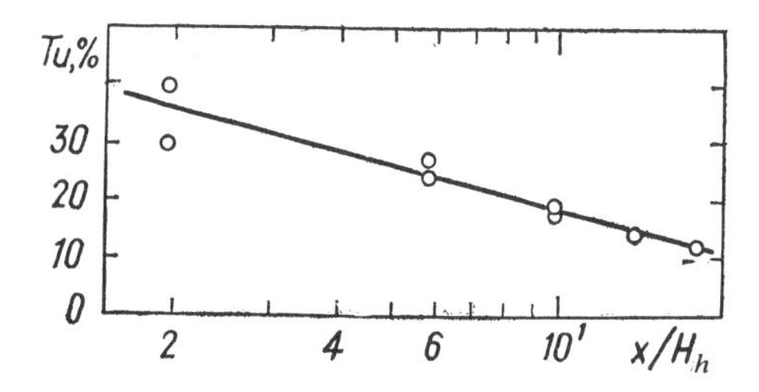


**FIG. 4.9** Variation in turbulence level versus relative length of channel 3 when blown down with cold air.

structural features of the turbulent boundary layer on a plate in an axial flow of air.*

These studies and also the estimates of $Pr_T$ obtained for high flow temperatures allow the assumption that the turbulence-level dependence of $Pr_T$ remains the same irrespective of the flow temperature conditions. This means that in the first approximation, $Pr_T$ of hot flows can be calculated from the expression

$$Pr_T = 0.86 \cdot \left(\frac{T_f}{T_w}\right)^{-0.71} \cdot [1 - 0.68 \, \text{th} \, (0.056 \, \text{Tu})] \tag{4.29}$$

which satisfactorily correlates the not too numerous estimates of $Pr_T$ (Fig. 4.8). Since it is impossible to directly measure the turbulence in high-temperature flows, it is recommended that measurements obtained upon cold-air blowdown of channels at the same values of $Re_f$ be used.

Using Eq. (4.29) and data listed in Fig. 4.9, the experimental results on convective heat transfer in the inlet region of the channel can be correlated by a single relationship based on the empirical description of heat transfer provided by Eqs. (4.14)–(4.26) (Fig. 4.10).

The above analysis should be regarded as a first attempt at correlating relationships governing convective heat transfer of turbulized flows, which are quite characteristic of the working conditions of various heat-engineering devices.

## 4.5. EXPERIMENTAL RESULTS ON CONVECTIVE HEAT TRANSFER IN STABILIZED FLOW

The relationships governing turbulent heat transfer in stabilized pipe flows of gases with variable physical properties with heat transferred from the flowing gas are best investigated experimentally, since some differences exist between results of calculations using familiar methods and experimental data [165]. This difference can be attributed to the fact that in all the

* These as yet unpublished data are included in Figs. 4.5 and 4.8 by permission of the authors.

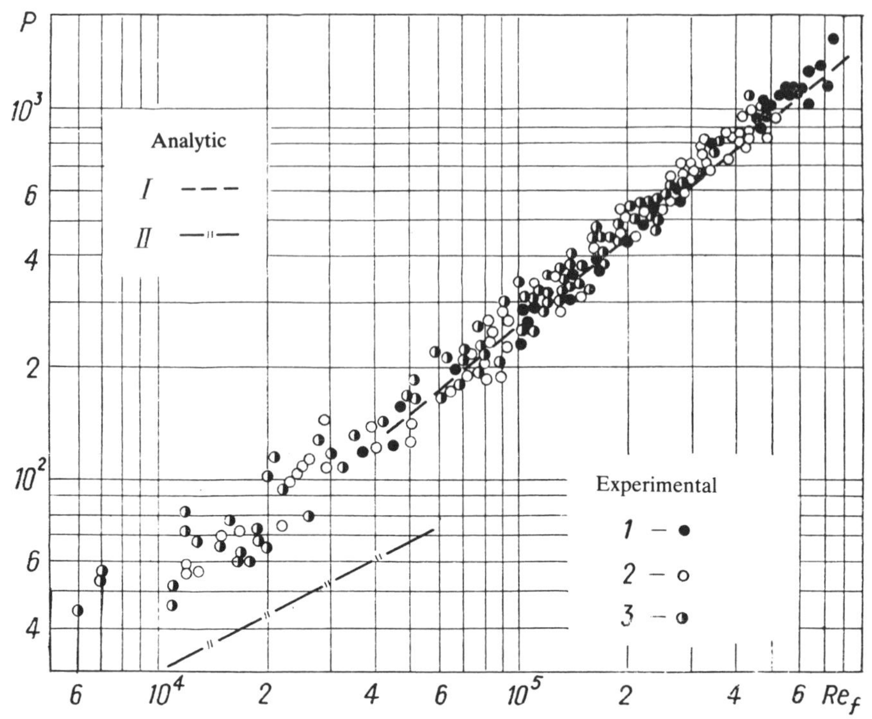


**FIG. 4.10** Local turbulent heat transfer and $T_f$ as the correlation temperature and the temperature dependence of $Pr_T$ given by Eq. (4.29). (I) Analytic curve for turbulent heat transfer, Eq. (4.21); (II) analytic curve for laminar heat transfer, Eq. (4.2). (1) Plate flow of combustion products; (2) channel flow of combustion products; (3) channel flow of air; $P = Nu^*_T/\varphi$, here $\varphi = \frac{Nu_{lim}\,\varphi\,/\,\varphi_T}{Nu^0_{lim}}$, and the value of $Nu^0_{lim}$ was determined at $Pr_T = 0.73$.

standard analytic studies the value of $Pr_T$ for gases was always taken to be close to unity. A detailed analysis of the features of stabilized heat transfer from combustion products over a wide range of velocities and temperatures was performed by the present author and his co-workers [134]. These results, concerning heat transfer of combustion products of a propane–oxygen mixture in a tube with an inner diameter of 11.92 mm at relative tube lengths from $x/H_h = 20$ to 60, were correlated by a nondimensional expression, which incorporates both the diffusive–convective energy transport and the transport of the kinetic energy of the flow.

$$Nu^*_{f,H} = Nu_{f,H} \cdot \left[1 + (\Lambda - 1) \cdot \frac{\Delta h_{chem}}{\Delta h} + (D - 1) \cdot \frac{\Delta h_{kin}}{\Delta h}\right] \tag{4.30}$$

Note that in the case of stabilized flow the values of $Nu_{f,H}$ and $Re_{f,H}$ are determined on the basis of the hydraulic diameter of the channel $H_h$.

Equation (4.30) was obtained on the basis of the heat flux balance at the wall with allowance for the total enthalpy difference $\Delta h$, including not only the difference $\Delta h_{chem}$ of dissociation energy, but also of the kinetic energy of the flow $\Delta h_{kin}$.

In the case of stabilized turbulent flow it was found that the proportionality factor $\Lambda$ between diffusive and convective energy transports over the range of parameters under study is smaller than unity and can be obtained from the empirical expression

$$\Lambda = \frac{0{,}68}{1+0.000217\,\exp\,(20080/T_f)} \tag{4.31}$$

An analogous proportionality factor between the transport of kinetic energy in compressible flow with convective energy transfer $D$ was estimated from experimental results of other investigators, and it was found that the value of $D = 0.53$ yields satisfactory results for the conditions under study.

The Nussel number $\mathrm{Nu}_f$, defining convective heat transfer with allowance for variability of the physical properties of the gas in the boundary layer, is given by the expression

$$\mathrm{Nu}_{f,\,H} = c\,\mathrm{Re}_{f,\,H}^{0.8} \cdot \mathrm{Pr}_f^{0.43} \cdot (T_f/T_w)^m \tag{4.32}$$

The present author and his co-workers [134] found that $c = 0.0208$ and $m = -0.05$. The slight difference between this value of $c$ and that obtained by Mikheyev can be attributed to differences in experimental conditions. Studies showed that the variations in the physical properties of the gas from which energy is transferred have little effect on heat transfer.

Since studies in [134] were performed only for a small-diameter tube, radiative energy transport was insignificant. The interaction between radiative and convective heat transfer under these conditions was investigated with the ADA-5 experimental stand, specially constructed for this purpose. Studies performed with this stand of heat transfer from flow of hot air and also from combustion products highly diluted by air or nitrogen confirmed the previously obtained equation (4.32). Due to the insignificant flow temperatures and velocities, the transfer of dissociation energy and of the kinetic energy of the flow can be neglected. Since new studies were performed for heat transfer with smaller relative channel lengths, the results were worked up with allowance for their dependence on $x/H_h$ (Fig. 4.11). Some departures from the previously obtained results can be attributed to specific conditions at the inlet to the test channel. As shown by turbulence measurements when blowing down the test channel, at its inlet $\mathrm{Tu} \simeq 15\%$; however, at a distance of 15–16 equivalent diameters from the inlet it drops to 5–7% and remains at this level to the end of the channel. It was thus established that the flow turbulence in the test channel of the ADA-5 stand is elevated.

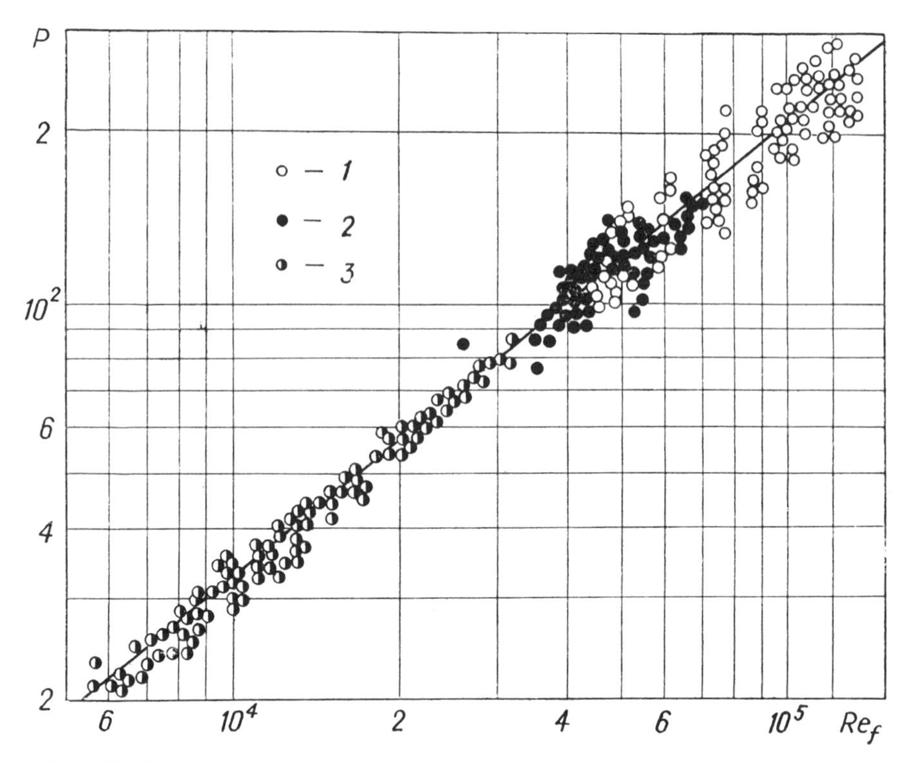


**FIG. 4.11.** Heat transfer in stabilized pipe flow. (1), (2) Heat transfer from hot air and from combustion products highly diluted by nitrogen in channel 4, $P = \mathrm{Nu}_{f,H}/\{\mathrm{Pr}_f^{0.43}(T_f/T_w)^{-0.05}[1-(x/H_h)^{-0.7}]\}$; (3) heat transfer in a 11.2-mm-diameter cylindrical channel [134], $P_2 = \mathrm{Nu}_{f,H}/\{\mathrm{Pr}_f^{0.43}\,(T_f/T_w)^{-0.05}[1 + (\Lambda - 2)\,\Delta h + (\mathrm{D} - 1)\Delta h_{\mathrm{chem}}/\Delta h]\}$.

The experimental results on heat transfer in stabilized flow at $x/H$ 7–24 can be expressed by the equation

$$\mathrm{Nu}_{f,H} = 0.0208 \cdot \mathrm{Re}_{f,H}^{0.8} \cdot \mathrm{Pr}_f^{0.43} \cdot (T_f/T_w)^{-0.05} \cdot \left[1 + \left(\frac{x}{H_h}\right)^{-0.7}\right] \quad (4.33)$$

The above expressions for convective heat transfer were used in analyzing relationships governing combined heat transfer.

Detailed data of experiments on convective heat transfer are presented in Appedixes 2 and 3.

CHAPTER

# FIVE

# RADIATIVE HEAT TRANSFER IN GASEOUS MEDIA

Design of heat engineering devices [furnaces, boilers, MHD channels, etc.] in which radiative energy transfer plays an important role is very work-consuming due to the integral nature of the processes. Significant simplific-ations of calculations are frequently achieved by using certain integral optical properties of the medium and enclosures [2–5, 7–9, 47, 48, 86, 87, 93, 94, 98, 166,]. However, in this case it is impossible to properly make allowance for the spectral features of the radiating medium and enclo-sures, whereas these play an important role under certain conditions. In particular, Mikk [167–169] showed that the effective radiation temperature of nonisothermal media is not the same for different wavelengths, i.e., it is impossible to determine it from the integral optical properties of the medium.

A detailed accounting of the spectral features of the medium is currently possible only in simplest cases, for example for a volume of gas between two infinite plates, for an infinitely long cylindrical volume, etc. When the geometric shape of the radiating system is more complex one usually has to use simplified computational techniques.

This chapter presents results of a study of governing relationships of radiative heat flux in volumes of different geometric shape at differnt distributions of the medium's variables. The problem is examined in the fundamental formulation, i.e., on the assumption that the temperature and concentration fields in the medium are known.

The calculations presented here correspond to conditions close to those actually encountered in heat engineering. The results can be used for estimating the effect of various factors on the radiative heat flux.

## 5.1. EFFECT OF THE GEOMETRY OF THE ISOTHERMAL MEDIUM

The simplest geometric shape in calculating radiative heat transfer is the hemisphere. In the absence of enclosures, the expression for radiative heat transfer for a hemisphere [Eq. (2.18)] can be integrated analytically. It can be easily shown that the radiative heat flux for a hemisphere determined in this manner is related rather simply to the total emissivity of gases:

$$q_{r,h} = \varepsilon \sigma T^4 \tag{5.1}$$

where $\varepsilon$ is obtained from Eq. (1.56) or (1.60).

In determining the radiative heat flux for other geometric shapes, use is frequently made of the fact that volumes with the same characteristic dimensions $R_0 = 4V/F$, where $V$ is the volume and $F$ is the area of the enclosures, have similar values to the total emissivity [3]. Vinogradov and Detkov [166] suggested a method for comparing radiation in volumes of different geometric shape, in which the standard was the total emissivity of a plane-parallel layer. Analysis of the difference between emissivity $\varepsilon_{vol}$ of various volumes of elementary configuration and a plane-parallel layer made it possible to obtain a general relationship for defining it, based on calculating the total emissivity $\varepsilon_{pl}$ of a plane-parallel layer at three different values of its thickness:

$$\varepsilon_{vol} = 0.0628 \cdot \varepsilon_{pl}\,(8.8H) + 0.4444 \cdot \varepsilon_{pl}\,(2H) + 0.4928 \cdot \varepsilon_{pl}\,(1.125H) \tag{5.2}$$

It is pointed out by Vinogradov and Detkov [166] that the error in determining the total emissivity $\varepsilon_{vol}$ of a volume of any shape for media with a band spectrum is significantly smaller than the error for a gray medium. Equation (5.2) reflects to some degree the dependence of the radiative heat flux on the length of beams; however, its accuracy has been checked out only for gaseous carbon dioxide and water vapor from scanty data of calculation of radiation in various volumes by zonal methods.

A more complete picture of the effect of geometry on the radiant heat flux can be obtained from numerical calculations; however, these are made difficult by the need of integrating Eq. (2.18) over the solid angle. In connection with this it is important to know the principal differences between radiation in volumes of different shape and the hemispheric radiation.

We shall consider in more detail this problem for the planar and cylindrical infinite gas layers.

The solution of the problem of radiation in an infinite plane-parallel or infinite cylindrical volume is very important from not only the theoretical but also the practical point of view, since such conditions are frequently encountered in various devices employed in heat engineering.

Radiation in a homogeneous layer between two infinite plates can be calculated exactly using equations for the one-dimensional approximation [Eqs. (2.109)–(2.116)]. When solving the same problem for an infinite cylindrical layer one may use Eqs. (2.117)–(2.142).

In analyzing relationships governing radiation in the planar or cylindrical layer it is important to first determine the difference between this and hemispherical radiation. It was assumed in comparing the radiative heat fluxes in the planar and hemispherical layers that there are no enclosures, i.e., that $I_H^{\dagger} = I_0^{-} = 0$. In some cases such an assumption may be very far from the actual conditions; however, it is frequently used by many investigators on account of its convenience in comparing the results of calculation of the natural radiation of a gaseous medium by various methods. Allowance for enclosures radiation from Eq. (2.111) or Eq. (2.116) does not involve principal difficulties.

Calculations performed for planar and cylindrical layers of a gray gas using Eqs. (2.109)–(2.142) produced the familiar relationships shown in Fig. 5.1 and served as tests for checking the accuracy of programs combined on the basis of the above equations.

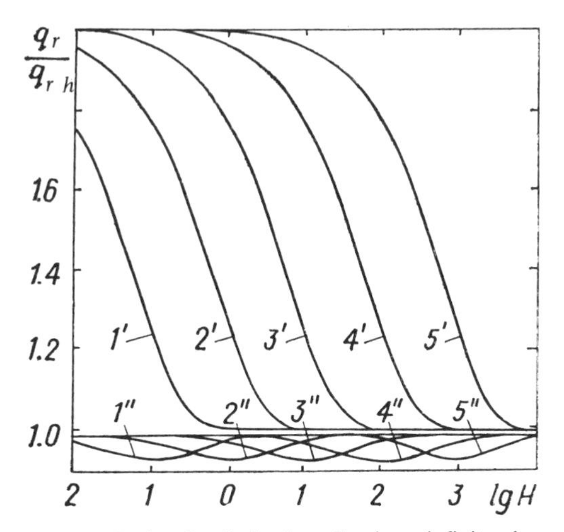


**FIG. 5.1** Ratio of radiative heat flux in an infinite plane-parallel layer (primed numbers) and cylindrical layers (double-primed numbers) to the hemispherical radiative heat flux in a gray gas as a function of layer thickness $H$ at different values of the absorption coefficient $\varkappa$, cm$^{-1}$: (1) 10; (2) 1; (3) 0.1; (4) 0.01; (5) 0.001.

It was established that the behavior of the ratio $q_r/q_{r,h}$ depends on the kind, temperature, pressure, and thickness of the layer of molecular gases (Figs. 5.2 and 5.3).

As the thickness of the infinite plane-parallel layer of gas increases, the value of $q_r/q_{r,h}$ approaches unity, whereas with a reduction in this ratio it approaches 2. The nature of the curves depends both on the kind of gas and on other conditions. Note that the expression for $q_r/q_{r,h}$ for molecular gases cannot be described by a unique dependence on the optical thickness of the layer, as this is done in studies employing the gray-gas approximation for optical properties of the medium.

The values of $q_r/q_{r,h}$ for an finite cylinder ranges from 0.9 to 1.0 (Figs. 5.2 and 5.3). In spite of the small range of variation in this quantity, its expression for molecular gases is quite different than for curves for a gray gas.

Calculations show that neglect of the geometric features of a radiating volume may result in significant errors. In estimating these features for parallel and cylindrical volumes of molecular gases, one can use results of calculations plotted in Figs. 5.2 and 5.3.

In calculating the radiation of a gas volume with a more complex geometry, it is apparently best to use the simplifications obtained by introducing the effective radiation length $R_e$, defined by Eq. (2.147). In parti-

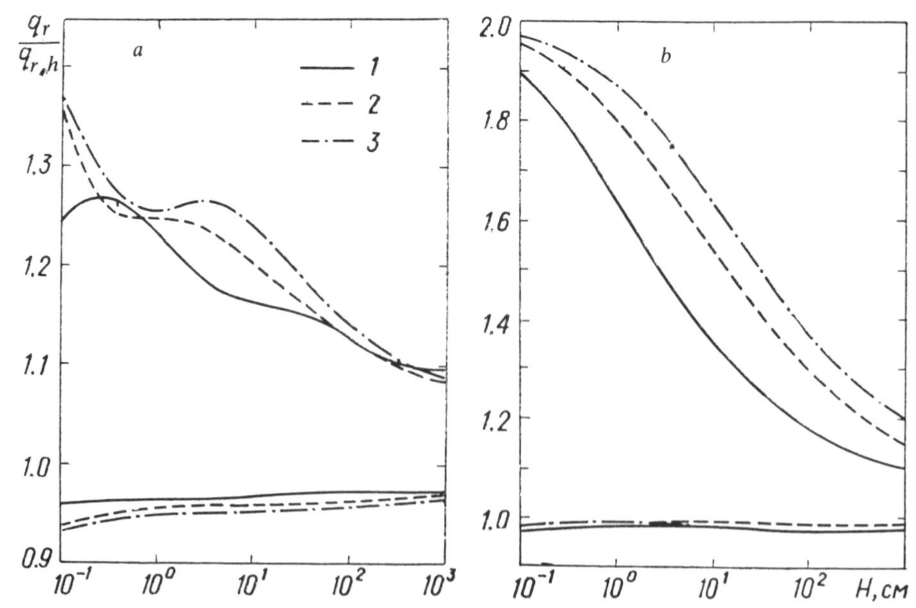


**FIG. 5.2** Ratio of radiative heat flux in infinite plane-parallel and cylindrical layers to the hemispherical radiative heat flux for different gases at atmospheric pressure as a function of layer thickness $H$. (a) Gaseous $CO_2$; (b) water vapor. (1) 1000 K; (2) 2000 K; (3) 3000 K.

cular, it is recommended that the radiative heat flux in volumes with a complex geometry be calculated from Eq. (2.148), which, as Eq. (5.2), incorporates the dependence of the radiant heat flux on the length of beams.

## 5.2. RADIATION IN A NONISOTHERMAL PLANAR LAYER OF GAS

In addition to geometric factors, the radiant flux in a planar layer is significantly affected by temperature conditions. A case of practical important is the radiation in a planar layer with linear temperature distribution. Radiation under such conditions was first investigated by Kavaderov [170], and was then comprehensively investigated by Nevskiy and Chukhanova [171] and Detkov [172]. The results obtained by these investigators differ significantly, which can be attributed in part to the use of different values of optical properties of the medium in the calculations. The results of calculations of radiation using the technique suggested in this paper for isothermal conditions are in satisfactory agreement with analytic data obtained by Kavaderov [170] for minimum and large thicknesses of the layer (Fig. 5.4.), but they are significantly higher in the case of medium thicknesses. The results of Nevskiy and Chukhanova [171] for isothermal conditions for large layer thickness are in satisfactory agreement with the results of the present study; however, they become significantly lower with reduc-

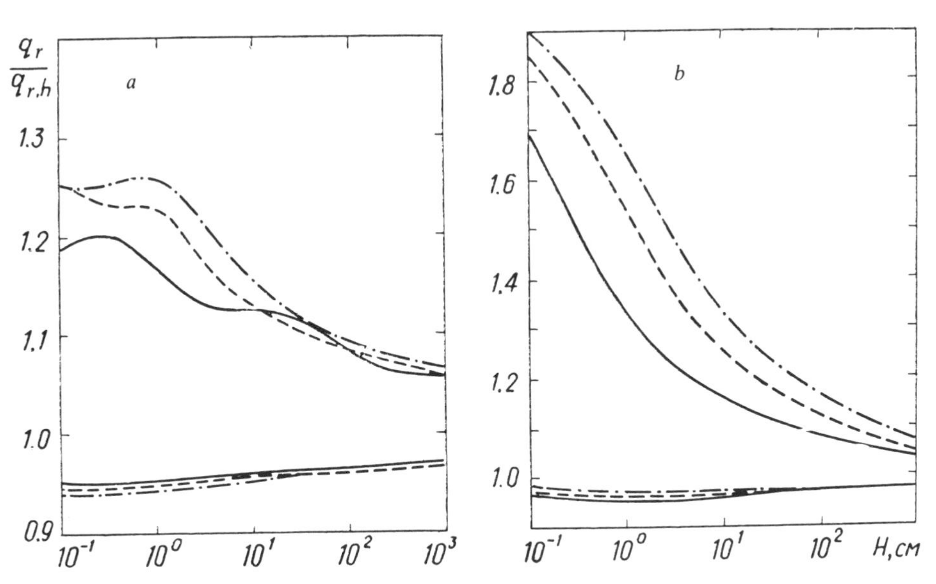


**FIG. 5.3** Ratio of radiative heat flux in an infinite plane-parallel layer and cylindrical layers to the hemispherical radiative heat flux for different gases at 5 atm. For the remaining legend, see Fig. 5.2.

tion in the layer thickness. The radiative heat flux at a high-temperature wall for small layer thicknesses corresponds to that calculated in [170], but with increasing $H$ the calculations yield higher values of $q_r$, which agree best with the results given in [171, 172]. The calculated radiant heat flux at a wall with a lower temperature is in satisfactory agreement with the results obtained by Nevskiy and Chukhanova, and Detkov, whereas the data from the study by Kavaderov for large thicknesses are significantly lower. This significant lack of agreement between results obtained by different investigators is apparently due not only to differences in the optical properties of the medium assumed in the calculations but also to different computational techniques.

The effect of nonisothermicity of the flow in a layer with different temperatures of black-body enclosures has been analyzed for the case of radiation of $CO_2$, $H_2O$, and combustion products of natural gas (Fig. 5.5). At small optical thicknesses the radiation is basically controlled by radiation from the enclosures. The radiant heat flux decreases with increasing optical thickness of the layer. A nonuniform temperature in the layer has a greater effect on the heat flux at the cooler wall. This is in agreement with the results obtained in Chapter 1 concerning differences in the effect of temperature conditions on the optical thickness of the layer upon transmission of radia-

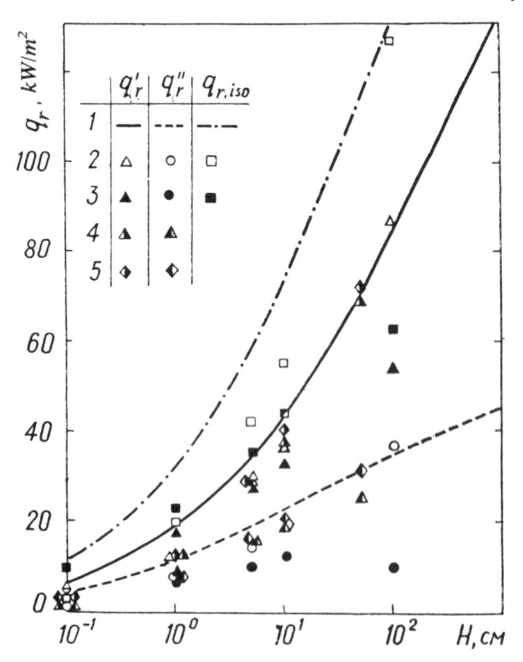


**FIG. 5.4** Analytic data for a layer of gaseous $CO_2$ with linear temperature distribution. (1) Present study; (2) [171]; (3) [170]; (4) [172]; $q_{r,iso}$ is the radiative heat flux for an isothermal layer; $T$=2000 K; $q_r'$ and $q_r''$ are the radiative heat fluxes at the low- and high-temperature boundaries, respectively.

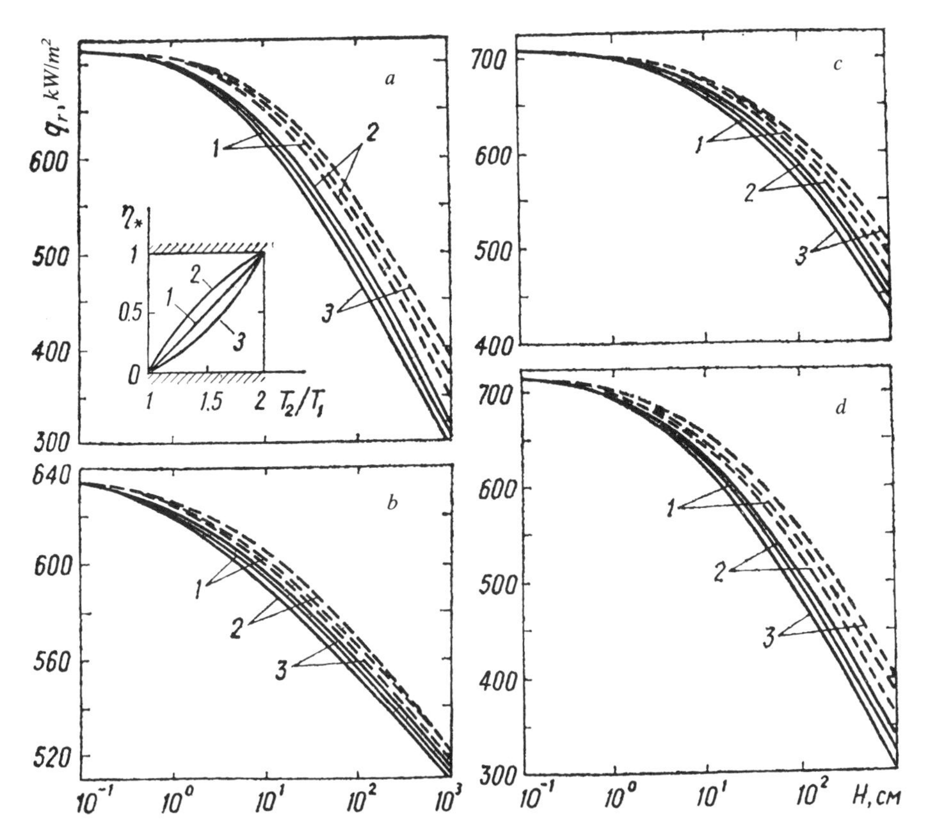


**FIG. 5.5** Radiative heat flux between two black-body walls at 1 atm and different temperature distributions in the gaseous medium. (1) $T = T_1(1+\eta_*)$; (2) $T = T_1\{3-2[\exp(-\eta_* \ln 2)]\}$; (3) $T=T_1 \exp(\eta_* \ln 2)$. The solid curves represent the heat fluxes at the wall with $T_1 = 1000$ K, the dashed curves are for the wall with $T_2 = 2000$ K. The insert represents a characteristic temperature distribution within a layer: (*a*) water vapor; (*b*) gaseous $CO_2$; (*c*) combustion products ($P_{CO_2} = 0.09$, $P_{H_2O} = 0.18$); (*d*) combustion products ($P_{CO_2} = 0.33$, $P_{H_2O} = 0.67$).

tion through hotter or cooler gas layers. In Fig. 5.5 the convex curve (2) of the temperature distribution in the layer corresponds to some rise in the radiative heat flux on both the cooler and hotter wall, whereas the concave curve (3) of the temperature distribution in a planar layer corresponds to a reduction in heat fluxes at both walls. These effects manifest themselves stronger with increasing layer thickness, i.e., they are a function of the radiation and absorption within the gas layer. The magnitudes of radiative fluxes due to natural radiation of the gas layer (Fig. 5.6) confirm the presence of the previously mentioned radiation features. The relationships governing radiation under the temperature conditions analyzed here were investigated by Detkov and his co-workers on the basis of zonal methods [172–175].

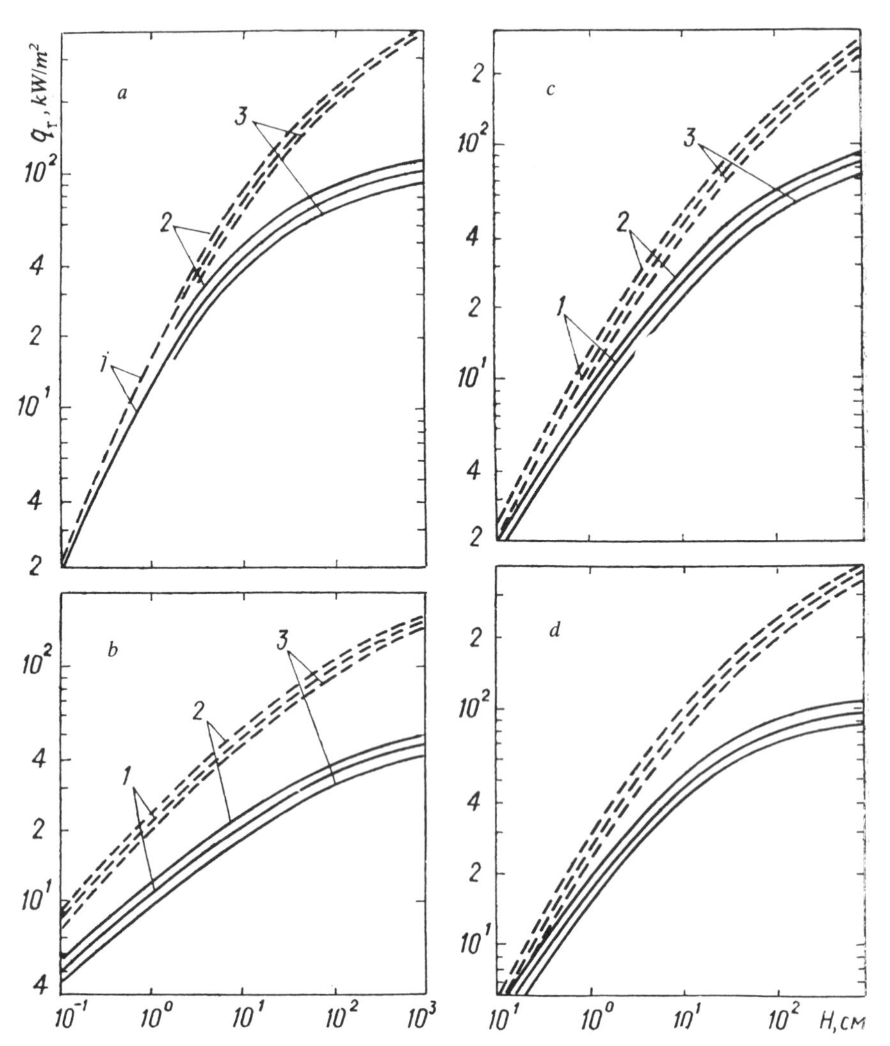


**FIG. 5.6** Radiative heat flux of a gas layer in the absence of enclosures. For the remaining legend, see Fig. 5.5.

Departure of the emissivity of enclosures from unity may result in significant changes in the radiative heat flux. The effect of the emissive and reflective powers of walls on the radiative energy transfer in this paper was analyzed for a linear temperature distribution between two walls with temperatures of 1000 and 2000 K (Figs. 5.7 and 5.8). A reduction in the emissivity of the lower wall (with the lower temperature) results in a significant drop in the radiative heat flux at both walls. The nature of this change depends little on the kind of gas and other radiation conditions. A reduction in the emissivity of the upper wall (at higher temperature) changes similarly the radiative heat flux of this wall, whereas the radiative heat flux of the lower wall changes much less. In this case the effect of the gaseous medium is greater.

There is another case of radiative energy transfer in a planar layer of importance, when the conditions are close to those of flow conditions in a channel, and the temperature changes over the entire he ht of the channel. In this case it is interesting to analyze the effect of nonisothermicity on the overall transfer of radiant energy at different temperatures $T_f$ of the medium in the center of a planar duct, but at the same bulk temperatures $T_m$ as the flow. The analysis is performed with the temperature distribution given by the equation

$$T = T_w + A\cos\left(\frac{\pi\eta}{2}\right) + (T_f - T_w - A)\cos^2\left(\frac{\pi\eta}{2}\right) \tag{5.3}$$

where
$$A = \frac{T_f + T_w - 2T_m}{1 - 4/\pi} \tag{5.4}$$

Equation (5.3) satisfies the boundary conditions

$$T = T_w \quad \text{at} \quad \eta = \pm 1 \quad \text{and} \quad T = T_f \quad \text{at} \quad \eta = 0 \tag{5.5}$$

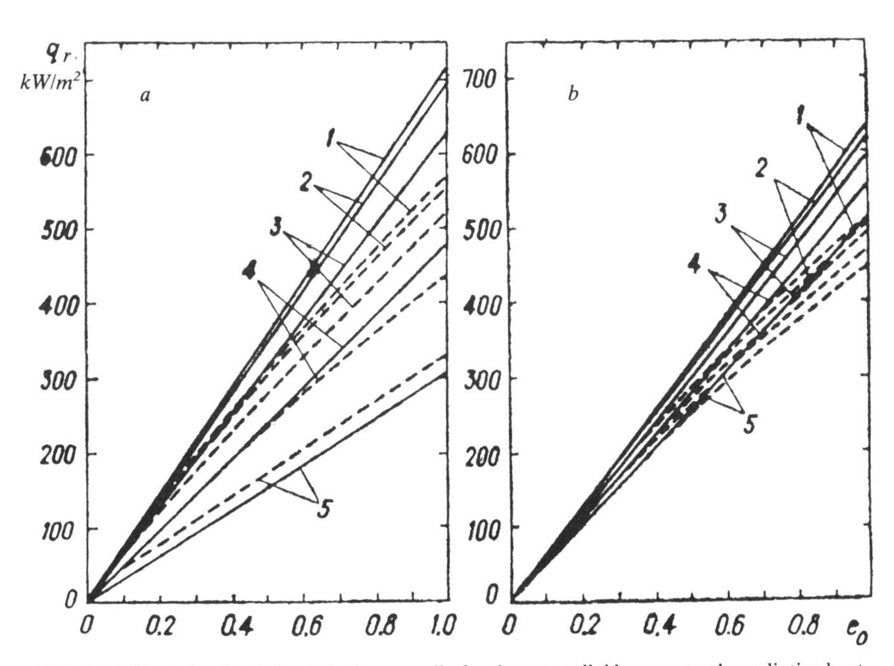


**FIG. 5.7** Effect of reflectivity of the lower wall of a plane-parallel layer $e_0$ on the radiative heat flux at $T_1$ = 1000 K, $T_2$ = 2000 K and linear temperature field at atmospheric pressure. (*a*) Water vapor; (*b*) gaseous $CO_2$. The solid curves represent calculations for the reflectivity of the upper wall $e_H$ = 1, the dashed curves are for $e_H$ = 0.08. *H* values: (1) 0.01 cm; (2) 1 cm; (3) 10 cm; (4) 100 cm (5) 1000 cm.

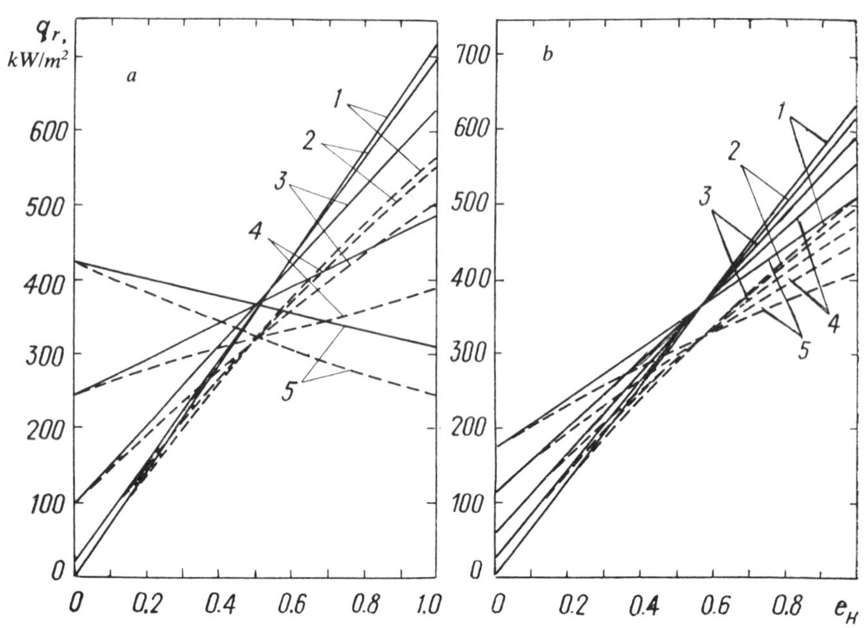


**FIG. 5.8** Effect of reflectivity of upper wall of a plane-parallel layer $e_H$ on radiative heat flux for different gases at atmospheric pressure. The solid curves represent calculations for the reflectivity of the lower wall $e_0 = 1$; the dashed curves are for $e_0 = 0.8$. For the remaining legend, see Fig. 5.7.

and makes it possible to obtain a family of temperature distribution curves in a planar layer (Fig. 5.9) as a function of $T_f$ at constant value of the bulk temperature of the layer

$$T_m = \frac{1}{2} \int_{-1}^{+1} T\, d\eta \tag{5.6}$$

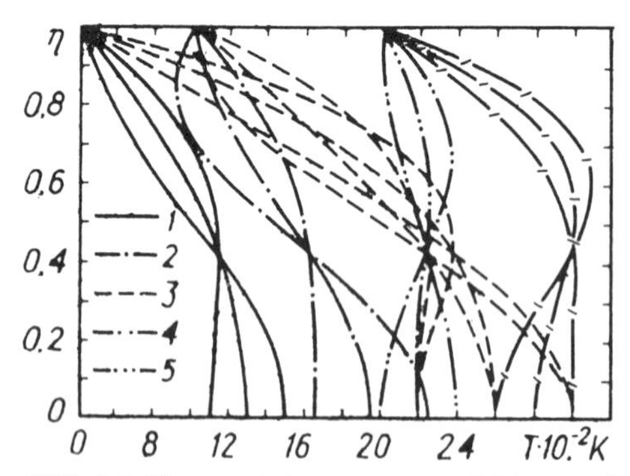


**FIG. 5.9** Characteristic temperature fields according to Eqs. (5.3) and (5.4) under different conditions for which numerical calculations of the effect of nonisothermicity in a planar layer were calculated numerically. $T_m$ values (K): (1) 1000; (2) 1500; (3) 2000; (4) 2200 (5) 2500.

For molecular gases such calculations were performed at $T_m$ of 1000, 1500, 2000, 2200 and 2800 K at wall temperatures $T_w$ of 400, 1000 and 2000 K (Figs. 5.10–5.15)

In the flow of hot gases in a channel with cold walls, the effect of the temperature field on radiative heat flux depends basically on the channel height $H$ (Figs. 5.10 and 5.11) With increasing $H$ and pressure of the medium, the radiative heat flux as a rule decreases as compared with the radiation of an isothermal medium that has the same bulk temperature. At channel heights from 0.1 to 10 cm and pressures from 1 to 5 atm, this difference ranges in the majority of cases within the limits of ±15%. As $H$ is increased further, these effects are significantly amplified, and under certain conditions the radiative heat flux may decrease two-fold or even more. In such cases the temperature curves of $q_r/q_{r,\text{iso}}$ have an extremum at which the radiative heat flux is at a minimum.

In practice, in structures with a geometry close to a planar duct (in furnaces, ovens, flues, etc.), in the majority of cases the wall temperatures are significantly higher than in the example considered here. However, the behavior of the curves, as seen from Figs. 5.12, 5.13 and 5.15, is the same as at low wall temperatures. A significant deviation from the behavior examined here was obtained in cases when the temperature of the radiating gases is close to the wall temperature. The behavior of the radiative heat flux under such condition is given by Fig. 5.14, which shows the results of calculations for a planar duct at $T_w$ = 2000 K and $T_m$ = 2200 K. In this case, depending on the temperature distribution, the radiative heat flux may either increase or decrease significantly as compared with the radiative heat flux under isothermal conditions. Conditions close to those shown in Figs. 5.14 and 5.15 occur in channels of MHD devices [176].

The above analysis allows estimating the effect of nonisothermicity of flow on radiant energy transfer in a planar layer. In cases to which the present calculations do not apply, one must use numerical methods.

## 5.3. RADIATION OF A NONISOTHERMAL MEDIUM IN A CYLINDRICAL CHANNEL

Radiation in an infinite cylindrical channel is close in magnitude to that in a hemispherical layer; however, in spite of this, in this case the effects of nonisothermicity may be no less strong than in a planar layer.

The effect of nonisothermicity on the total radiant energy transfer in flows in cylindrical channels with different temperature distributions was analyzed for bulk flow temperatures of 1000, 1500 and 2000 K at channel wall temperatures of 400 and 1000 K.

As in the case of a planar layer, the temperature distribution obeys Eq.

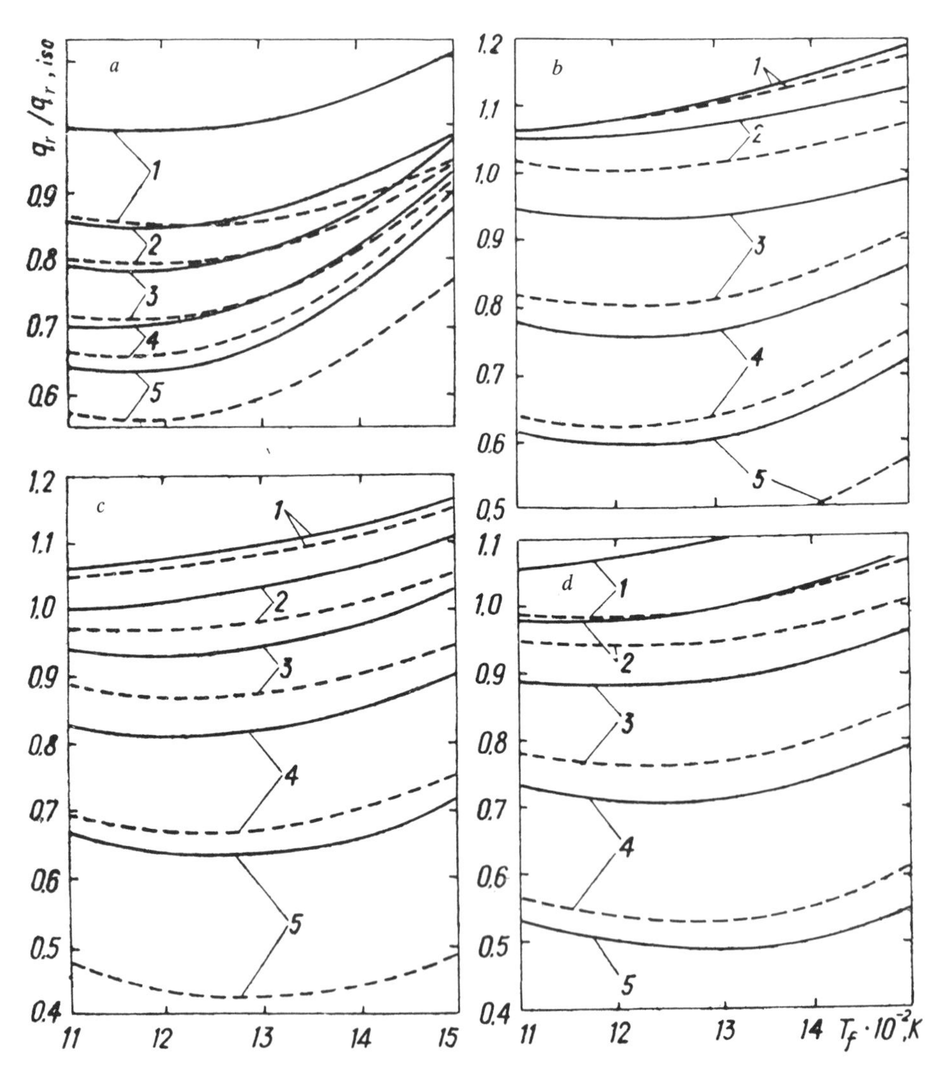


**FIG. 5.10** Radio of radiative heat flux $q_r$, calculated for a temperature distribution given by Eqs. (5.3) and (5.4), to the radiative heat fluxes $q_{r,\text{iso}}$ of an isothermal gas in an infinite plane parallel volume at the same bulk temperature of the gas $T_m$ = 1000 K and wall temperature $T_w$ = 400 K as a function of temperature $T_f$ in the center of the flow. The solid curves are for $p$=1 atm (absolute), the dashed for $p$= 5 atm (absolute). $H$ values (cm): (1) 0.1; (2) 1; (3) 10; (4) 100; (5) 1000. (*a*) Gaseous carbon dioxide; (*b*) water vapor; (*c*) products of combustion of methane with air ($P_{CO_2}P$ = 0.09, $P_{H_2O}/P$ = 0.18); (*d*) products of combustion of methane with oxygen ($P_{CO_2}/P$ = 0.33, $P_{H_2O}/P$ = 0.67).

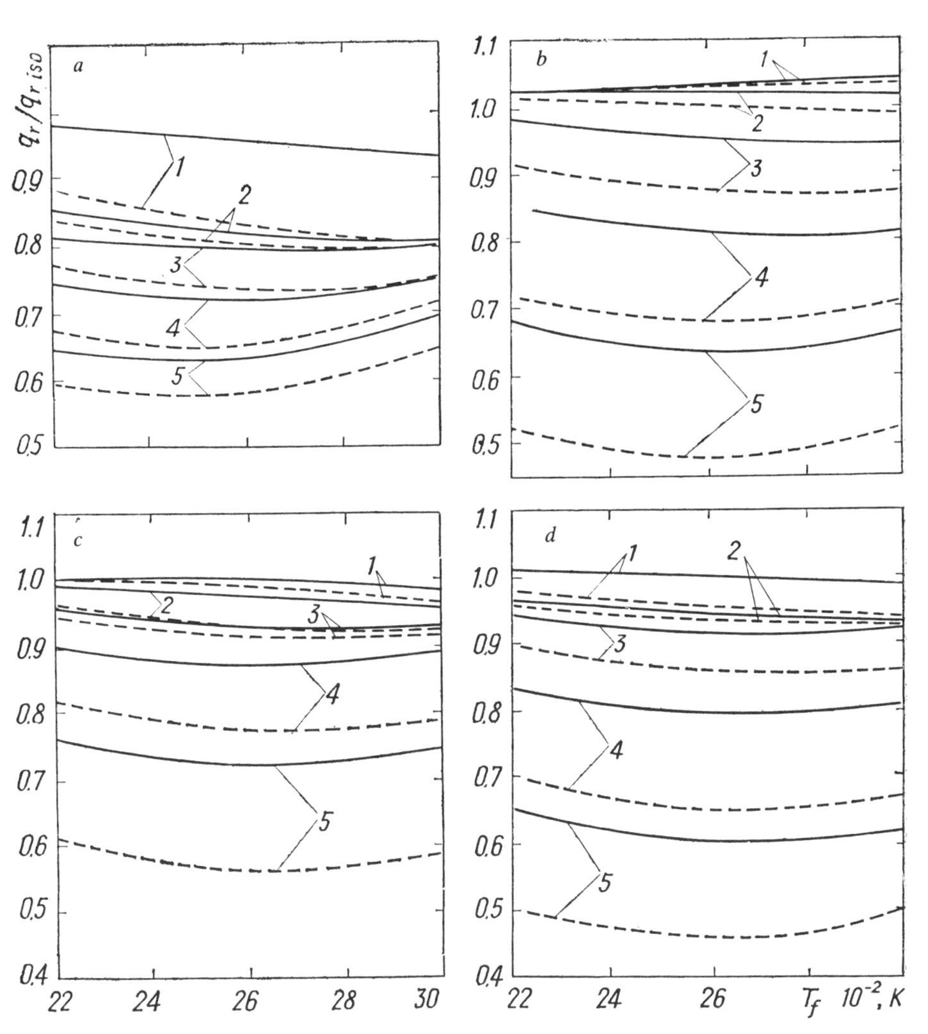


**FIG. 5.11** Effect of nonisothermicity of a plane-parallel volume for different molecular gases at $T_m$ = 2000 K and $T_w$ = 400 K as a function of $T_f$. For the remaining legend, see Fig. 5.10.

(5.3); however, in order to satisfy the condition

$$T_m = 2 \int_0^1 T\eta \, d\eta = \text{const} \tag{5.7}$$

constant $A$ is expressed by the equation

$$A = \frac{2\pi^2 T_m - T_w(\pi^2 + 4) - T_f(\pi^2 - 4)}{8\pi - \pi^2 - 12} \tag{5.8}$$

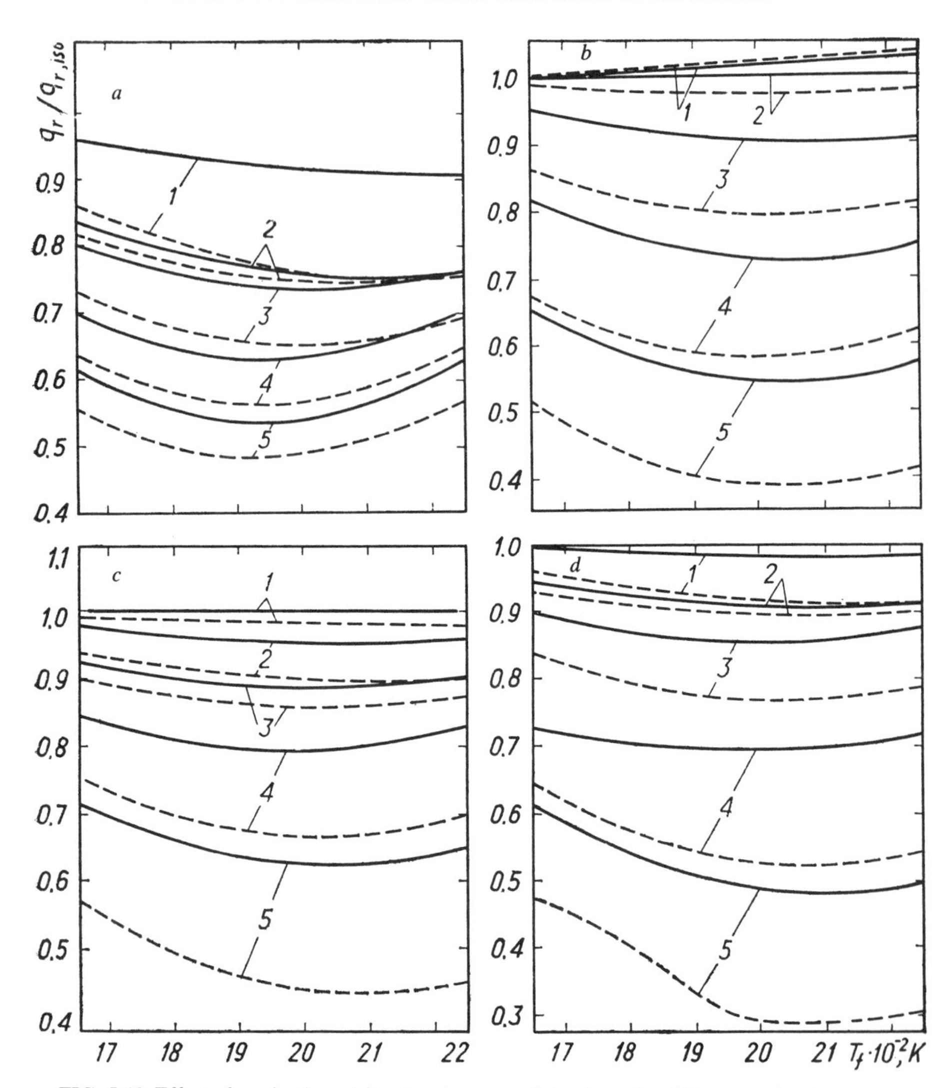


**FIG. 5.12** Effect of nonisothermicity of a plane-parallel volume for different molecular gases at $T_m$ = 1500 K and $T_w$ = 1000 K as a function of $T_f$. For the remaining legend, see Fig. 5.10.

The characteristic temperature fields calculated from Eqs. (5.3) and (5.8) are shown in Fig. 5.16.

According to calculations, the effect of nonisothermicity on the magnitude of the radiant heat flux on the cylinder wall is analogous to the case of a planar layer (Figs. 5.17–5.20).

At $T_w$ = 400 K, $T_m$ = 1000 K and large layer thickness (in the given case equal to the channel diameter), nonisothermicity results in a more significant reduction of radiation in the channel than in a planar layer (Figs.

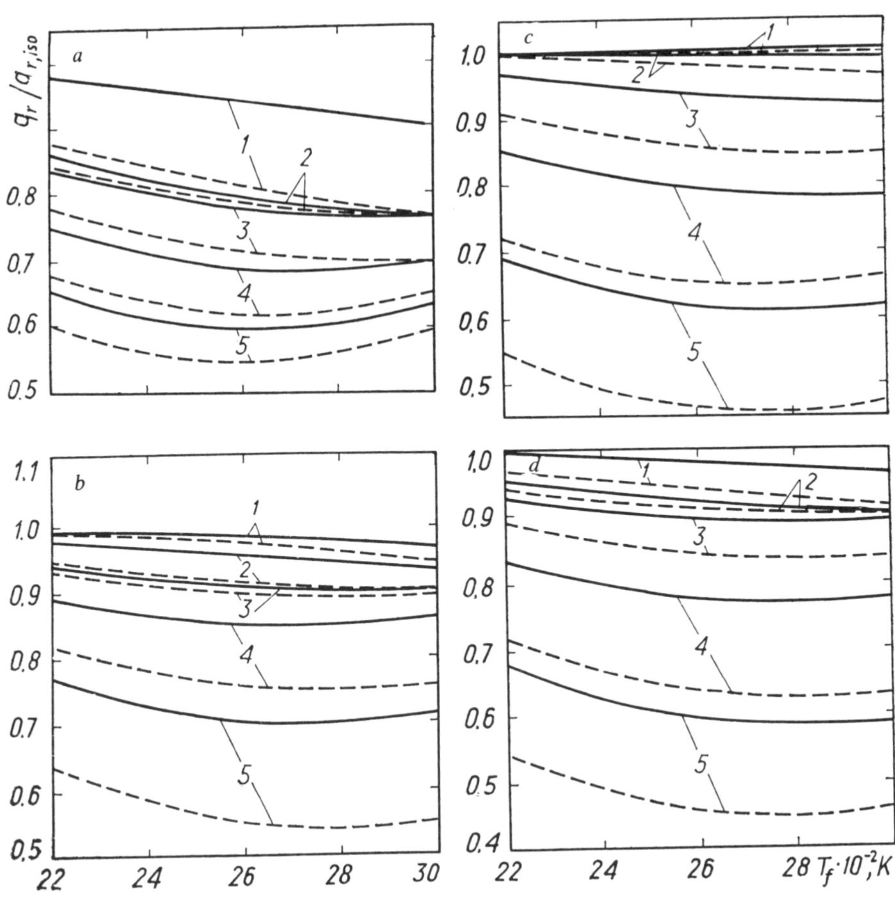


**FIG. 5.13** Effect of nonisothermicity of a plane-parallel volume for different molecular gases with $T_m$ = 2000 K and $T_w$ = 1000 K as a function of $T_f$. For the remaining legend, see Fig. 5.10.

5.10 and 5.17). Unlike a planar layer, in temperature distributions calculated for a channel there are no extreme values of the relative radiative heat flux, which possibly do exist but at different temperature distributions, for which the present calculations were not performed. An analogous situation is observed at the same wall temperature and $T_m$=2000 K (Fig. 5.18).

The greatest deviation from the case of a planar layer occurs at $T_w$ = 1000 K for a cylindrical channel (Figs. 5.19 and 5.20), since the effect of nonisothermicity, analogous to that occurring in a planar layer, is retained only at moderate channel diameters. With increasing channel diameter, the radiative heat flux may either increase or decrease. As a rule, the most significant changes in heat flux occur at its moderate absolute magnitude under isothermal conditions. Cases are frequent when the radiative heat flux at the wall changes sign owing to a different temperature distribution in the layer

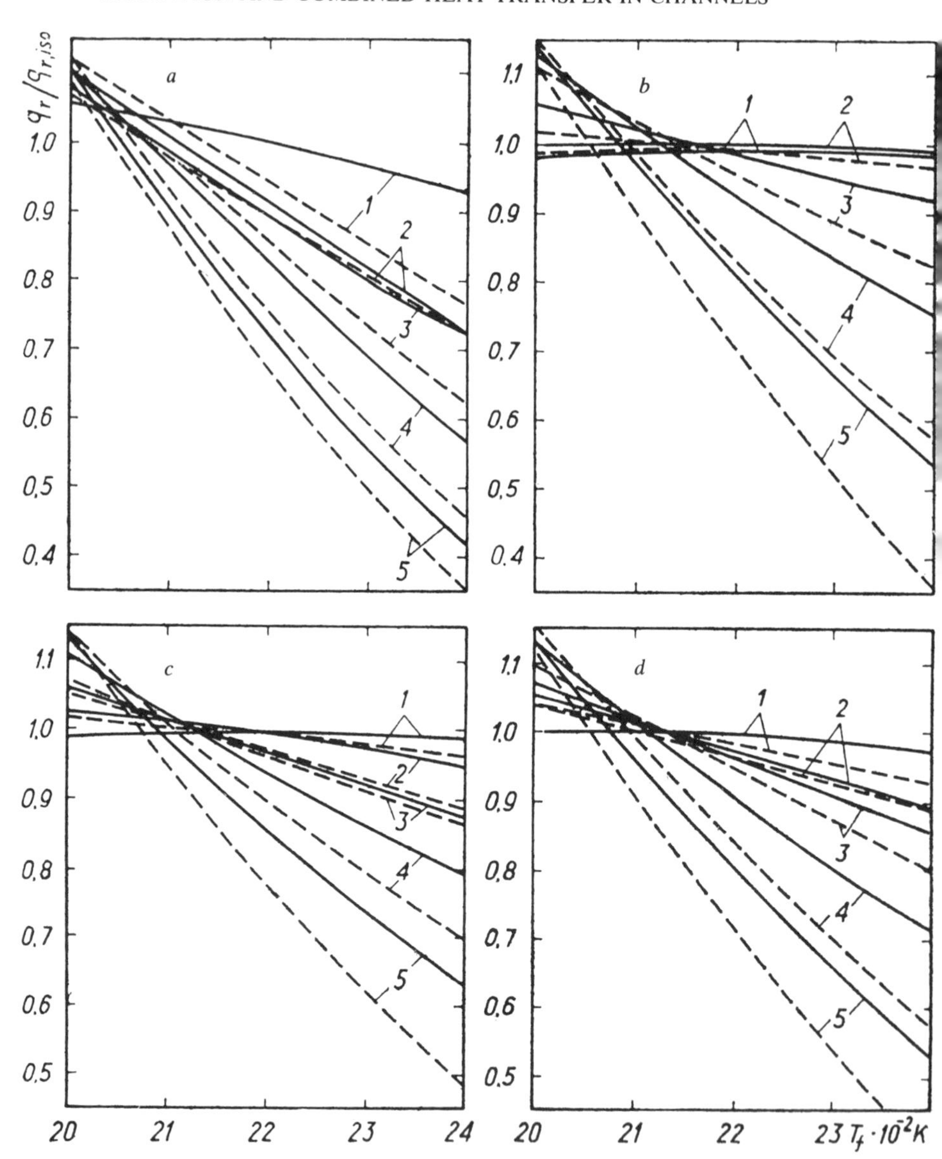


**FIG. 5.14** Effect of nonisothermicity of a plane-parallel volume for different molecular gases at $T_m$ = 2200 K and $T_w$ = 2000 K as a function of $T_f$. For the remaining legend, see Fig. 5.10.

(Figs. 2.19 and 5.20). Hence in estimating the effect of nonisothermicity in cylindrical volumes, calculations should be preferred to estimates from the graphical data presented here.

We note in conclusion that the calculations presented in this chapter were performed in the Curtis–Godson approximation, in which the nonisothermal layer was subdivided uniformly into 20 sections over the channel

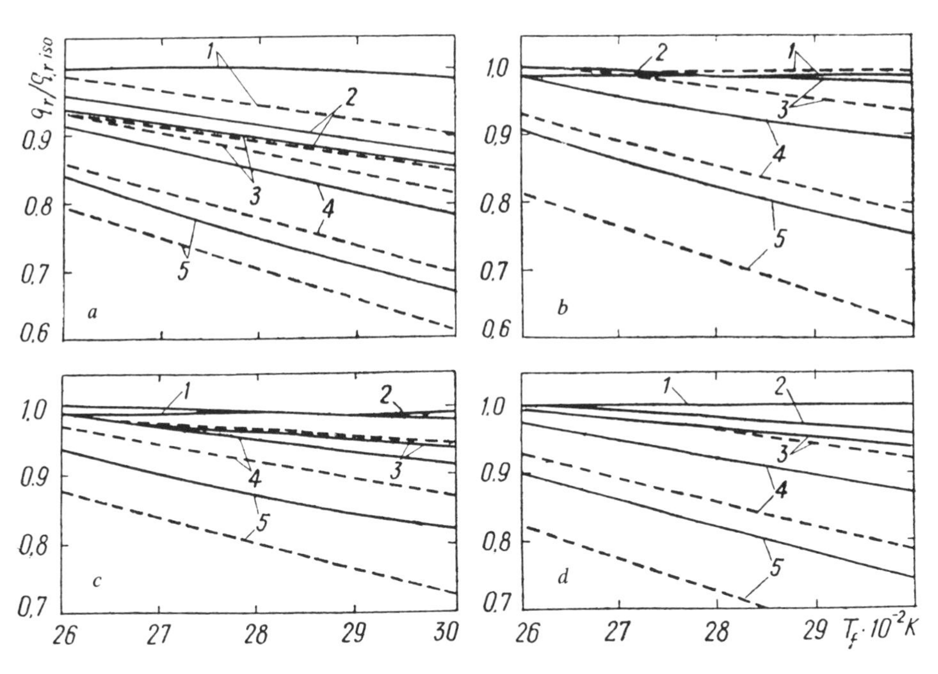


**FIG. 5.15** Effect of nonisothermicity of a plane-parallel volume for different molecular gases at $T_m$ = 2800 K and $T_w$ = 2000 K as a function of $T_f$. For the remaining legend, see Fig. 5.10

height (over the radius in the case of cylindrical channels). Given the limits on subdividing the layer into thin slices, the computational errors, particularly in the case of thick layers, may increase. For this reason these relationships governing radiation under nonisothermal conditions must be checked experimentally, although this verification, as pointed out in Chapter 3, involves great difficulties.

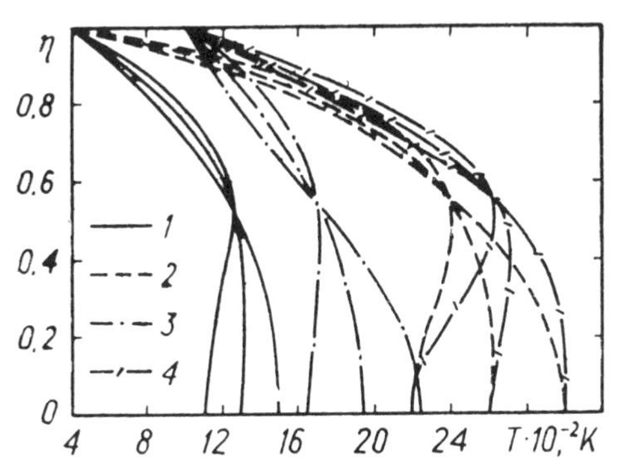


**FIG. 5.16** Characteristic temperature fields according to Eqs. (5.2) and (5.7) at different temperature conditions for which numerical calculations of the effect of nonisothermicity in a cylindrical volume were performed. $T_m$ values (K): (1) 1000; (2) 1500; (3) 1500; (4) 2000.

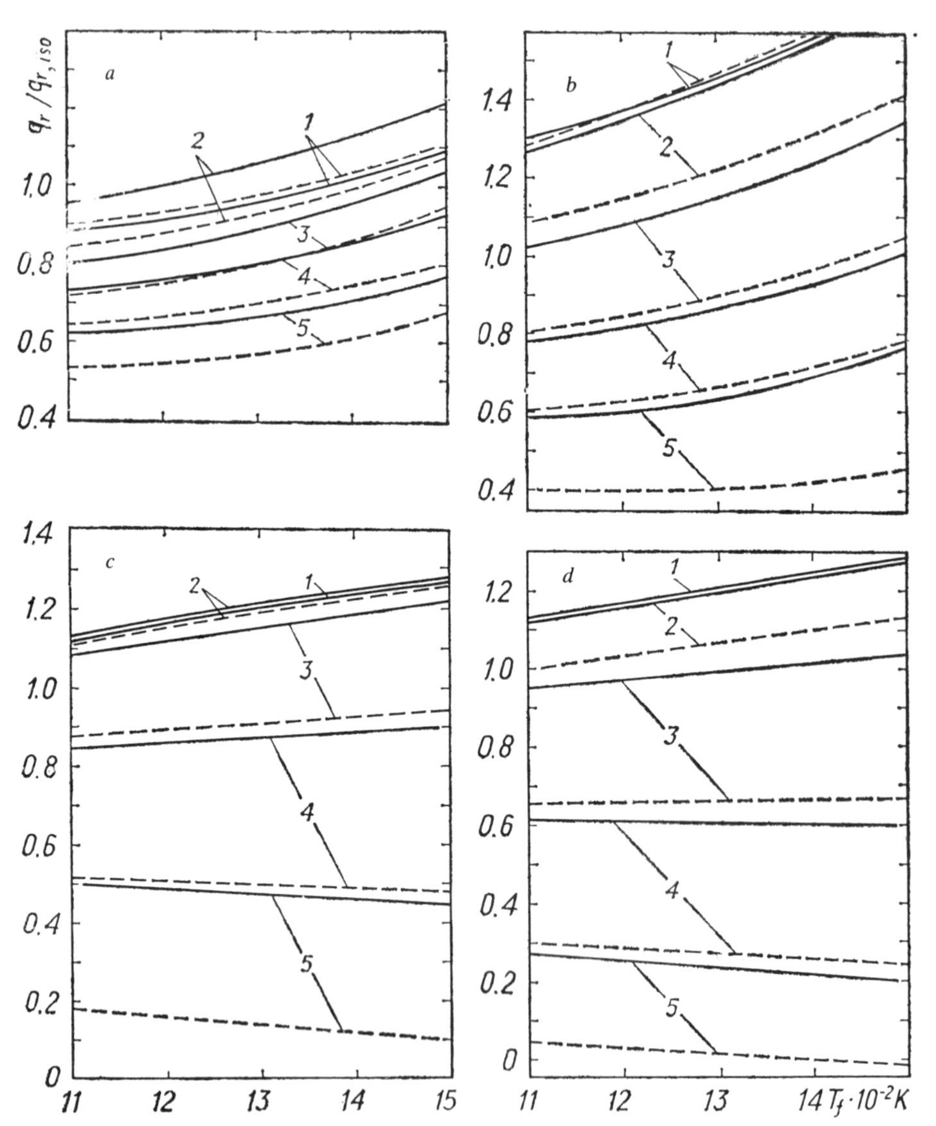


**FIG. 5.17** Ratio of radiative heat flux $q_r$, calculated for a temperature distribution given by Eqs. (5.2) and (5.7), to the radiant heat flux of a nonisothermal gas $q_{r,\text{iso}}$ in an infinite cylindrical volume at the same bulk temperature of the flow $T_m = 1000$ K and wall temperature $T_w = 400$ K. For the remaining legend, see Fig. 5.10.

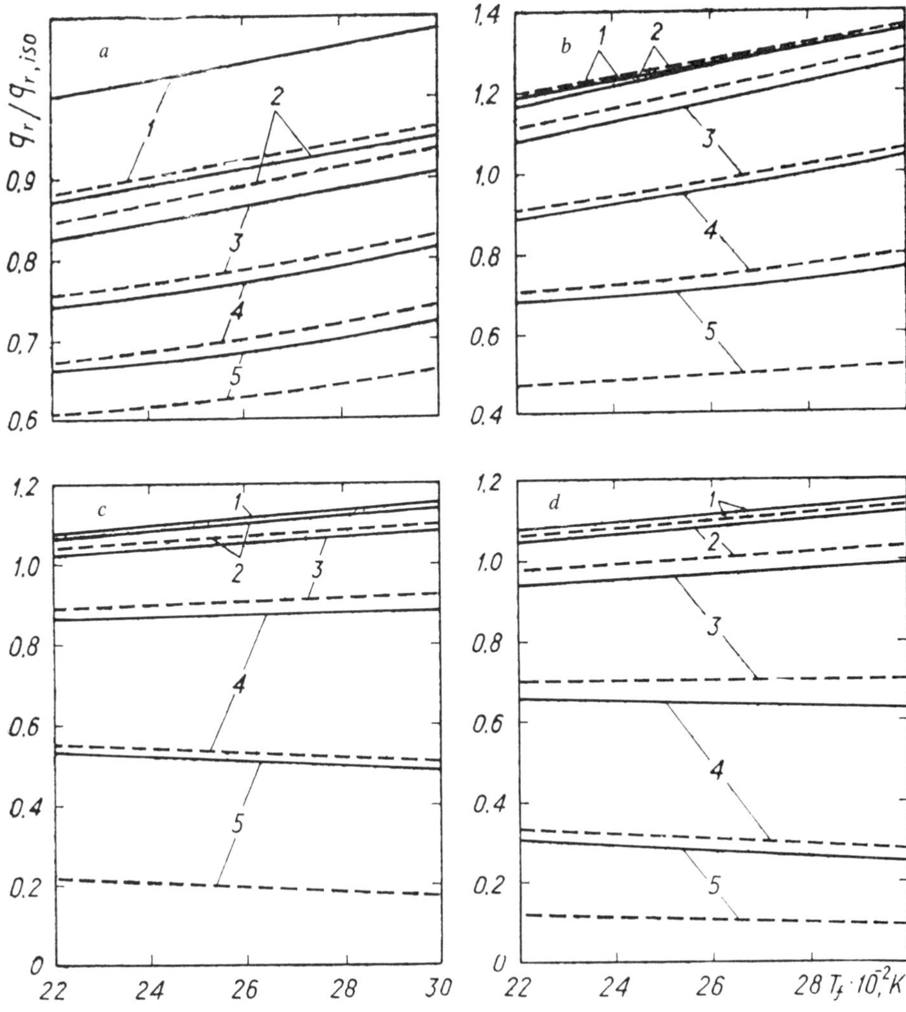


**FIG. 5.18** Effect of nonisothermicity of an infinite cylindrical volume for different molecular gases at $T_m$ = 2000 K and $T_w$ = 400 K as a function of $T_f$. For the remaining legend, see Fig. 5.17.

## 5.4. EXPERIMENTAL STUDY OF RADIATIVE HEAT FLUX

Usually experimental studies of radiation are performed for two purposes:

1. Determination of optical properties of gases. In this case one must know the temperature, concentration and pressure distributions of the

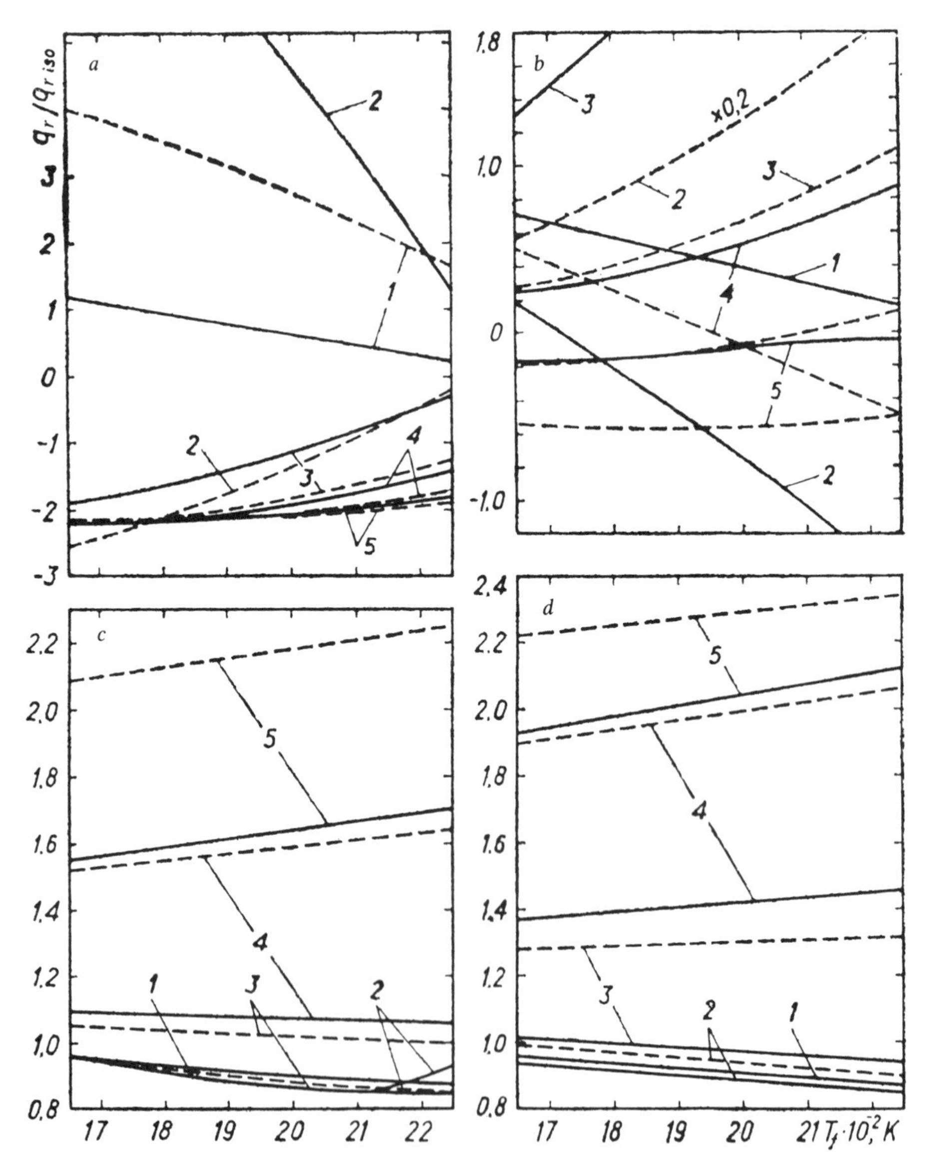


**FIG. 5.19** Effect of nonisothermicity of an infinite cylindrical volume for different molecular gases at $T_m$ = 1500 K and $T_w$ = 1000 K as a function of $T_f$. For the remaining legend, see Fig. 5.17.

radiating medium over the pertinent directions, and also the optical properties of the enclosures.

2. Determination of radiative heat flux. In this case the information on the magnitude of the radiative heat flux can be used for comparing analytic calculations with experiments, if the conditions necessary for item 1 above

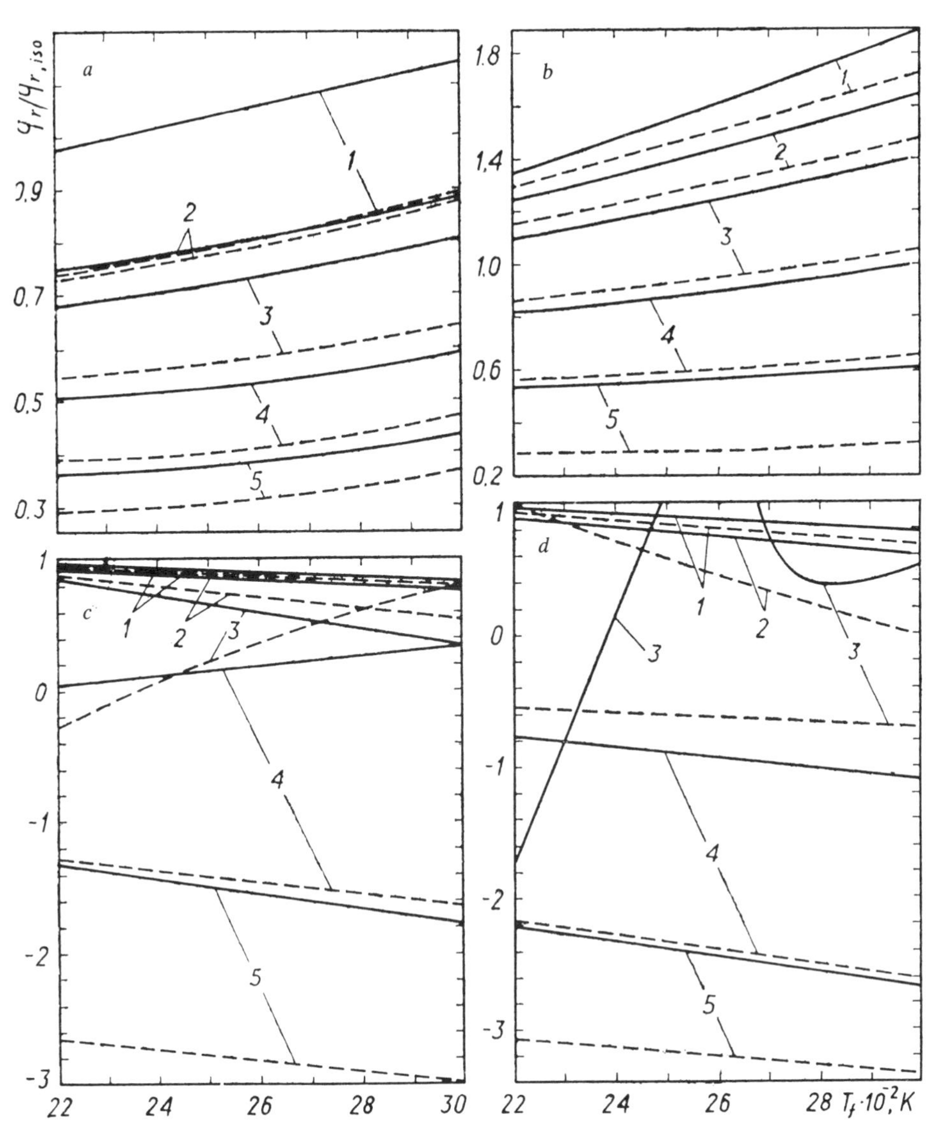


**FIG. 5.20** Effect of nonisothermicity of an infinite cylindrical volume for different molecular gases at $T_m = 2000$ K and $T_w = 1000$ K as a function of $T_f$. For the remaining legend, see Fig. 5.17.

have been satisfied. If, however, this information is incomplete, then the measurements of radiative heat flux can be used only for estimating certain variables of state of the medium.

As a rule, in experimental study of high-temperature fluxes, information about the state of these fluxes is quite limited. In such a case only a qualitative comparison of experimental and analytic results is possible.

The bulk of experiments on investigating radiative heat transfer of combustion products in cooled channels was performed by means of narrow-angle radiometers. Thus, two narrow-beam radiometers (Fig. 3.6) were installed in each measuring location of the rectangular test channel 3; these radiometers allowed measuring the radiation intensity in the channel cross-section in two directions. Since narrow-angle radiometers are calibrated against an ideal black body, the magnitude of their signal makes it rather easy to determine a quantity close to the total emissivity $\varepsilon_{\text{exp}}$ of the gas. The value of $\varepsilon_{\text{exp}}$ was determined upon incorporating a correction, making allowance for the difference between the walls of the channel and the sensing element of the radiometer using Eq. (3.4). Actually, the value of the $\varepsilon_{\text{exp}}$ thus determined contains within it the radiation intensity of the opposite wall; however, under experimental conditions the radiation of the channel walls was low as compared with that of the flow. The experimentally determined values of $q_{r,\text{exp}}$ or $\varepsilon_{\text{exp}}$ were compared with those calculated under the same conditions (Figs. 5.21 and 5.22). Calculations of the total emissivity of the gas were performed on the basis of the bulk temperature of the gas, determined from the measured temperature distributions of the channel or from the results of heat-balance calculations. Satisfactory agreement between the analytic and experimental results was obtained only under conditions of complete combustion. A significant deviation between the analytic values of $\varepsilon$ and $\varepsilon_{\text{exp}}$ was obtained at relatively low flow temperatures, when the oxidant was air diluted with a small quantity of oxygen, and $\alpha \leqslant 1$. Under these conditions there is a good possibility of incomplete combustion with formation of soot, which is responsible for the high values

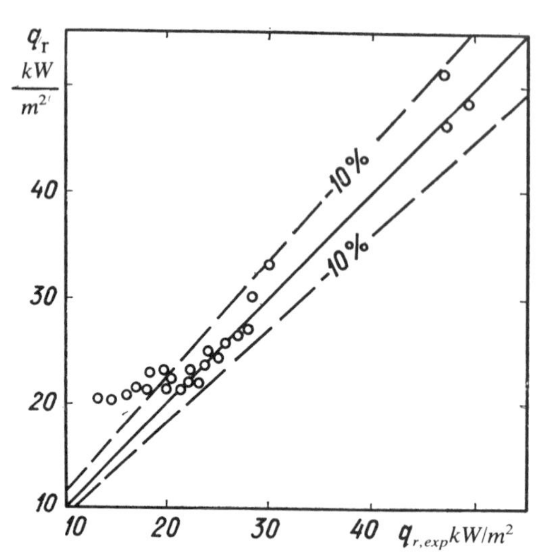


**FIG. 5.21** Comparison of measured radiative heat fluxes with the fluxes calculated analytically for test channel 4.

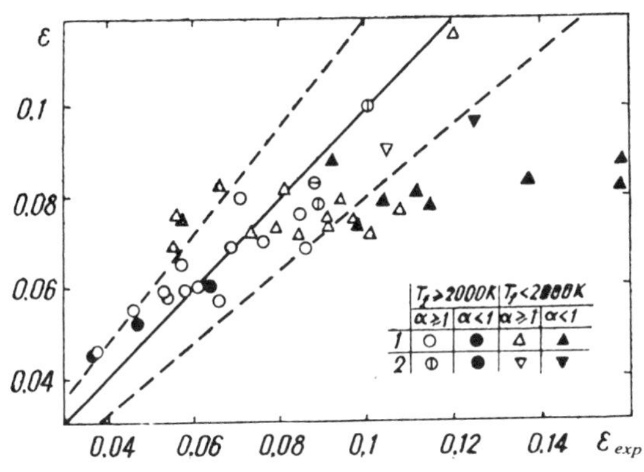


**FIG 5.22** Comparison of experimental and analytic values of the total emissivity of gas in channel 3 at different flow temperatures $T_f$ and oxidant excess ratio α. (1) In horizontal direction at $H=14.7$ cm; (2) in vertical direction at $H=10.2$ cm.

of $\varepsilon_{exp}$. The scatter in ε within ±20% as compared with $\varepsilon_{exp}$ should be regarded as satisfactory, if we consider the error in [determining] the optical properties and the inability to monitor the temperature and concentration fields in each of the experiments.

The greatest disagreement between analytic and experimental results was obtained for conditions of incomplete combustion, i.e., for luminous combustion products (Fig. 5.22). Naturally, in this case the value of ε must be calculated with allowance for the radiation of soot particles.

We have thus far considered the radiation of only the gaseous components of the medium. Actually, however, the combustion products contain to a lesser or greater degree soot particles that eradiate in the continuous spectrum. Calculations of radiation in such "luminous" flames is complicated by difficulties in preliminary determination of the concentration of soot particles and their optical properties. Detailed studies of methods of calculation of radiative heat flux in such dispersed media have been performed by Blokh [2, 145].

A more detailed analysis of the effect of dispersed particles on radiative heat transfer is not presented here, and comparison of analytic with experimental results on radiative heat flux is given for cases in which the radiation from solid particles can be neglected.

Since the analytic and experimental results on the total emissivity of gases under the complete combustion conditions presented in Fig. 5.22 agree within the limits of accuracy of determining the optical properties of gases, they were used for determining the radiative heat flux on the channel wall. The radiative heat flux was calculated using Eq. (2.148), whose accuracy was first checked in calculating the radiative heat flux in an infinite plane-parallel layer between two black enclosures. As seen from Figs. 5.23 and 5.24, data on the radiative heat flux calculated from Eq. (2.148) for

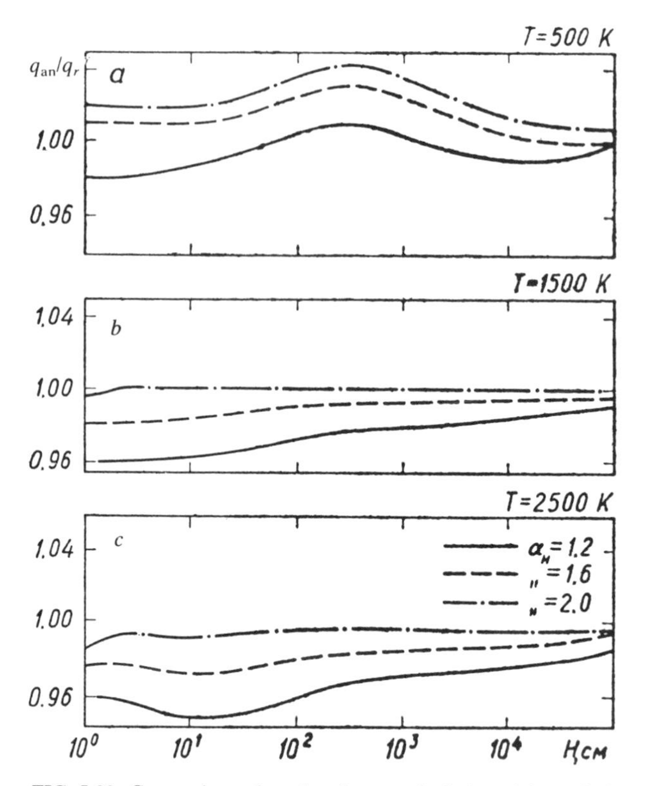


**FIG. 5.23** Comparison of results of exact calculation of the radiative heat flux of a planar layer $q_{an}$ using Eqs. (2.107)–(2.114) with approximate calculation of $q_r$ from Eq. (2.148) as a function of the layer thickness $H$ for gaseous $CO_2$ at different temperatures $T$ of the medium and ratios $\alpha_H = {}_2/H_1$.

water vapor and gaseous $CO_2$ over a wide range of parameters are in satisfactory agreement with results of exact calculations using Eqs. (2.107)–(2.116). For water vapor and gaseous $CO_2$, the error increases somewhat with reduction in the optical thickness of the layer, without, however, exceeding $\pm 10\%$. This method allows more exact incorporation of the geometric features of the medium as compared with other known methods, since it contains the radiation intensity at two different beam lengths $R_1$ and $R_2$ and also the effective radiation length $R_e$ of the given volume.

The radiation intensities $I(R_1)$ and $I(R_2)$ were calculated from expressions (2.149)–(2.152) for a hemispherical layer, which allow for radiation of the enclosures. Beam lengths $R_1$ and $R_2$ were calculated in such a manner that $R_1$ would be equal to the height $H$ of the plane-parallel layer, whereas $R_2 = \alpha_H H$. Note that the results of calculations depend only insignificantly on the selected value of $\alpha_H$. The best agreement is obtained at $\alpha_H = 2$.

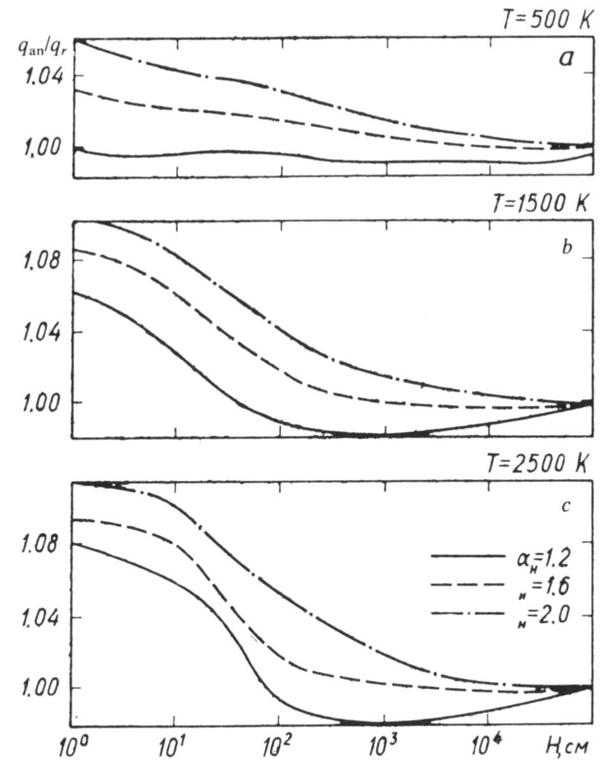


**FIG. 5.24** Comparison of results of exact calculation of the radiative heat flux of a planar layer $q_{an}$ using Eqs. (2.107)–(2.114) with approximate calculation of $q_r$ from Eq. (2.148) as a function of the layer thickness $H$ for water vapor at different temperatures $T$ of the medium and ratios $\alpha_H = H_2/H_1$.

An important advantage of this method is the fact that it suffices to calculate the effective beam length for a volume of any geometry only once. Its value was calculated not only for the channels employed in the experiments, but also for more general cases. The effective beam length calculated from Eq. (2.147) by the Monte Carlo method [130] (Fig. 5.25) is referred to the lower side surface in the case of rectangular channels or the linear dimension of the length of the cylindrical channel. This means that in order to calculate the total radiative heat flux, one should calculate from Eq. (2.148) in sequence the radiative heat fluxes for all the walls of the channel and sum up the results. The experimental values of radiative heat flux $q_{r,\mathrm{exp}}$ for channel 3 (Fig. 5.26) agree satisfactorily with the value of $q_r$ determined by calculating the radiation intensities from Eq. (2.148) under the same conditions. Calculations of $q_r$ were performed with allowance for the radiation of water vapor, gaseous $CO_2$, CO, and NO, on the basis of the equilibrium composition of combustion products.

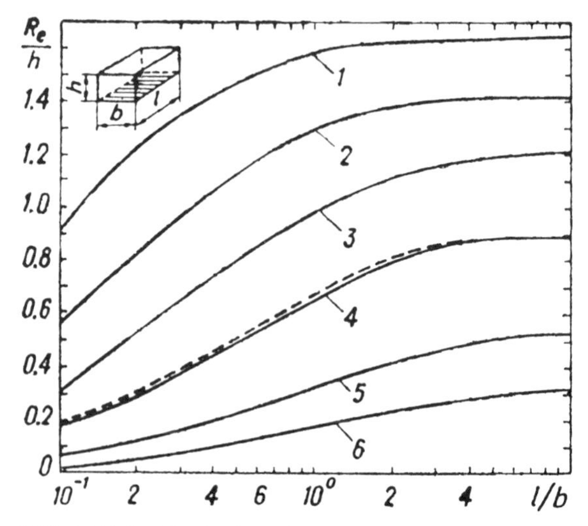


**FIG. 5.25** The effective beam length obtained from Eq. (2.147) for calculating the radiant heat flux to the lower wall of rectangular ducts. The dashed curve is for a cylindrical channel with diameter $h$. Values of $h/b$: (1) 0.1; (2) 0.25; (3) 0.5; (4) 1; (5) 2; (6) 5.

The experimental study of radiative heat flux performed by us confirmed the suitability of the selected values of the spectral optical properties of gases for calculating the radiation of combustion products, at least within the accuracy of determination of optical properties.

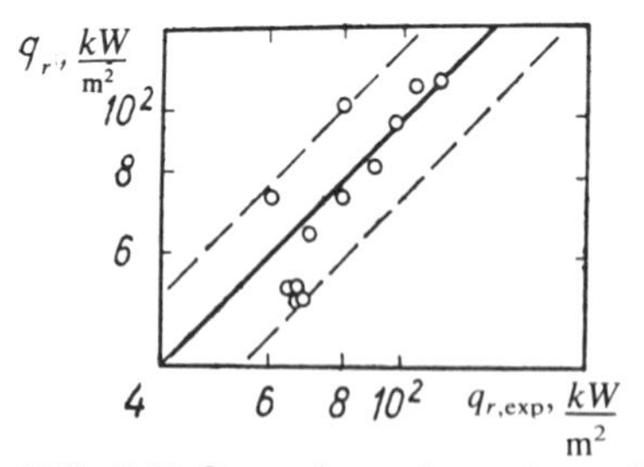


**FIG. 5.26** Comparison of experimental and analytic data of radiative heat fluxes under complete combustion conditions for channel 3.

CHAPTER

# SIX

# COMBINED HEAT TRANSFER

During the past two decades the problem of interaction between radiation and other forms of energy transfer in combined heat transfer has been attracting a great deal of attention. Accumulation of information in this field allows continuous refinement of the methods of investigation being used and of numerical methods of calculation.

Interest in these studies is due to the important role of combined heat transfer both in power generation and in industry. Combined heat transfer must be considered in the thermal design of furnace chambers and exhaust passages of boilers, industrial furnaces and gas turbines. Combined heat transfer is particularly important in power facilities with MHD generators and other high-temperature devices. Combined heat transfer is responsible for the effective operation of various kinds of melting and heating furnaces in metallurgy, glass-making furnaces, and many pieces of equipment employed in the chemical industry.

A survey of studies on radiant and combined heat transfer performed during 1966–1968 can be found in investigations by Viskanta and Grosh [88, 177], Cess and Sparrow [121, 178], Pai [80], Adrianov [86], and Filimonov and Khrustalev [179]. The studies performed during that period characteristically employed the gray-gas approximation as the main approach. It was found subsequently that selectivity in radiation of gases plays a very important role in radiant energy transport and that it is virtually impossible to use relationships governing gray-gas radiation.

Subsequent studies were directed toward investigating the governing relationships for radiant and combined heat transfer in selectively absorbing and emitting media. Surveys of these studies are given by Cess and Tiwari [180] and Biberman [181, 182]. An important role in present studies of combined heat transfer is played by numerical methods of calculation of radiation in real gases. Modern high-speed computers already allow the incorporation of quite real optical properties of gases. There appear studies which, in addition to radiation, also incorporate certain aspects of the motion of the radiating medium [30–33]; however, as a whole such studies are just about starting.

The investigation of this problem is hampered by the unavailability of reliable methods of experimental study. Thus, the frequently employed air blow-down and two-radiometer methods, measurement of molecular temperature of the medium in the vicinity of the heat-sensing surface, and others [183] have a number of shortcomings. In particular, the methods of air blow-down can, in the opinion of this author, be used for determining the radiant component only upon detailed accounting for the physical properties of the air, radiating medium, and turbulent flow features, which, as noted in Chapter 4, have a significant effect on convective heat transfer. For this reason, particular attention in the study of combined heat transfer must be given to analytic study of the interaction between radiation and other forms of energy transport, whose results can then be used as a mathematical model in analyzing experimental data on combined heat transfer.

## 6.1. SIMILITUDE PARAMETERS FOR COMBINED HEAT TRANSFER

The principal similitude parameters of a radiating medium were examined by Shorin [98] and Pai [80]. Given the integral nature of heat transfer in radiating media, it is difficult to hold a large number of similitude parameters invariant. Simulation of processes for selectively radiating media with finite optical thickness is virtually impossible, and conformance to certain similitude principles is possible only in the case of gray, optically thick media. In spite of this, experimental results for real media are frequently correlated on the basis of criteria of radiation gas dynamics, introduced for gray gases [184–189]. A wider use of such criteria for real media is hampered by lack of concrete recommendations on estimating the individual parameters contained in the criteria being used.

The governing relationships for interaction between different forms of energy transfer are now successfully expressed by relative simulation principles, when the reference parameters are ratios of certain quantities defined without allowance for interaction. Such simultation principles were used in radiation gas dynamics problems by Biberman [182] for describing combined heat transfer at the stagnation point of blunt-nosed bodies.

Let us consider the application of the principles of similitude theory for combined heat transfer in the more general formulation of the problem.

The substance of interaction between radiation and other forms of energy transport consists in the fact that the heat flux $q_\Sigma$ for combined heat transfer is not identical to the sum $q_\Sigma^0 = q_{\text{conv}}^0 + q_{\text{rad}}^0$,determined without allowance for their interaction. It is assumed here that quantity $q_\Sigma^0$ has been defined at the same boundary conditions, but was calculated without consideration for radiation, whereas the radiative flux $q_r^0$, contained in $q^0{}_\Sigma$ was determined from temperature fields due solely to convection, conduction, etc. This means that $q_{\text{conv}}^0$ can be determined from familiar expressions for convective heat transfer, whereas $q_r^0$ should be determined in the fundamental formulation of the problem.

All energy transfer involves a change in the total enthalpy of the medium, which, for combined heat transfer, can be written as:

$$\Delta h^* = \Delta h + \Delta h_r \tag{6.1}$$

where $\Delta h_r$ is the change in radiation enthalpy, $\Delta h$ is the change in the enthalpy of the medium owing to its thermal state, kinetic energy, energy of chemical transformations, etc. The term $\Delta h$ can be written as

$$\Delta h = c_{p,\,m}\,\Delta T + \frac{U^2}{2} + \Delta h_{\text{chem}} + \dots \tag{6.2}$$

According to Biberman [182] the total heat flux $q_\Sigma$ for combined heat transfer can be expanded in a Taylor series with respect to heat flux $q_\Sigma^0$ in the absence of interaction between radiation and other forms of energy transfer. If we remember that this interaction will vanish as $\Delta h^*$ approaches $\Delta h$, the series expansion of $q_\Sigma$ can be written in the form

$$q_\Sigma = q_\Sigma^0 + \left.\frac{dq_\Sigma}{dh^*}\right|_{\Delta h^* \approx \Delta h} \cdot (\Delta h^* - \Delta h) \tag{6.3}$$

and if we use similitude numbers for heat transfer, then

$$\mathrm{Nu}_\Sigma = \mathrm{Nu}_\Sigma^0 + \left.\frac{d\mathrm{Nu}_\Sigma}{dh^*}\right|_{\Delta h^* \approx \Delta h} \cdot (\Delta h^* - \Delta h) \tag{6.4}$$

When $\Delta h^* \to \Delta h$, the above equations can be significantly simplified:

$$\frac{q_\Sigma}{q_\Sigma^0} = 1 + \frac{\Delta h}{q_\Sigma^0} \cdot \frac{dq_\Sigma^0}{dh} \cdot \frac{\Delta h_r}{\Delta h} \tag{6.5}$$

or

$$\frac{\mathrm{Nu}_\Sigma}{\mathrm{Nu}_\Sigma^0} = 1 + \frac{\Delta h}{\mathrm{Nu}_\Sigma^0} \cdot \frac{d\mathrm{Nu}_\Sigma^0}{dh} \cdot \frac{\Delta h_r}{\Delta h} \tag{6.6}$$

This means that ratio $q_\Sigma/q_\Sigma^0$ is a function of the nondimensional parameter

$$\chi \equiv \frac{\Delta h}{q_\Sigma^0}\cdot\frac{dq_\Sigma^0}{dh}\cdot\frac{\Delta h_r}{\Delta h} \tag{6.7}$$

or

$$\chi \equiv \frac{\Delta h}{\mathrm{Nu}_\Sigma^0}\cdot\frac{d\mathrm{Nu}_\Sigma^0}{dh}\cdot\frac{\Delta h_r}{\Delta h} \tag{6.8}$$

Actually the linearity of $q_\Sigma/q_\Sigma^0$ or $\mathrm{Nu}_\Sigma/\mathrm{Nu}_\Sigma^0$ as a function of $\chi$ can prevail only at low values of the latter. Hence the real dependence of these ratios on $\chi$ should be determined from analytic or experimental results on combined heat transfer. This means that $\chi$, defined in terms of known quantities, can serve as a similitude criterion in combined heat transfer.

In determining the value of $\chi$, the ratio of enthalpy differences $\Delta h_p/\Delta h$ is best defined in terms of heat flux density. For this purpose we shall consider the balance of increments of heat fluxes, which in general can be represented in the form of the equation [98]

$$\mathrm{div}\,\vec{q}_{\mathrm{M},\,d}+\mathrm{div}\,\vec{q}_{\mathrm{T},\,d}+\mathrm{div}\,\vec{q}_r+\mathrm{div}\,(\rho\,\vec{U}\,\Delta h)=q_{\mathrm{sc}} \tag{6.9}$$

where the heat flux density due to molecular diffusion is

$$\vec{q}_{\mathrm{M},\,d}=\rho\,\vec{V}_{\mathrm{M},\,d}\,\Delta h \tag{6.10}$$

the heat flux vector due to eddy diffusion is

$$\vec{q}_{\mathrm{T},\,d}=\rho\vec{V}_{\mathrm{T},\,d}\,\Delta h \tag{6.11}$$

and the heat flux vector due to radiant energy transfer is

$$\vec{q}_r=\rho\vec{c}_0\,\Delta h_r \tag{6.12}$$

The remaining terms in Eq. (6.9) designate divergence of convective energy transfer and strength of internal energy sources of the medium. Depending on the predominace of these and other processes, the balance of increments of heat fluxes can be significantly simplified. Let us consider the most important particular cases.

1. In laminar flow of a radiating medium without internal energy sources the balance of increments of heat flux vectors is written in the form:

$$\mathrm{div}\,\vec{q}_{\mathrm{M},\,d}+\mathrm{div}\,\vec{q}_r+\mathrm{div}\,(\rho\vec{U}\,\Delta h)=0 \tag{6.13}$$

Using expressions for the heat flux vectors, Eq. (6.13) can be represented in the form:

$$\operatorname{div}\left[\vec{q}_r\left(1+\frac{\rho \vec{V}_{M,d}\,\Delta h}{\rho \vec{c}_0\,\Delta h_r}\right)\right] = -\operatorname{div}(\rho \vec{U}\,\Delta h) \tag{6.14}$$

Since $$(\rho V_{M,d}\,\Delta h)/(\rho \vec{c}_0,\,\Delta h_r) >> 1$$

and dropping symbolic operations over individual quantities while retaining only dimensional quantities, we can obtain the similitude ratio

$$\frac{q_r^0}{H}\cdot\frac{\rho V_{M,d}\,\Delta h}{\rho c_0\,\Delta h_r} \sim \frac{\rho U \Delta h}{H} \quad \text{or} \quad \frac{\Delta h_p}{\Delta h} \sim \frac{q_r^0}{\rho U \Delta h} \tag{6.15}$$

where it is assumed that $V_{M,d}/c_0 \approx$ constant.

This means that for laminar flow of a radiating medium without internal energy sources, the similitude parameter $\chi$ becomes

$$\chi_I \equiv \frac{\Delta h}{q_\Sigma^0}\cdot\frac{dq_\Sigma^0}{dh}\cdot\frac{q_r^0}{\rho U \Delta h} \tag{6.16}$$

The change in total enthalpy in supersonic gas flows over bodies is basically due to its kinetic energy $U^2/2$. In this case,

$$\chi_I \equiv \frac{\Delta h}{q_\Sigma^0}\cdot\frac{dq_\Sigma^0}{dh}\cdot\frac{2q_r^0}{\rho U^3} \tag{6.17}$$

An analogous parameter

$$\chi_I \equiv \frac{d\ln q_\Sigma^0}{d\ln h}\cdot\frac{4q_r^0}{\rho U^3} \tag{6.18}$$

which differs from Eq. (6.17) only by a constant factor, was used by Biberman [182].

If the change in the total energy of the medium is controlled mainly by its thermal state, then

$$\chi_I \equiv \frac{(\Theta_0-1)}{q_\Sigma^0}\cdot\frac{dq_\Sigma^0}{d\Theta_0}\cdot\frac{q_r^0}{\rho U\cdot c_{p,m}\cdot(T_f-T_w)} \tag{6.19}$$

where $$\Theta_0 = T_f / T_w \tag{6.20}$$

On the other hand, an analogous analysis of Eq. (6.13) for hydrodynamically stabilized flow yields

$$\rho U \Delta h \sim q_\Sigma^0 \tag{6.21}$$

Then Eqs. (6.15) and (6.16) yield

$$\chi_{\mathrm{I}} \equiv \frac{(\Theta_0 - 1)}{\mathrm{Nu}_\Sigma^0} \cdot \frac{d\,\mathrm{Nu}_\Sigma^0}{d\Theta_0} \cdot \frac{\mathrm{Nu}_r^0}{\mathrm{Nu}_\Sigma^0} \tag{6.22}$$

where the similitude parameter for combined heat transfer in stabilized laminar flow is expressed in terms of the Nusselt number.

2. In turbulent flow of a medium without internal energy sources, the balance of heat flux vector increments is written as

$$\mathrm{div}\,\vec{q}_{\mathrm{T},d} + \mathrm{div}\,\vec{q}_r + \mathrm{div}\,(\rho\vec{U}\,\Delta h) = 0 \tag{6.23}$$

Using expressions for heat flux vectors, Eq. (6.23) can be represented in the form

$$\mathrm{div}\left[\vec{q}_r \cdot \left(1 + \frac{\rho\vec{V}_{\mathrm{T},d}\,\Delta h}{\rho\vec{c}_0\,\Delta h_r}\right)\right] = -\,\mathrm{div}\,(\rho\,\vec{U}\,\Delta h) \tag{6.24}$$

Since

$$\frac{\rho\vec{V}_{\mathrm{T},d}\,\Delta h}{\vec{\Delta h}_r} >> 1$$

and using similitude analysis, we obtain

$$\frac{q_r^0}{H} \cdot \frac{\rho V_{\mathrm{T},d}\,\Delta h}{\rho c_0\,\Delta h_r} \sim \frac{\rho U \Delta h}{H} \tag{6.25}$$

If it is remembered that for turbulent flow $V_{r,d}/U \approx$ constant, then Eq. (6.25) yields

$$\frac{\Delta h_r}{\Delta h} \sim \frac{q_r^0}{\rho\Delta h} \equiv \mathrm{St}_r^0 \tag{6.26}$$

In the denominator of Eq. (6.26) we dropped the speed of light $c_0$ since it is a constant, but, in spite of this, this ratio can be regarded as a Stanton number of a kind for radiative heat transfer.

Then similitude parameter $\chi$ for turbulent flow of a radiating medium without internal energy sources becomes

$$\chi_{\mathrm{II}} \equiv \frac{\Delta h}{\mathrm{Nu}_\Sigma^0} \cdot \frac{d\mathrm{Nu}_\Sigma^0}{dh} \cdot \mathrm{St}_r^0 \tag{6.27}$$

Depending on the predominance of individual modes of energy transfer, parameter $\chi_{\mathrm{II}}$ may have different forms. If the change in the total enthalpy is defined by its thermal state, then $\Delta h = c_{p,m}(T_f - T_w)$ and

$$\chi_{\mathrm{II}} \equiv \frac{(\Theta_0 - 1)}{\mathrm{Nu}_\Sigma^0} \cdot \frac{d\mathrm{Nu}_\Sigma^0}{d\Theta_0} \cdot \mathrm{St}_r^0 \tag{6.28}$$

3. For a medium with internal energy sources, the expressions for the similitude parameters may differ depending on the nature of the medium's flow. For a nonmoving medium the balance of increments of heat flux vectors is

$$\operatorname{div}(\vec{q}_{\mathrm{M},d}) + \operatorname{div}(\vec{q}_r) = q_{\mathrm{sc}} \tag{6.29}$$

It can be shown by analogous transformations of Eq. (6.29) that in this case

$$\frac{\Delta h_r}{\Delta h} \sim \frac{q_r^0}{q_{\mathrm{sc}}\, H} \tag{6.30}$$

where it is also assumed that $V_{M,d}/c_0 \approx$ constant. Using Eq. (6.30), the similitude parameter of combined heat transfer in a nonmoving medium with internal heat sources, expressed in terms of heat fluxes, is written as

$$\chi_{\mathrm{III}} \equiv \frac{\Delta h}{q_\Sigma^0} \cdot \frac{dq_\Sigma^0}{dh} \cdot \frac{q_r^0}{q_{\mathrm{sc}}\, H} \tag{6.31}$$

4. In laminar flow of a medium with internal energy sources, the balance of heat flux increment vectors is written as

$$\operatorname{div}(\vec{q}_{\mathrm{M},d}) + \operatorname{div}(\vec{q}_r) + \operatorname{div}(\rho\, \vec{U}\, \Delta h) = q_{\mathrm{sc}} \tag{6.32}$$

In this case the expression for the similitude parameter of combined heat transfer will depend on the ratio $q_{\mathrm{is}}\, H/\varrho U \Delta h$. For the given problem at $q_{\mathrm{is}}\, H/\varrho U \Delta h << 1$, parameter $\chi$ can be obtained from Eq. (6.16). When $q_{\mathrm{is}} H/\varrho U \Delta h >> 1$, the divergence of convective energy transport in Eq. (6.32) can be neglected and $\chi$ can be expressed by Eq. (6.31).

5. The balance of heat flux vector increments for turbulent flow is

$$\operatorname{div}(\vec{q}_{\mathrm{T},d}) + \operatorname{div}(\vec{q}_r) + \operatorname{div}(\rho\, \vec{U}\, \Delta h) = q_{\mathrm{sc}} \tag{6.33}$$

As in the case of laminar flow, there are two cases possible, depending on the strength of the energy source. At $q_{\mathrm{is}} H/\varrho U \Delta h << 1$, the similitude parameter for this problem can be obtained from Eq. (6.28). At $q_{\mathrm{is}} H/\varrho U \Delta h >> 1$, neglecting the divergence of convective energy transport, we obtain

$$\frac{\Delta h_r}{\Delta h} \sim \frac{q_r^0 U_f}{q_{\mathrm{sc}}\, H}$$

where the speed of light $c_0$ has been dropped and it is assumed, as in the previously examined problem, that the turbulent diffusion rate $V_{T,d}$ is proportional to the free-stream velocity $U_f$. Using Eq. (6.32), parameter $\chi$ can be represented in the form

$$\chi_{\mathrm{IV}} \equiv \frac{\Delta h}{q_\Sigma^0} \cdot \frac{dq_\Sigma^0}{dh} \cdot \frac{q_r^0 U_f}{q_{\mathrm{sc}}\, H} \tag{6.34}$$

Similitude parameters for combined heat transfer for other specific problems can be obtained analogously.

In general the derivatives in Eqs. (6.16), (6.22), (6.27)–(6.28), (6.31) and (6.34) can always be determined numerically upon an insignificant change in the argument. This quantity in general allows for the rate of rise in heat transfer as a function of the change in the total enthalpy. The derivatives of heat flux with respect to total enthalpy are most conveniently defined by the nondimensional expressions

$$\mathrm{Nu}_c^0 = \mathrm{Nu}_{c,0}^0 \cdot \left(\frac{T_f}{T_w}\right)^{m_c} \tag{6.35}$$

and

$$\mathrm{Nu}_r^0 = \mathrm{Nu}_{r,0}^0 \cdot \left(\frac{T_f}{T_w}\right)^{m_r} \tag{6.36}$$

Using the fact that $\Delta h = c_{p,m}(T_f - T_w)$, the derivative contained in the expression for $\chi$ can be written as

$$\frac{\Delta h}{\mathrm{Nu}_\Sigma^0} \cdot \frac{d\mathrm{Nu}_\Sigma^0}{dh} = \frac{m_K \, \mathrm{Nu}_c^0 + m_r \, \mathrm{Nu}_r^0}{\mathrm{Nu}_\Sigma^0} \cdot \frac{\Theta_0 - 1}{\Theta_0} \tag{6.37}$$

This means that in defining $\chi$ it becomes possible to extensively use the well-explored relationships governing convective heat transfer in non-radiating media.

The practical application of the approximate similitude parameters will be shown in more detail in solving specific problems of combined heat transfer.

## 6.2. CONDUCTIVE–RADIATIVE HEAT TRANSFER IN A PLANAR LAYER

Investigation of the relationships governing energy transfer and the temperature field in an absorbing and radiating medium between two infinite parallel plates is of great cognitive and practical significance. In many cases the temperature distribution in semitransparent materials such as glass, semiconductors, glass melt, artificial crystals, etc., plays an important role in the technology of their production. Frequently the calculation of heating of materials used in furnace technology is reduced to solution of such problems. Depending on the nature of the practical application, problems of this class can have highly varied boundary conditions. Many problems reduce to the following elementary cases: (1) when there is no release of heat in the layer, and only the surface temperatures or heat flux densities at the layer enclosures are specified, i.e., the heat is released only within the walls; and (2) when the distribution of heat sources in a layer of radiating gas is specified.

The problem of conductive–radiative heat transfer is expressed by the equation

$$\frac{\partial}{\partial y}\left(\lambda \frac{\partial T}{\partial y}\right)+\frac{\partial q_r}{\partial y}+S(y)=0 \tag{6.38}$$

which is a particular case of energy equation (2.11) for a nonmoving medium with internal heat sources.

Solutions of Eq. (6.38) for a gray medium with constant physical properties were obtained for the first time by Viskanta and Grosh [88, 177]. Some limiting cases of solution of the problem of radiation together with heat conduction are examined by Wang and Tien [192]. The combined effect of radiation and conduction on temperature- and frequency-dependent optical properties of gases is examined by Grief [193], who assumed that the thermal conductivity and absorption coefficients are a power-law function of temperature.

A number of problems of radiative–conductive heat transfer were solved by Rubtsov and his co-workers [194, 195]. These studies were intended for investigating rather complex boundary conditions of radiative energy transfer.

The most detailed analysis of interaction between conduction and radiation in a layer of molecular gases ($CO$, $CO_2$, $H_2O$ and $CH_4$) was performed by Cess and Tiwari [196]. The optical properties of molecular gases were were described by the wide-band Edwards model, and the enclosures were assumed to be black. The temperature dependence of thermal conductivity was not considered. The authors limited themselves basically to parametric calculation of temperature distributions in a layer for different values of nondimensional parameters characterizing the problem.

We now consider in more detail the problem of conductive–radiative transport without internal energy sources:

$$\frac{\partial}{\partial y}\left(\lambda \frac{\partial T}{\partial y}\right)=-\frac{\partial q_r}{\partial y} \tag{6.39}$$

for specified surface temperatures

$$T=T_1 \quad \text{at} \quad y=0 \tag{6.40}$$

$$T=T_2 \quad \text{at} \quad y=H \tag{6.41}$$

This problem will be solved numerically using the technique presented in Chapter 2.

Using the nondimensional variables

$$\eta_* = \frac{y}{H} \tag{6.42}$$

$$\Theta = \frac{T}{T_1} \tag{6.43}$$

$$\bar{q}_r = \frac{q_r \cdot H}{\lambda_1 T_1} \tag{6.44}$$

and auxiliary function $S_\Theta$, defined by the differential equation

$$\frac{\partial S_\Theta}{\partial \eta_*} = \frac{\lambda}{\lambda_1} \cdot \frac{\partial \Theta}{\partial \eta_*} \tag{6.45}$$

we write Eq. (6.39) in the form

$$\frac{\partial^2 S_\Theta}{\partial \eta_*^2} + S_\Theta = S_\Theta - \frac{\partial \bar{q}_r}{\partial \eta_*} \tag{6.46}$$

with boundary conditions

$$\Theta = 1 \quad \text{at} \quad \eta_* = 0 \tag{6.47}$$

$$\Theta = \Theta_2 = \frac{T_2}{T_1} \quad \text{at} \quad \eta_* = 1 \tag{6.48}$$

For convenience of calculations function $S_\Theta$ has been added to both sides of Eq. (6.46). In iterative calculations the right-hand side of Eq. (6.46) will be treated as a function of coordinate $\eta_*$ (it can be determined from the preceding approximation of the temperature field). Then the general solution of the nonhomogeneous equation (6.46) can be written as

$$S_\Theta(\eta_*) = C_1(\eta_*) \cdot \sin \eta_* + C_2(\eta_*) \cdot \cos \eta_* \tag{6.49}$$

The method of variation of constants allows representing the general solution of Eq. (6.49) in the form

$$S_\Theta(\eta_*) = \sin \eta_* \cdot \int_0^{\eta_*} \cos \eta_1 \cdot \left(S_\Theta^0 - \frac{\partial \bar{q}_r}{\partial \eta_1}\right) d\eta_1 + c_1 \cdot \sin \eta_* -$$

$$- \cos \eta_* \cdot \int_0^{\eta_*} \sin \eta_1 \cdot \left(S_\Theta^0 - \frac{\partial \bar{q}_r}{\partial \eta_1}\right) d\eta_1 + c_2 \cdot \cos \eta_* \tag{6.50}$$

where quantity $S_\Theta^0$, as function of $\eta^*$, is defined from temperature fields of the preceding approximation.

Integrating Eq. (6.50) with respect to $\eta_*$ and substituting the derivative $S_\Theta$ into Eq. (6.49), we obtain

$$\frac{\partial \Theta}{\partial \eta_*} = G(\eta_*) + c_1 \cdot \frac{\lambda_1}{\lambda} \cos \eta_* - c_2 \cdot \frac{\lambda_1}{\lambda} \cdot \sin \eta_* \tag{6.51}$$

where

$$G(\eta_*)=\frac{\lambda_1}{\lambda}\cdot\left[\cos\eta_*\cdot\int_0^{\eta_*} S_\Theta^0\cdot\cos\eta_1\,d\eta_1+\sin\eta_*\int_0^{\eta_*} S_\Theta^0\cdot\sin\eta_1\,d\eta_1+\right.$$

$$+\sin\eta_*\cdot\int_0^{\eta_*}\bar{q}_r\cdot\cos\eta_1\,d\eta_1-\cos\eta_*\cdot\int_0^{\eta_*}\bar{q}_r\cdot\sin\eta_1\,d\eta_1+$$

$$\left.+\bar{q}_r(0)\cdot\cos\eta_*-\bar{q}_r\right] \tag{6.52}$$

This means that in defining function $G(\eta_*)$ it is necessary to calculate, for each step of iteration, the field of the nondimensional radiative heat flux $\bar{q}_r$ and function $S_\Theta^0$. The nondimensional radiative heat flux $\bar{q}_r$ was defined from equations for radiation in a planar layer, performed in Chapter 2, with allowance for the values of the temperature field of the preceding iteration. Function $S_\Theta^0$ was calculated by numerical integration of Eq. (6.45):

$$S_\Theta^0=\int_0^{\eta_*}\frac{\lambda}{\lambda_1}\cdot\frac{\partial\Theta}{\partial\eta_*}\,d\eta_* \tag{6.53}$$

where the integrand was also determined from data of temperature fields of the preceding approximation.

Integration of Eq. (6.52) allows one to obtain the following expression for the temperature field with two unknown constants in

$$\Theta=1+\int_0^{\eta_*}G(\eta_1)\cdot d\eta_1+c_1\cdot\int_0^{\eta_*}\frac{\lambda_1}{\lambda}\cdot\cos\eta_1\,d\eta_1-$$

$$-c_2\cdot\int_0^{\eta_*}\frac{\lambda_1}{\lambda}\cdot\sin\eta_1\,d\eta_1 \tag{6.54}$$

integrating which, in boundary condition (6.47) has already been incorporated. The second boundary condition (6.48) yields the expression

$$\Theta_2=1+\int_0^1 G(\eta_*)\,d\eta_*+c_1\int_0^1\frac{\lambda_1}{\lambda}\cos\eta_*\,d\eta_*-$$

$$-c_2\int_0^1\frac{\lambda_1}{\lambda}\sin\eta_*\,d\eta_* \tag{6.55}$$

The additional condition used was that the process is steady, which is expressed by equating the total heat fluxes at both walls:

$$q_{w,1}=q_{w,2} \tag{6.56}$$

or

$$-\frac{\partial\Theta}{\partial\eta_*}\bigg|_{\eta_*=0}+\bar{q}_r(0)=-\frac{\lambda_2}{\lambda_1}\cdot\frac{\partial\Theta}{\partial\eta_*}\bigg|_{\eta_*=1}+\bar{q}_r(1) \tag{6.57}$$

It must be taken into account in Eq. (6.57) that

$$\frac{\partial\Theta}{\partial\eta_*}\bigg|_{\eta_*=0}=c_1, \tag{6.58}$$

and

$$\frac{\partial\Theta}{\partial\eta_*}\bigg|_{\eta_*=1}=G(1)+c_1\cdot\frac{\lambda_1}{\lambda_2}\cdot\cos(1)-c_2\cdot\frac{\lambda_1}{\lambda_2}\cdot\sin(1) \tag{6.59}$$

Equations (6.58) and (6.59) were obtained from Eq. (6.51) at the corresponding values of coordinate $\eta_*$. Equations (6.55) and (6.57)–(6.59) were used in defining constants of integration $c_1$ and $c_2$.

The problem was solved in the zero approximation by specifying a linear temperature field. Then the field of values of thermal conductivity $\lambda(\eta_*)$ was calculated and auxiliary funtion $S_\Theta^0(\eta_*)$ was determined from Eq. (6.53) by numerical integration by the method of trapezoids or parabolas.

The values of thermophysical properties of combustion products and individual gaseous components of these products were determined in accordance with the recommendations of the present author and his coworkers [153]. Numerical integration was performed with the coordinate $\eta_*$ subdivided uniformly into 200–300 parts. The field of radiative heat flux was determined by subdividing the layer thickness into 20 slices. The width of these sections was selected either in accordance with the cosine law, or by subdividing the layer into parts with the same temperature difference. At the remaining points the radiative heat flux was determined by linear interpolation.

After determining the field of radiative heat flux $\bar{q}_r(\eta_*)$, auxiliary function $G(\eta_*)$ was calculated from Eq. (6.52), and then all the expressions contained in Eqs. (6.55)–(6.59) were defined. This made it possible to determine constants of integration $c_1$ and $c_2$ for each iteration step, whereupon Eq. (6.51) was used for calculating the field temperature derivatives $\Theta'_{\eta^*}$, and Eq. (6.54) for calculating the temperature field $\Theta$. For the convergence of the solutions, the new temperature field was determined with allowance for the damping factor $\delta_n$ from the equation

$$\Theta^n=\Theta^{n-1}+(\Theta-\Theta^{n-1})\cdot\delta_n \tag{6.60}$$

where $n$ is the number of iteration. The damping factor was determined by requiring that the new temperature field $\Theta^n$ differ from the preceding by not more than 10%. The criterion for convergence was a change in the temperature derivative at $\eta_* = 0$. To attain third decimal point accuracy of this derivative required from 10 to 30 iterations.

This problem was solved numerically for different conditions. Characteristic temperature fields for layers of gaseous $CO_2$ and water vapor at different optical thicknesses are plotted in Figs. 6.1 and 6.2. An increase in optical thickness of the layer results in a reduction in temperature about the hotter wall and in a rise in temperature near the cooler wall. These changes depend little on the kind of gas.

The effect of the temperature of enclosures is illustrated by Fig. 6.3, which represents temperature fields at different values of $\Theta_2$ for a layer of gaseous $CO_2$. Calculations performed at different pressures of the medium show that raising the pressure to twice its previous value has a different effect on the temperature field than an analogous increase in the layer thickness (Fig. 6.4).

The above analysis of interaction between conductive and radiative heat transfer can yield only a more general idea about the nature of these processes. According to results of calculations, radiation has a significant effect on the distribution of temperature fields within the gas layer. However, over the range of parameters under study, radiation was not found to have a significant effect on combined heat transfer (Table 6.1). As a result of distortion of the temperature field, the change in conductive heat transfer is virtually balanced by changes in the radiative heat flux.

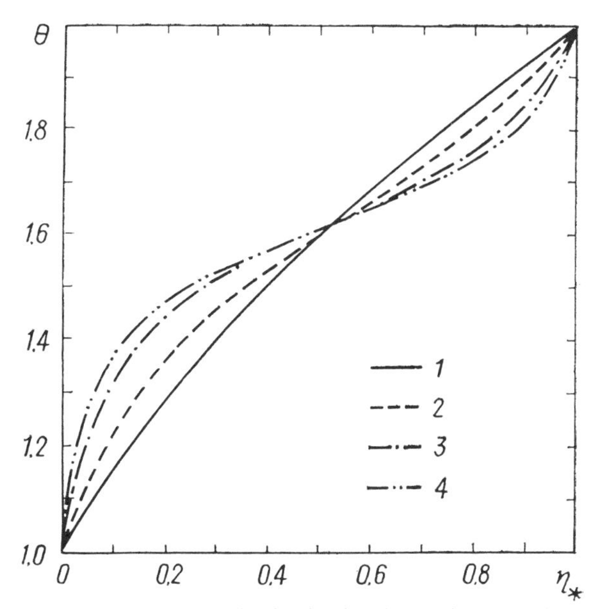


**FIG. 6.1** Temperature distribution in a layer of gaseous $CO_2$, bounded by ideal black-body plane-parallel enclosures with temperatures $T_1 = 1000$ K and $T_2 = 2000$ K. Layer thickness (cm): (1) 0.1; (2) 2; (3) 5; (4) 10. Here the relative layer thickness $\eta_*$ is obtained from Eq. (6.42) and the relative temperature $\Theta$ from Eq. (6.43).

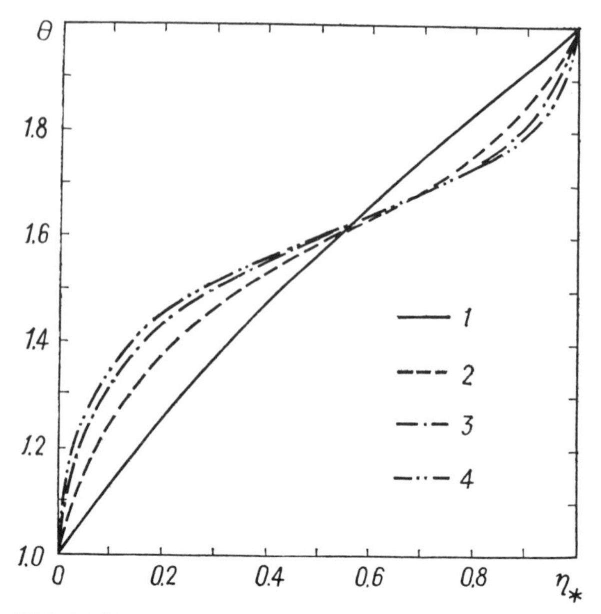


**FIG. 6.2** Temperature distribution in a layer of water vapor, bounded by ideal black-body plane-parallel enclosures with temperatures $T_1 = 1000$ K and $T_2 = 2000$ K. For the remaining legend, see Fig. 6.1

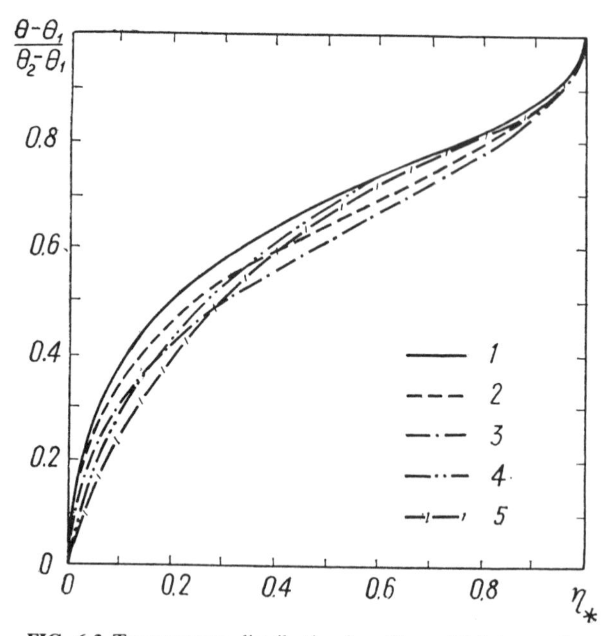


**FIG. 6.3** Temperature distribution in a 10-cm-thick layer of gaseous $CO_2$, bounded by ideal black-body plane-parallel enclosures at temperatures $T_1 = 500$ K and different $T_2$. $\theta_2$: (1) 2; (2) 1.8; (3) 1.6; (4) 1.4; (5) 1.2.

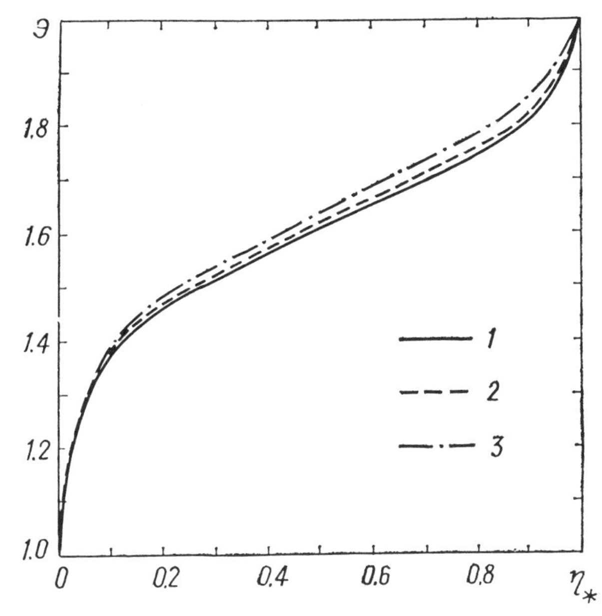


**FIG. 6.4.** Effect of pressure of medium on the temperature distribution in a 10-cm-thick layer of gaseous $CO_2$, bounded by ideal black-body plane-parallel enclosures with temperature $T_1 = 1000$ K and $T_2 = 2000$ K. Pressure of medium $p$(atm): (1) 1; (2) 2.5; (3) 10.

The governing relationships for combined heat transfer under these conditions can be analyzed by similitude parameter $\chi_{\rm III}$ [Eq. (6.31)], obtained for a nonmoving medium with internal heat sources. Since in the case at hand the energy sources are concentrated within the walls, their strength can be defined by the equation

$$q_{\rm sc} = \frac{\sigma(T_2^4 - T_1^4)}{H} \tag{6.61}$$

Since for the problem under study the temperatures of the enclosures are known, it can be assumed that $\Delta h \approx c_{p,m}(T_2 - T_1)$.Then similitude parameter $\chi$ for combined heat transfer is

$$\chi_{\rm III} \equiv \frac{\Theta_2 - 1}{q_\Sigma^0} \cdot \frac{dq_\Sigma^0}{d\Theta_2} \cdot \frac{q_r^0}{\sigma(T_2^4 - T_1^4)} \tag{6.62}$$

It was established that, in spite of the wide range of investigated values of $\chi_{\rm III}$, combined heat transfer increases on the average only by 2% (Table 6.1). In this analysis the conductive heat flux $q_{\rm cond}^0$ was determined by solving the purely conductive problem, and the radiant heat flux $q_r^0$ was calculated from the teperature field of conductive heat transfer.

One could expect changes in the relationships governing combined conductive–radiative heat transfer at significantly greater thickness of the

**TABLE 6.1** Results of Calculation of Conductive–Radiative Heat Transfer in an Infinite Plane-Parallel Layer

| Gas | Height $H$, cm | $p$, atm | $T_1$, K | $T_2$, K | Predicted value $y$, cm | $q_r^\circ$, kW/m² | $q_K^\circ$ and kW/m² | $q_r$, kW/m² | $q$ and, kW/m² | $\frac{dq_\Sigma^\circ}{d\Theta_2}$ | $\frac{q_\Sigma}{q_\Sigma^\circ}$ | $\chi_{II}$ |
|---|---|---|---|---|---|---|---|---|---|---|---|---|
| $CO_2$ | 0.1 | 1 | 1000 | 2000 | 0 | 632.6 | 95.4 | 632.6 | 98.0 | 1160 | 1.004 | 1.39 |
| | | | | | 0.1 | 633.5 | 95.4 | 633.5 | 97.2 | 1163 | 1.006 | 1.39 |
| $CO_2$ | 2 | 1 | 1000 | 2000 | 0 | 591.2 | 4.77 | 595.7 | 12.6 | 993.5 | 1.021 | 1.65 |
| | | | | | 2 | 597.7 | 4.77 | 596.9 | 11.4 | 1004 | 1.009 | 1.65 |
| $CO_2$ | 5 | 1 | 1000 | 2000 | 0 | 561.4 | 1.91 | 571.4 | 8.83 | 947.6 | 1.030 | 1.71 |
| | | | | | 5 | 571.4 | 1.91 | 571.2 | 9.10 | 963.9 | 1.012 | 1.68 |
| $CO_2$ | 10 | 1 | 1000 | 2000 | 0 | 532.0 | 0.95 | 547.4 | 6.47 | 903.3 | 1.039 | 1.69 |
| | | | | | 10 | 544.7 | 0.95 | 545.6 | 8.25 | 924.5 | 1.015 | 1.69 |
| $CO_2$ | 10 | 2.5 | 1000 | 2000 | 0 | 479.3 | 0.95 | 503.9 | 7.64 | 833.3 | 1.015 | 1.73 |
| | | | | | 10 | 499.2 | 0.95 | 500.6 | 10.9 | 867.9 | 1.023 | 1.73 |
| $CO_2$ | 10 | 10 | 1000 | 2000 | 0 | 373.9 | 0.95 | 411.6 | 9.23 | 668.4 | 1.123 | 1.78 |
| | | | | | 10 | 404.4 | 0.95 | 406.2 | 14.6 | 728.4 | 1.014 | 1.79 |
| $CO_2$ | 10 | 1 | 500 | 600 | 0 | 2.569 | 0.04 | 2.595 | 0.18 | 31.09 | 1.065 | 2.35 |
| | | | | | 10 | 2.615 | 0.04 | 2.715 | 0.06 | 31.90 | 1.046 | 2.37 |
| $CO_2$ | 10 | 1 | 500 | 700 | 0 | 6.846 | 0.08 | 7.049 | 0.36 | 49.88 | 1.071 | 2.85 |
| | | | | | 10 | 7.061 | 0.08 | 7.181 | 0.23 | 52.12 | 1.038 | 2.89 |
| $CO_2$ | 10 | 1 | 500 | 800 | 0 | 13.51 | 0.13 | 14.18 | 0.51 | 75.63 | 1.077 | 3.29 |
| | | | | | 10 | 14.09 | 0.13 | 13.99 | 0.70 | 79.95 | 1.033 | 3.34 |
| $CO_2$ | 10 | 1 | 500 | 900 | 0 | 23.39 | 0.19 | 24.60 | 0.87 | 109.5 | 1.078 | 3.68 |
| | | | | | 10 | 24.59 | 0.19 | 24.42 | 1.01 | 116.5 | 1.029 | 3.73 |
| $CO_2$ | 10 | 1 | 500 | 1000 | 0 | 37.39 | 0.25 | 38.83 | 1.68 | 172.1 | 1.076 | 4.54 |
| | | | | | 10 | 39.52 | 0.25 | 39.81 | 0.69 | 162.2 | 1.018 | 4.05 |
| $H_2O$ | 2 | 1 | 1000 | 2000 | 0 | 689.7 | 8.56 | 692.5 | 16.7 | 1230 | 1.013 | 1.74 |
| | | | | | 2 | 1665 | 8.56 | 1663 | 14.2 | 3679 | 1.002 | 2.19 |
| $H_2O$ | 2 | 5 | 1000 | 2000 | 0 | 634.3 | 8.56 | 641.5 | 26.9 | 1163 | 1.040 | 1.79 |
| | | | | | 5 | 1558 | 8.56 | 1561 | 17.0 | 3521 | 1.007 | 2.24 |
| $H_2O$ | 10 | 1 | 1000 | 2000 | 0 | 639.1 | 1.71 | 651.4 | 8.90 | 1131 | 1.030 | 1.76 |
| | | | | | 10 | 1563 | 1.71 | 1561 | 8.59 | 3480 | 1.003 | 2.22 |
| $H_2O$ | 10 | 5 | 1000 | 2000 | 0 | 527.9 | 1.71 | 557.3 | 11.7 | 1979 | 1.074 | 3.72 |
| | | | | | 10 | 1325 | 1.71 | 1328 | 14.7 | 3033 | 1.012 | 2.28 |

radiating layer. However, such calculations are made difficult by the appearance of instabilities in solutions, which are difficult to overcome when it is impossible to significantly increase the number of subdivisions $M$ into zones in determining the field of the radiative heat flux from Eq. (2.114). In addition, under such conditions there form zones with large temperature gradients, which may result in hydraulic instability of the medium. In this case it becomes meaningless to solve only one radiation–conduction problem.

The method for calculating conductive–radiative heat transfer in a planar layer developed here can be used not only for gaseous radiating media but also in solving a wide circle of similar problems.

## 6.3. LAMINAR CONVECTIVE–RADIATIVE HEAT TRANSFER IN A PLANAR DUCT

As shown in Chapter 2, interaction between convective and radiative heat transfer is in general described by a set of integrodifferential equations. As of now, only attempts at obtaining particular solutions of these equations are known.

The problem of flow of a gray gas in a planar duct of two infinite parallel diffusely radiating plates was solved for the first time by Viskanta [197]. It was assumed in solving the problem that the velocity field is known and corresponds to that in stabilized laminar flow of a nonradiating gas in the duct,

$$\frac{u}{U_m} = \frac{3}{2} - \frac{3}{2}\eta^2 \tag{6.63}$$

Subsequent studies [30, 31] incorporated not only the spectral optical properties of the medium on the basis of the wide-band Edwards model but also the flow mode of the medium.

In general, stabilized flow of a radiating gas in a planar duct is described by Eqs. (2.104)–(2.106). Let us consider the interaction between convective and radiative heat transfer for selectively radiating gas in a formulation analogous to that used by Viskanta [197]. If it is remembered that the velocity of the medium is given by Eq. (6.63), then it is no longer necessary to solve equations of motion (2.104). Energy equation (2.105) for a plane laminar layer simplifies significantly and can be written in the form

$$\frac{\partial^2 S_\Theta}{\partial \eta^2} = \Phi_T(\eta) \tag{6.64}$$

where $\Phi_T$ is obtained from Eq. (2.106) at $k = 0$, whereas auxiliary function $S_\Theta$ was found from the differential equation

$$\frac{\partial S_\Theta}{\partial \eta} = \frac{\lambda}{\lambda_w} \cdot \frac{\partial \Theta}{\partial \eta} \tag{6.65}$$

With the addition, as in the case of solving the problem of conductive–radiative heat transfer, of $S_\Theta$ to both sides of Eq. (6.64), we shall treat the right-hand side of the equation as a function of coordinate $\eta$:

$$\frac{\partial^2 S_\Theta}{\partial \eta^2} + S_\Theta = S_\Theta^0 (\eta) + \Phi_T (\eta) \tag{6.66}$$

Iteration solution of Eq. (6.66) can be obtained by the same technique as used for solving the problem of conductive–radiative heat flux (6.46). Here the boundary conditions are the following:

$$\Theta = 1 \quad \text{at} \quad \eta = 1 \tag{6.67}$$

$$\frac{\partial \Theta}{\partial \eta} = 0 \quad \text{at} \quad \eta = 0 \tag{6.68}$$

Boundary condition (6.68) expresses symmetricity of the flow about the axis of the channel ($\eta = 0$) and, according to Eq. (2.98), it can be replaced by the condition

$$\frac{\partial S_\Theta}{\partial \eta} = 0 \quad \text{at} \quad \eta = 0 \tag{6.69}$$

By analogy with the solution of the problem of conductive–radiative heat transfer (6.50), the solution of Eq. (6.66) can be written as

$$S_\Theta (\eta) = \sin \eta \cdot \int_0^\eta \cos \eta_1 \cdot (S_\Theta^0 + \Phi_T) \, d\eta_1 + c_1 \cdot \sin \eta -$$

$$- \cos \eta \cdot \int_0^\eta \sin \eta_1 \cdot (S_\Theta^0 + \Phi_T) \, d\eta_1 + c_2 \cdot \cos \eta \tag{6.70}$$

The derivative of $S_\Theta$ with respect to $\eta$ is

$$\frac{dS_\Theta}{d\eta} = \cos \eta \cdot \int_0^\eta \cos \eta_1 \cdot (S_\Theta^0 + \Phi_T) \, d\eta_1 + c_1 \cdot \cos \eta +$$

$$+ \sin \eta \cdot \int_0^\eta \sin \eta_1 \cdot (S_\Theta^0 + \Phi_T) \, d\eta_1 - c_1 \cdot \sin \eta \tag{6.71}$$

Using symmetricity conditions (6.68), we find

$$c_1 = 0 \tag{6.72}$$

Then the derivative of the temperature field for any iteration can be expressed in the form

$$\frac{\partial \Theta}{\partial \eta} = - G (\eta) \cdot \frac{\bar{q}_w}{1 - \Theta_m} + H (\eta) - c_2 \cdot \frac{\lambda_w}{\lambda} \cdot \sin \eta \tag{6.73}$$

where

$$G(\eta)=\frac{\lambda_w}{\lambda}\cdot\left[\frac{\sin\eta}{Q_m}\cdot\int_0^{\eta}(1-\Theta)\cdot\rho c_p u\cdot\sin\eta_1\,d\eta_1+\right.$$

$$\left.+\frac{\cos\eta}{Q_m}\cdot\int_0^{\eta}(1-\Theta)\cdot\rho c_p u\cdot\cos\eta_1\,d\eta_1\right] \tag{6.74}$$

$$H(\eta)=\frac{\lambda_w}{\lambda}\cdot\left[\sin\eta\cdot\int_0^{\eta}S_\Theta^0\cdot\sin\eta_1\,d\eta_1+\cos\eta\cdot\int_0^{\eta}S_\Theta^0\cdot\cos\eta_1\,d\eta_1+\right.$$

$$+\sin\eta\cdot\int_0^{\eta}\bar{q}_p(\eta_1)\cdot\cos\eta_1\,d\eta_1-\cos\eta\cdot\int_0^{\eta}\bar{q}_p(\eta_1)\times$$

$$\left.\times\sin\eta_1\,d\eta_1-\bar{q}_p(\eta)\right] \tag{6.75}$$

$$Q_m=\int_0^{\eta}\rho c_p u\,d\eta \tag{6.76}$$

Integration of Eq. (6.73) with respect to $\eta$, using boundary condition (6.67), yields

$$\Theta=1-\frac{\bar{q}_w}{1-\Theta_m}\cdot\int_0^{\eta}G(\eta_1)\cdot d\eta_1+\int_0^{\eta}H(\eta_1)\,d\eta_1-c_2\times$$

$$\times\int_0^{\eta}\frac{\lambda_w}{\lambda}\sin\eta_1\,d\eta_1. \tag{6.77}$$

It is seen that conditions (6.67) and (6.68) do not suffice for unique solution of the problem. One must additionally know either the temperature in the center of the flow, or the bulk temperature, or the value of $\bar{q}_w$, the heat flux at the wall. In this problem we specified the bulk temperature of the flow

$$\Theta_m=\frac{1}{Q_m}\cdot\int_0^{1}\rho c_p u\Theta\,d\eta \tag{6.78}$$

Constant of integration $c_2$ satisfying this condition was obtained from the equation

$$c_2=\left[1-\Theta_m-\frac{\bar{q}_w(0)}{1-\Theta_m}\cdot A_1+A_2\right]\Big/A_3 \tag{6.79}$$

where

$$A_1 = \frac{1}{Q_m} \cdot \int_0^1 \rho c_p u \int_0^\eta G(\eta_1)\, d\eta_1\, d\eta \tag{6.80}$$

$$A_2 = \frac{1}{Q_m} \cdot \int_0^1 \rho c_p u \int_0^\eta H(\eta_1)\, d\eta_1\, d\eta \tag{6.81}$$

$$A_3 = \frac{1}{Q_m} \cdot \int_0^1 \rho c_p u \int_0^\eta \frac{\lambda_w}{\lambda} \cdot \sin \eta_1\, d\eta_1\, d\eta \tag{6.82}$$

Equation (6.77) was solved with the aid of the equality

$$\left.\frac{\partial \Theta}{\partial \eta}\right|_{\eta=1} = -G(1) \cdot \frac{\bar{q}_w(0)}{1-\Theta_m} + H(1) - c_2 \cdot \sin 1 \tag{6.83}$$

obtained from Eq. (6.73).

The zero approximation was obtained by specifying a temperature field obtained in solving the same problem for a nonradiating gas. Numerical integration was performed by uniformly subdividing coordinate $\eta$ into 200–300 segments. As in the case of conductive–radiative heat transfer, the radiant flux was determined by slicing the layer into 20 layers with subsequent linear integration. The convergence of this computational scheme is rather good. Obtaining an accuracy to the third decimal point requires 15–20 iterations. For convergence of the solutions for optically thick media, the temperature field was determined employing the damping factor from Eq. (6.60).

In the case of the constant physical properties, the solution of the energy equation yields Nu = 7.541, which is in satisfactory agreement with the value obtained by another method [197].

Calculations show that in laminar stabilized flow the radiation produces a significant equalization of the temperature field (Figs. 6.5–6.8). In conjunction with this, the temperature gradient about the channel walls rises, i.e., the interaction between radiative and convective energy transports results in enhancement of convective heat transfer. At high radiative heat flux densities, the convective heat transfer may increase several fold as compared with the convective flow determined for the same gas without allowance for radiation. However, such an increase in convective heat transfer was obtained in this case only when the fraction of convective heat transfer in the overall heat flux was moderate (of the order of 10%) (Table 6.2).

The radiative heat flux calculated for a temperature distribution in a nonradiating gas under analogous conditions is, as a rule, higher than the flux obtained under conditions of interaction of radiative and convective energy transports. This means that equalization of the temperature field in combined heat transfer results in some reduction in the radiative heat flux. For this reason, as a result of interaction, the total heat flux increases by

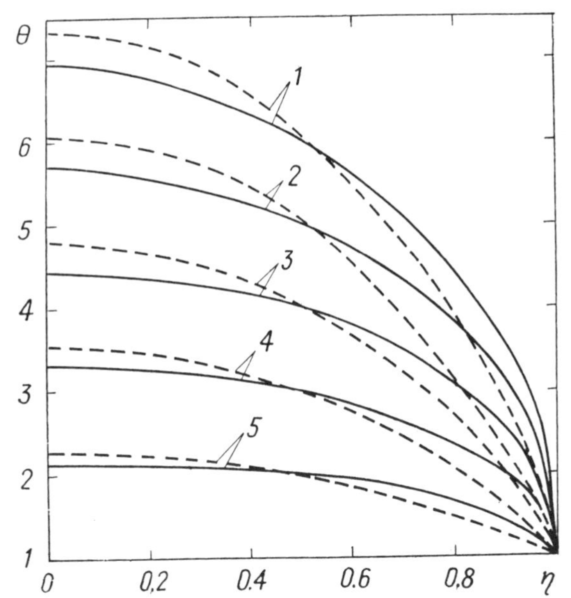


**FIG. 6.5** Temperature distribution in stabilized laminar flow of gaseous $CO_2$ in a 10–cm-high planar duct as a function of bulk temperature $T_m$. $T_m$ (K): (1) 3000; (2) 2500; (3) 2000; (4) 1500; (5) 1000. The ideal black-body channel walls were at temperature $T_w = 500$ K. The solid curves were obtained with allowance for radiant energy transfer, and the dashed curves without incorporating radiation.

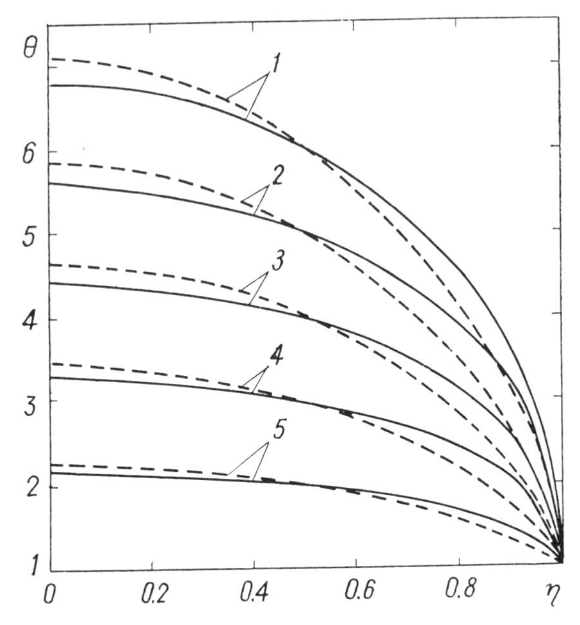


**FIG. 6.6** Temperature distribution in stabilized laminar flow of water vapor. For the remaining legend, see Fig. 6.5.

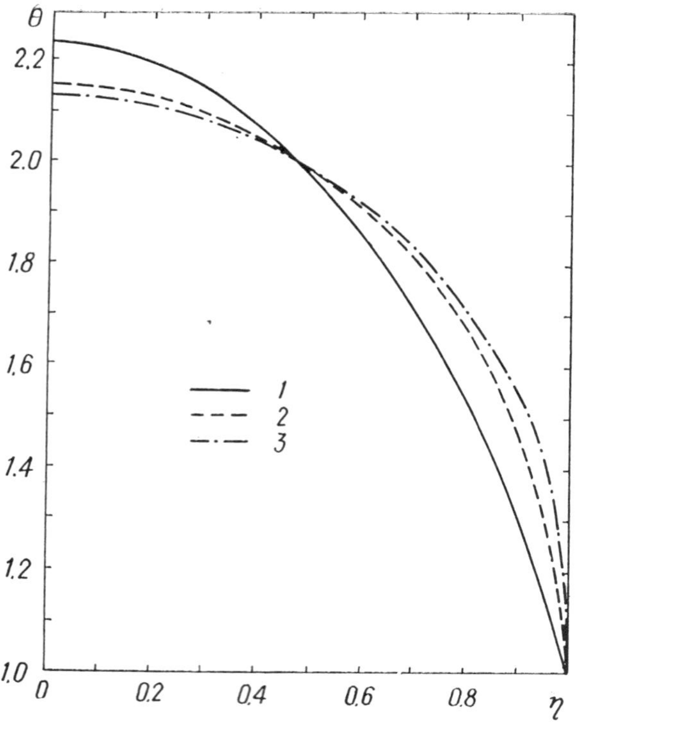


**FIG. 6.7** Effect of pressure on the temperature distribution in stabilized laminar flow of water vapor with bulk temperature $T_m = 1000$ K in a 10-cm-high planar duct. (1) Without allowance for radiation at $p = 1$ atm; (2) with allowance for radiation at $p = 1$ atm; (3) same as (2) at $p = 10$ atm.

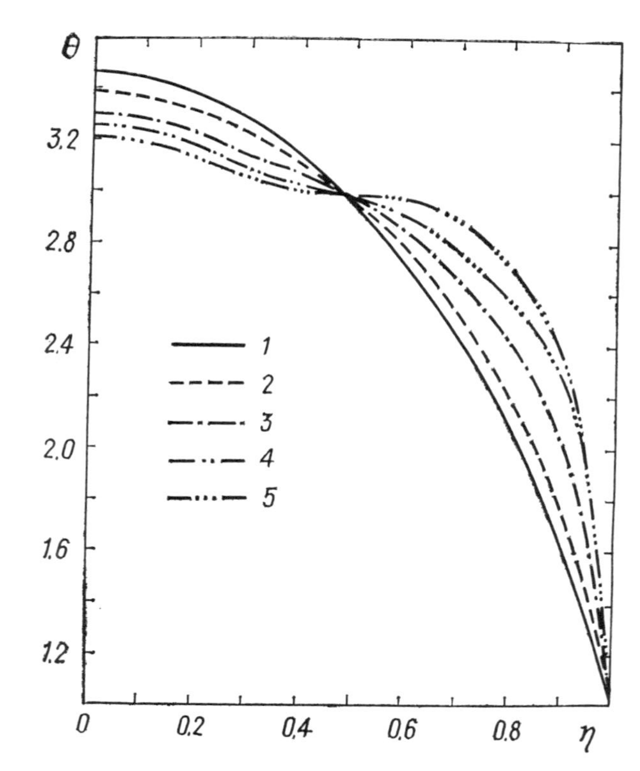


**FIG. 6.8** Effect of thickness of water-vapor layer at bulk temperature $T_m = 1500$ K on the temperature distribution in stabilized laminar flow in a planar duct as a function of duct height. $H$ (cm): (1) 0.1; (2) 5; (3) 15; (4) 100; (5) 500.

**TABLE 6.2** Results of Calculation of Convective–Radiative Heat Transfer in Stabilized Laminar Flow of Gas in a Planar Duct

| Gas | Height $H$, cm | $p$, atm | $T_m$, $K$ | $Nu^0_K$ | $Nu^0_D$ | $Nu^0_\Sigma$ | $\frac{\partial Nu^0_\Sigma}{\partial \Theta}$ | $Nu_D$ | $Nu_K$ | $Nu_\Sigma$ | $\frac{Nu_\Sigma}{Nu^0_\Sigma}$ | $\chi_{II}$ |
|---|---|---|---|---|---|---|---|---|---|---|---|---|
| $CO_2$ | 1 | 1.0 | 1000 | 12.9 | 2.7 | 15.6 | 7.7 | 2.8 | 13.8 | 16.6 | 1.067 | 0.086 |
| $CO_2$ | 1 | 1.0 | 1500 | 17.5 | 5.3 | 22.8 | 6.8 | 5.6 | 19.1 | 24.7 | 1.079 | 0.140 |
| $CO_2$ | 1 | 1.0 | 2000 | 21.7 | 7.5 | 29.2 | 5.8 | 6.9 | 23.0 | 29.9 | 1.026 | 0.154 |
| $CO_2$ | 5 | 1.0 | 1000 | 12.9 | 20.1 | 32.9 | 27.2 | 20.6 | 19.3 | 39.9 | 1.211 | 0.503 |
| $CO_2$ | 5 | 10.0 | 1500 | 17.5 | 44.1 | 61.6 | 28.4 | 45.7 | 29.1 | 74.9 | 1.216 | 0.661 |
| $CO_2$ | 5 | 1.0 | 2000 | 21.7 | 66.2 | 87.9 | 24.1 | 69.7 | 36.4 | 106.1 | 1.208 | 0.621 |
| $CO_2$ | 1 | 1.0 | 750 | 11.3 | 2.7 | 14.0 | 10.7 | 2.7 | 11.6 | 14.2 | 1.020 | 0.073 |
| $H_2O$ | 1 | 1.0 | 1000 | 15.5 | 3.9 | 19.4 | 11.8 | 3.9 | 15.9 | 19.8 | 1.023 | 0.123 |
| $H_2O$ | 1 | 1.0 | 1500 | 24.3 | 6.2 | 30.5 | 11.0 | 6.2 | 24.8 | 31.0 | 1.017 | 0.147 |
| $H_2O$ | 5 | 1.0 | 750 | 11.3 | 25.3 | 36.6 | 37.2 | 25.8 | 16.1 | 42.0 | 1.148 | 0.352 |
| $H_2O$ | 5 | 1.0 | 2000 | 33.0 | 111.9 | 144.9 | 40.4 | 113.7 | 47.6 | 161.3 | 1.113 | 0.646 |
| $H_2O$ | 5 | 1.0 | 3000 | 49.5 | 163.6 | 213.1 | 28.9 | 169.6 | 64.8 | 234.3 | 1.099 | 0.520 |
| $H_2O$ | 5 | 5.0 | 1000 | 15.5 | 70.7 | 86.2 | 85.5 | 78.6 | 31.9 | 110.5 | 1.282 | 0.815 |
| $H_2O$ | 10 | 1.0 | 1000 | 15.5 | 101.4 | 116.8 | 102.7 | 110.3 | 33.8 | 144.1 | 1.233 | 0.763 |
| $H_2O$ | 10 | 1.0 | 1500 | 24.3 | 205.9 | 230.1 | 118.2 | 223.6 | 55.9 | 279.5 | 1.215 | 0.919 |
| $H_2O$ | 10 | 1.0 | 2000 | 33.0 | 312.5 | 345.5 | 111.1 | 337.0 | 73.7 | 410.7 | 1.189 | 0.873 |
| $H_2O$ | 10 | 1.0 | 2500 | 41.5 | 407.9 | 449.3 | 97.1 | 439.6 | 86.9 | 526.5 | 1.172 | 0.785 |
| $H_2O$ | 10 | 1.0 | 3000 | 49.5 | 489.2 | 538.7 | 83.2 | 521.0 | 98.7 | 619.7 | 1.150 | 0.701 |
| $H_2O$ | 10 | 1.5 | 1000 | 15.5 | 118.4 | 133.8 | 126.0 | 129.3 | 39.5 | 168.8 | 1.261 | 0.833 |
| $H_2O$ | 10 | 2.5 | 1000 | 15.5 | 136.1 | 151.6 | 153.2 | 150.8 | 44.7 | 195.5 | 1.290 | 0.908 |
| $H_2O$ | 10 | 5.0 | 1000 | 15.5 | 153.1 | 168.5 | 180.6 | 171.6 | 51.3 | 222.8 | 1.322 | 0.973 |
| $H_2O$ | 10 | 7.5 | 1000 | 15.5 | 160.2 | 175.6 | 191.4 | 181.7 | 52.8 | 234.5 | 1.335 | 0.994 |
| $H_2O$ | 10 | 10.0 | 1000 | 15.5 | 165.5 | 180.6 | 197.8 | 186.2 | 54.6 | 240.8 | 1.338 | 1.005 |
| $H_2O$ | 15 | 1.0 | 750 | 11.3 | 96.2 | 107.5 | 132.3 | 104.1 | 28.1 | 132.2 | 1.229 | 0.551 |
| $H_2O$ | 15 | 1.0 | 1000 | 15.5 | 167.9 | 183.4 | 171.2 | 182.1 | 43.6 | 225.8 | 1.231 | 0.855 |
| $H_2O$ | 15 | 1.0 | 1500 | 24.3 | 354.4 | 378.7 | 208.5 | 384.4 | 74.9 | 459.3 | 1.213 | 1.031 |
| $H_2O$ | 15 | 1.0 | 2000 | 33.0 | 554.0 | 587.0 | 204.0 | 595.6 | 100.8 | 696.4 | 1.186 | 0.984 |
| $H_2O$ | 20 | 1.0 | 750 | 11.3 | 134.2 | 145.4 | 185.4 | 136.7 | 40.4 | 177.1 | 1.218 | 0.588 |

approximately 30% as compared with the overall heat flux, determined from the magnitudes of heat fluxes without allowance for their interaction. The results of calculation of combined heat transfer are correlated on the basis of parameter $\chi_I$ [Eq. (6.19)] (Fig. 6.9). Over the range of parameters under study one can use the linear relationship

$$\frac{q_\Sigma}{q^0_\Sigma} = 1 + 0{,}255\,\chi_I \tag{6.84}$$

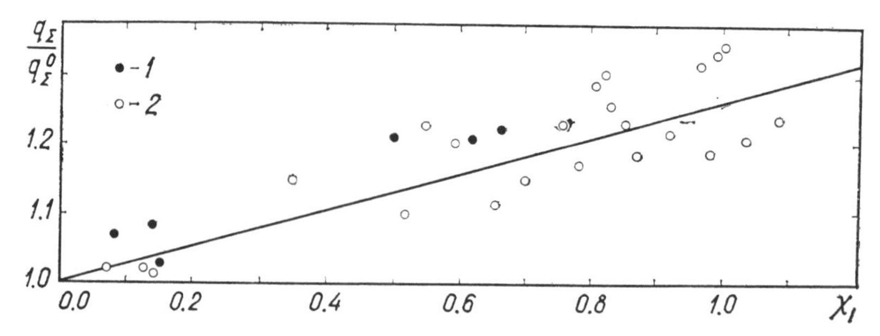


**FIG. 6.9** Combined heat transfer in laminar stabilized flow of radiating gas in a planar duct as a function of parameter $\chi_{III}$. (1) Gaseous $CO_2$; (2) water vapor. Detailed data on calculations are given in Table 6.2.

which describes within ±10% all the analytic results in laminar flow in a planar duct. Note that the scatter of analytic results is significantly affected by the derivative in the expression for $\chi_I$.

Since it utilizes general principles of the theory of similitude, Eq. (6.84) is recommended for estimating the interaction between radiative and convective heat transfer and also other cases of laminar flow.

## 6.4 TURBULENT CONVECTIVE–RADIATIVE HEAT TRANSFER IN A CYLINDRICAL CHANNEL

Since transfer of heat by radiation and convection in a cylindrical space is frequently encountered in practice, it is very important to know its magnitude when designing combustion chambers. This problem has been studied by many investigators [31, 185-187, 189, 199]; however, owing to the difficulty of calculating radiation in an infinite cylindrical layer, its solution is obtained under simplifying assumptions. In this study we used expressions for radiative heat flux listed in Chapter 2.

Gas-dynamically stabilized flow in a cylindrical channel at $k = 1$ is described by Eqs. (2.104)–(2.106), which are more complex than for a planar duct.

Let us consider turbulent stabilized flow in a pipe, which is described by equations of motion (2.104) and energy (2.105) with the following boundary conditions:

$$u=0, \quad \Theta=1 \quad \text{at} \quad \eta=1 \tag{6.85}$$

$$\frac{\partial u}{\partial \eta}=0, \quad \frac{\partial \Theta}{\partial \eta}=0 \quad \text{at} \quad \eta=0 \tag{6.86}$$

To construct an iterative scheme we add to both sides of Eqs. (2.104) and (2.105) the quantities $s_u$ and $S_\Theta$, respectively:

$$\frac{1}{\eta}\cdot\frac{\partial}{\partial\eta}\left(\eta\,\frac{\partial S_u}{\partial\eta}\right)+S_u=S_u+2\left.\frac{\partial u}{\partial\eta}\right|_{\eta=1} \tag{6.87}$$

$$\frac{1}{\eta}\cdot\frac{\partial}{\partial\eta}\left(\eta\,\frac{\partial S_\Theta}{\partial\eta}\right)+S_\Theta=S_\Theta+\Phi_T \tag{6.88}$$

We assume, as in problems for a planar duct, that the right-hand sides of Eqs. (6.87) and (6.88) were evaluated from the velocity and temperature fields of the preceding iterations. This assumption allows treating Eqs. (6.87) and (6.88) as zero-order nonhomogeneous Bessel equations. The general solution of the equation

$$\frac{1}{\eta}\cdot\frac{\partial}{\partial\eta}\left(\eta\,\frac{\partial S}{\partial\eta}\right)+S=f(\eta) \tag{6.89}$$

can be written in the form

$$S=C_1(\eta)\cdot J_0(\eta)+C_2(\eta)\cdot Y_0(\eta) \tag{6.90}$$

Having determined the unknown functions $C_1(\eta)$ and $C_2(\eta)$ by variation of constants, we obtain

$$S=-J_0(\eta)\cdot\int_0^\eta f(\eta_1)\cdot Y_0(\eta_1)\cdot Z(\eta_1)\,d\eta_1+c_1\cdot J_0(\eta)+$$

$$+Y_0(\eta)\cdot\int_0^\eta f(\eta_1)\cdot J_0(\eta_1)\cdot Z(\eta_1)\,d\eta_1+c_2\cdot Y_0(\eta) \tag{6.91}$$

where

$$Z(\eta)\equiv\frac{1}{J_1(\eta)\cdot Y_0(\eta)-J_0(\eta)\cdot Y_1(\eta)} \tag{6.92}$$

The derivative of the sought function is

$$\frac{\partial S}{\partial\eta}=J_1(\eta)\cdot\int_0^\eta f(\eta_1)\cdot Y_0(\eta_1)\cdot Z(\eta_1)\,d\eta_1-c_1\cdot J_1(\eta)-$$

$$-Y_1(\eta)\cdot\int_0^\eta f(\eta_1)\cdot J_0(\eta_1)\cdot Z(\eta_1)\,d\eta_1-c_2\cdot Y_1(\eta) \tag{6.93}$$

To conform to symmetry conditions (6.86) in the equation of energy and motion it is required that

$$\frac{\partial S}{\partial\eta}=0\quad\text{at}\quad\eta=0 \tag{6.94}$$

It follows from Eq. (6.93) that condition (6.94) is always satisfied in the case

$$c_2 = 0 \tag{6.95}$$

Using Eqs. (6.95) and (2.105), the solution of the equation of motion for any approximation becomes

$$\frac{\partial u}{\partial \eta} = -c_1 \cdot U_1(\eta) + 2 \cdot \frac{\partial u}{\partial \eta}\bigg|_{\eta=1} \cdot U_2(\eta) + U_3(\eta) \tag{6.96}$$

and the velocity field is

$$u = -c_1 \cdot \int_1^{\eta} U_1(\eta_1)\, d\eta_1 + 2 \cdot \frac{\partial u}{\partial \eta}\bigg|_{\eta=1} \cdot \int_1^{\eta} U_2(\eta_1)\, d\eta_1 +$$

$$+ \int_1^{\eta} U_3(\eta_1)\, d\eta_1 \tag{6.97}$$

where

$$U_1(\eta) = \frac{\mu_w}{\mu + \rho\varepsilon_\tau} \cdot J_1(\eta) \tag{6.98}$$

$$U_2(\eta) = \frac{\mu_w}{\mu + \rho\varepsilon_\tau} \cdot \left[ J_1(\eta) \cdot \int_0^{\eta} Y_0(\eta_1) \cdot Z(\eta_1)\, d\eta_1 - \right.$$

$$\left. - Y_1(\eta) \cdot \int_0^{\eta} J_0(\eta_1) \cdot Z(\eta_1) \cdot d\eta_1 \right] = \frac{\mu_w}{\mu + \rho\varepsilon_\tau} \cdot J_1(\eta) \tag{6.99}$$

$$U_3(\eta) = \frac{\mu_w}{\mu + \rho\varepsilon_\tau} \cdot \left[ J_1(\eta) \cdot \int_0^{\eta} S_u(\eta_1) \cdot Y_0(\eta_1) \cdot Z(\eta_1)\, d\eta_1 - \right.$$

$$\left. - Y_1(\eta) \cdot \int_0^{\eta} S_u(\eta_1) \cdot J_0(\eta_1) \cdot Z(\eta_1)\, d\eta_1 \right] \tag{6.100}$$

Equation (6.97) incorporates conditions (6.85) and (6.86). The obtaining of a unique solution requires the use of the additional stipulation that the flow rate be constant. According to Eq. (2.9), we introduce the bulk velocity of the flow

$$U_m = \frac{G}{\pi Y^2 \rho_w} = \frac{2}{\rho_w} \cdot \int_0^1 \rho u \eta\, d\eta \tag{6.101}$$

From Eq. (6.97), with the bulk velocity expressed by Eq. (6.101), we obtain an expression for the constant of integration $c_1$:

$$\frac{\rho_w U_m}{2} = -c_1 \cdot \int_0^1 \rho\eta \int_1^\eta U_1(\eta_1)\, d\eta_1\, d\eta + 2 \cdot \frac{\partial u}{\partial \eta}\Big|_{\eta=1} \times$$

$$\times \int_0^1 \rho\eta \int_1^\eta U_2(\eta_1)\, d\eta_1\, d\eta + \int_0^1 \rho\eta \int_1^\eta U_3(\eta_1)\, d\eta_1\, d\eta \qquad (6.102)$$

The value of $c_1$ is determined from Eq. (6.102) by also using its relationship to the velocity derivative at the wall, obtained from Eq. (6.96):

$$\frac{\partial u}{\partial \eta}\Big|_{\eta=1} = -c_1 \cdot U_1(1) + 2\frac{\partial u}{\partial \eta}\Big|_{\eta=1} \cdot U_2(1) + U_3(1) \qquad (6.103)$$

The solution of the equation of motion yields not only the velocity field, but also the friction velocity of the flow,

$$u_* = \sqrt{\nu_w \cdot \left|\frac{\partial u}{\partial \eta}\right|_{\eta=1}} \qquad (6.104)$$

The eddy viscosity in the Prandtl hypothesis approximation is described, according to Schlichting [78], by the expression

$$\varepsilon_\tau = N^2 \cdot Y \cdot \left|\frac{\partial u}{\partial \eta}\right| \cdot [0{,}14 - 0{,}08 \cdot \eta^2 - 0{,}06 \cdot \eta^4]^2 \qquad (6.105)$$

where the damping factor $N$ is determined from the Van Driest formula

$$N = 1 - \exp\left[-(1-\eta) \cdot \frac{Y \cdot u_*}{A^+ \cdot \nu}\right] \qquad (6.106)$$

at $A^+ = 26$.

The equation of energy for stabilized turbulent channel flow yields highly cumbersome expressions. According to Eqs. (2.102) and (6.93), the temperature derivative can be expressed as

$$\frac{\partial \Theta}{\partial \eta} = -c_1 \cdot T_1(\eta) - \frac{\bar{q}_w}{1-\Theta_m} \cdot T_2(\eta) + T_3(\eta) + T_4(\eta) \qquad (6.107)$$

where use is made of the notation:

$$T_1(\eta) = \frac{\lambda_w}{\lambda + \rho c_p \varepsilon_q} \cdot J_1(\eta) \qquad (6.108)$$

$$T_2(\eta) = \frac{\lambda_w}{\lambda + \rho c_p \varepsilon_q} \cdot \left[\frac{J_1(\eta)}{Q_m} \cdot \int_0^\eta \rho c_p u \cdot (1-\Theta) \cdot Y_0(\eta_1) \cdot Z(\eta_1)\, d\eta_1 - \right.$$

$$-\frac{Y_1(\eta)}{Q_m}\cdot\int_0^\eta \rho c_p u\cdot(1-\Theta)\cdot J_0(\eta_1)\cdot Z(\eta_1)\,d\eta_1\Big] \tag{6.109}$$

$$T_3(\eta)=\frac{\lambda_w}{\lambda+\rho c_p\,\varepsilon_q}\cdot\Big[J_1(\eta)\cdot\int_0^\eta S_\Theta(\eta_1)\cdot Y_0(\eta_1)\cdot Z(\eta_1)\,d\eta_1-$$

$$-Y_1(\eta)\cdot\int_0^\eta S_\Theta(\eta_1)\cdot J_0(\eta_1)\cdot Z(\eta_1)\,d\eta_1\Big] \tag{6.110}$$

$$T_4(\eta)=\frac{\lambda_w}{\lambda+\rho c_p\,\varepsilon_q}\cdot\Big[J_1(\eta)\cdot\int_0^\eta \frac{1}{\eta_1}\cdot\frac{\partial(\eta_1\bar{q}_r)}{\partial\eta_1}\cdot Y_0(\eta_1)\cdot Z(\eta_1)\,d\eta_1-$$

$$-Y_1(\eta)\cdot\int_0^\eta \frac{1}{\eta_1}\cdot\frac{\partial(\eta_1\bar{q}_r)}{\partial\eta_1}\cdot J_0(\eta_1)\cdot Z(\eta_1)\,d\eta_1\Big] \tag{6.111}$$

$$Q_m=\int_0^1 \rho c_p u\eta\,d\eta. \tag{6.112}$$

Equation (6.111) contains a derivative of the density of the nondimensional radiant heat flux and it is hence inconvenient for numerical calculations.

Since

$$\frac{\partial}{\partial\eta}\Big[\frac{1}{\eta}\cdot Y_0(\eta)\cdot Z(\eta)\Big]=-\frac{1}{\eta}\cdot Y_1(\eta)\cdot Z(\eta) \tag{6.113}$$

$$\frac{\partial}{\partial\eta}\Big[\frac{1}{\eta}\cdot J_0(\eta)\cdot Z(\eta)\Big]=-\frac{1}{\eta}\cdot J_1(\eta)\cdot Z(\eta) \tag{6.114}$$

Eq. (6.111) can be integrated by parts. Then

$$T_4(\eta)=\frac{\lambda_w}{\lambda+\rho c_p\,\varepsilon_q}\cdot\Big[J_1(\eta)\cdot\int_0^\eta \bar{q}_r(\eta_1)\cdot Y_1(\eta_1)\cdot Z(\eta_1)\,d\eta_1-$$

$$-Y_1(\eta)\cdot\int_0^\eta \bar{q}_r(\eta_1)\cdot J_1(\eta_1)\cdot Z(\eta_1)\cdot d\eta_1-\bar{q}_r(\eta)\Big] \tag{6.115}$$

Equation (6.115) was obtained on the assumption that at $\eta=0$, $\bar{q}_r=0$.

Integration of Eq. (6.107) with respect to $\eta$ yields

$$\Theta=1-\frac{\bar{q}_w}{1-\Theta_m}\cdot\int_1^\eta T_2(\eta_1)\,d\eta_1-c_1\cdot\int_1^\eta T_1(\eta_1)\,d\eta_1+$$

$$+\int_{1}^{\eta}[T_3(\eta_1)+T_4(\eta_1)]\,d\eta_1 \tag{6.116}$$

which incorporates boundary condition (6.85).

Determination of the constant of integration $c_1$ requires an additional assumption. The latter may vary, as in the case of a planar layer. In investigating interaction between convective and radiative heat transfer, it is most convenient to perform the comparison at the same bulk temperature. For this reason we shall assume that the bulk temperature of the flow is specified and constant.

In nondimensional form the bulk temperature defined by Eq. (2.103) in the case of cylindrical symmetry can be represented by the equation

$$\Theta_m=\frac{1}{Q_m}\cdot\int_{0}^{1}\rho uc_p\Theta\eta\,d\eta \tag{6.117}$$

where $Q_m$ is defined by Eq. (6.112).

Multiplying Eq. (6.117) by $\rho uc_p\eta/Q_m$ and then integrating with respect to $\eta$ from 0 to 1, we obtain the condition for determining constant $c_1$:

$$\Theta_m=1-\frac{c_1}{Q_m}\int_{0}^{1}\rho uc_p\eta\int_{1}^{\eta}T_1(\eta_1)\,d\eta_1\,d\eta-\frac{\bar{q}_w}{(1-\Theta_m)\cdot Q_m}\times$$

$$\times\int_{0}^{1}\rho uc_p\eta\int_{1}^{\eta}T_2(\eta_1)\,d\eta_1\,d\eta+\frac{1}{Q_m}\cdot\int_{0}^{1}\rho uc_p\eta\times$$

$$\times\int_{1}^{\eta}[T_3(\eta_1)+T_4(\eta_1)]\,d\eta_1\,d\eta \tag{6.118}$$

It must be remembered in defining $c_1$ that

$$\bar{q}_w=-\left.\frac{\partial\Theta}{\partial\eta}\right|_{\eta=1}+\bar{q}_r(1) \tag{6.119}$$

The relationship between $c_1$ and the temperature derivative at the wall can be obtained from Eq. (6.107):

$$\left.\frac{\partial\Theta}{\partial\eta}\right|_{\eta=1}=-c_1\cdot T_1(1)-\frac{\bar{q}_w}{1-\Theta_m}\cdot T_2(1)+T_3(1)+T_4(1) \tag{6.120}$$

Equations (6.118)–(6.120) suffice for defining $c_1$, which means that all the boundary conditions can be satisfied for each iteration.

In solving the problem we first specified the velocity and temperature fields of the zero approximation. Then new fields of these quantities were determined from Eqs. (6.97) and (6.116) and the calculations were repeated until the desired accuracy was attained.

In numerical solution of the problem the coordinate $\eta$ was subdivided into 200–300 nonequal parts, in accordance with the logarithmic law according to which the computational nodes at the wall are crowded closer together [31].

The solution of the above set of equations in the absence of radiation and eddy fission in a flow with constant physical properties yields Nu = 3.678, which is in agreement with familiar results for laminar flow.

If it is assumed that $\mathrm{Pr}_T = 1.0$, then the results of calculation of heat transfer in a turbulent flow of a nonradiating gas will be in satisfactory agreement with the familiar results cited in [95, 96, 149].

Our calculations show that the radiation from a stabilized turbulent flow modifies the temperature field, which differs significantly from analogous changes in laminar flow. In turbulent flow, radiation, as a rule, results in some reduction in temperature in the wall regions of the channel and rise in temperature in the center of the channel as compared with the temperature distribution of nonradiating gas under the same conditions. The magnitude of these changes is a function both of the thickness of the radiating layer and of the flow pattern (Fig. 6.10). The slight rise in temperature in the center of the channel is due to blocking of radiation by the wall layers of the gas. In turbulent flow the change in temperature fields is weaker than in laminar flow. The redistribution of temperatures in turbulent flow is responsible for a slight reduction in the combined heat transfer as compared with the sum of heat fluxes $q_\Sigma^0$ calculated without allowance for interaction between convection and radiation. A more perceptible effect of combined heat transfer can be expected in channels with larger diameter. Obviously, these effects will also be related to the turbulent structure of the flow. In the present calculations, performed at $\mathrm{Pr}_T = 1.0$, we assumed that the turbulence characteristics of the flow are independent of radiation. The validity of such assumptions can be checked only experimentally. For this reason a great deal of attention was paid to the experimental study of combined heat transfer in turbulent flows.

## 6.5. EXPERIMENTAL STUDY OF COMBINED HEAT TRANSFER IN CHANNELS

Combined heat transfer in the flow of combustion products of hydrocarbon fuels was investigated in test channels 3 and 4. In addition to the total heat transfer, determined by calorimetry, we measured the radiant heat transfer component by means of narrow-angle radiometers.

The results on combined heat transfer were correlated by using the previously obtained reference parameter for turbulent flow of the medium,

**TABLE 6.3** Results of Calculation of Convective–Radiative Heat Transfer in Stabilized Turbulent Flow of Water Vapour at Atmospheric Pressure in a Cylindrical Channel with $T_m$ = 500 K

| Height $H$, cm | $T_m$, K | $Re_w$ | $Re_f$ | $Nu^{\circ}_{c,f}$ | $Nu^{\circ}_{r,f}$ | $St^{\circ}_{p}$ | $\frac{d \ln Nu^{\circ}_{\Sigma}}{d \ln \Theta_0}$ | $\chi_{II}$ | $Nu_{c,f}$ | $Nu_{r,f}$ | $\frac{Nu_{\Sigma}}{Nu^{\circ}_{\Sigma}}$ |
|---|---|---|---|---|---|---|---|---|---|---|---|
| 15 | 1000 | 5×10⁴ | 2.28×10⁴ | 65.6 | 31.9 | 0.043 | −0.007 | −0.0003 | 65.5 | 31.8 | 0.998 |
| 15 | 1500 | 5×10⁴ | 1.54×10⁴ | 45.1 | 35.1 | 0.108 | 0.094 | 0.010 | 45.0 | 35.0 | 0.998 |
| 15 | 2000 | 5×10⁴ | 1.23×10⁴ | 36.8 | 36.7 | 0.191 | 0.142 | 0.027 | 36.5 | 36.6 | 0.995 |
| 15 | 1000 | 1×10⁵ | 4.55×10⁴ | 118.6 | 32.1 | 0.043 | −0.011 | −0.0005 | 118.5 | 32.1 | 0.999 |
| 15 | 1500 | 1×10⁵ | 3.09×10⁴ | 82.5 | 35.5 | 0.109 | 0.087 | 0.009 | 82.3 | 35.4 | 0.998 |
| 15 | 2000 | 1×10⁵ | 2.47×10⁴ | 67.2 | 37.2 | 0.194 | 0.158 | 0.031 | 66.9 | 37.2 | 0.997 |
| 30 | 1000 | 5×10⁴ | 2.28×10⁴ | 65.6 | 79.4 | 0.054 | 0.081 | 0.004 | 65.3 | 78.9 | 0.994 |
| 30 | 1500 | 5×10⁴ | 1.54×10⁴ | 45.1 | 92.9 | 0.143 | 0.190 | 0.027 | 45.1 | 92.3 | 0.995 |
| 30 | 2000 | 5×10⁴ | 1.23×10⁴ | 36.8 | 100.4 | 0.264 | 0.204 | 0.054 | 36.5 | 100.7 | 0.994 |

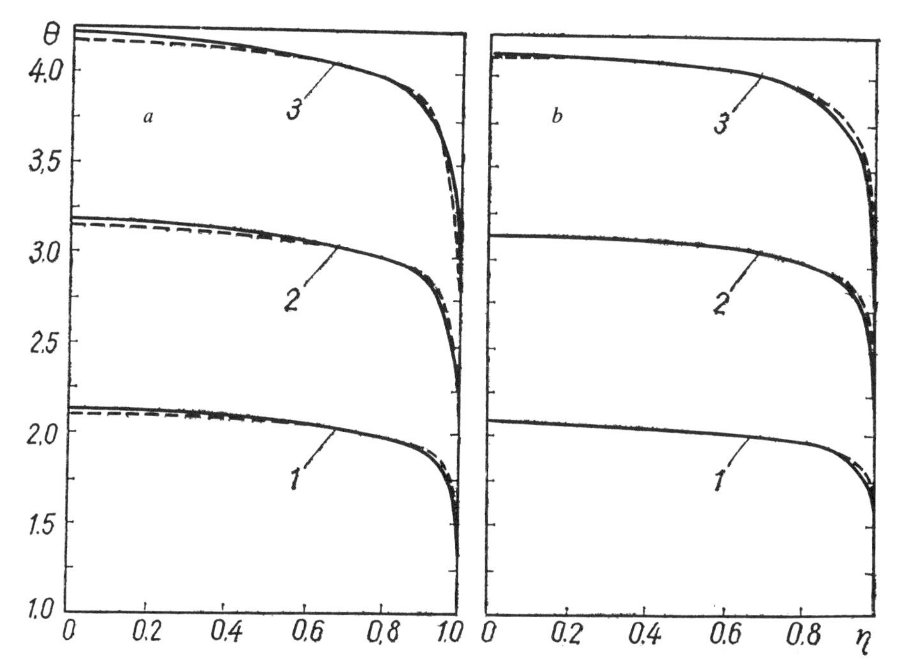


**FIG. 6.10** Temperture distribution in stabilized turbulent flow of water vapor in a 15-cm-diameter channel at $Re_w$ = (*a*) 5 × 10⁴ and (*b*) 1 × 10⁵. The temperature of the channel walls $T_w$ = 500 K and the bulk temperature of the flow is given by the curves. The solid curves were obtained with allowance for radiant energy transfer, and the dashed curves without incorporating radiation. $T_m$(K): (1) 1000; (2) 1500; (3) 2000.

$$\chi_{\rm II} \equiv \left(\frac{T_f - T_w}{T_f}\right)^{\frac{m_c \mathrm{Nu}_c^0 + m_r \mathrm{Nu}_r^0}{\mathrm{Nu}_\Sigma^0}} \mathrm{St}_r^0 \qquad (6.121)$$

The total heat transfer in the absence of interaction in convective radiative transfer $\mathrm{Nu}_\Sigma^0$, contained in similitude parameter (6.121), was determined from the additivity rule

$$\mathrm{Nu}_\Sigma^0 = \mathrm{Nu}_c^0 + \mathrm{Nu}_r^0 \qquad (6.122)$$

For heat transfer in the inlet region of the channel, the value of $\mathrm{Nu}_c^0$ was determined from expressions for convective heat transfer (4.14)–(4.26) with allowance for dissociation of the combustion products and also for the dependence of $\mathrm{Pr}_T$ on the temperature conditions and flow turbulence [Eq. (4.29)]. It was assumed that the flow turbulence for heated gases remains the same as when blowing down with cold air (Fig. 4.9), whereas temperature $T_f$ in the center of the flow along the channel does not change and is equal to the bulk temperature at the inlet to the channel.

Experimental determination of the radiant heat flux $q_r^0$ or $\mathrm{Nu}_r^0$ in the absence in interaction with convection is very difficult to determine experimentally. However, as shown in Chapter 5, when the thickness of the radiating gas is equal to the height of the channel, the value of $q_r^0$ depends rather insignificantly on the nature of the temperature distribution, which allows determining it analytically at bulk temperature of the flow in individual locations within the channel from Eq. (2.148). In conjunction with this, values of $R_e$ were precalculated for the individual walls of the channel. It was shown in Chapter 5 that the results of such calculations are in satisfactory agreement with experimental data.

The radiation Stanton number $\mathrm{St}_r^0 \equiv \dfrac{q_r^0}{\rho_f(h_f - h_w)}$ was determined on the basis of the total enthalpy difference, i.e., for dissociated products of combustion consideration was given to the difference in the energy of chemical transformations. The actual enthalpy difference was determined from the expression

$$h_f - h_w = c_{p,m}^* \cdot (T_f - T_w) \qquad (6.123)$$

in which, in calculating the effective specific heat $c_{p,m}^*$, the contribution of the energy of chemical transformations is incorporated. Detailed data on the thermophysical properties of combustion products are given by Makarevicius [100].

The power exponent $m_c$ in the case of convective turbulent heat transfer was calculated from Eqs. (4.14)–(4.26) with allowance for the temperature dependence (4.29) of $\mathrm{Pr}_T$. Calculations show that the value of $m_c$ over the range of parameters under study depends basically on $\mathrm{Re}_f$ and very little on other factors (Fig. 6.11).

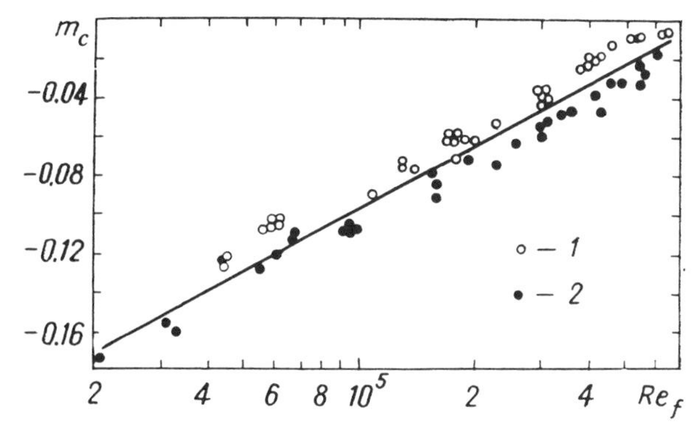


**FIG. 6.11.** For determining the power exponent $m_c$ of the temperature dependence of convective heat transfer in channel 3 as a function of $\mathrm{Re}_f$. (1) $T_m < 2999$ K; (2)$T_m > 2000$ K.

In determining the power exponent for the temperature dependence of radiative heat transfer $m_r$, use is made of the fact that the radiative heat flux is approximately a power-law function of the temperature (Fig. 6.12). Over the range of temperatures in the experimental studies we can assume $m_r = 3.2$ for channel 3 and $m_r = 2.9$ for channel 4.

The experimental results on combined heat transfer are analyzed using the expression

$$\frac{\mathrm{Nu}_\Sigma}{\mathrm{Nu}_\Sigma^0} = 1 - 3{,}76\chi_{\mathrm{II}}^{1{,}47} \tag{6.124}$$

Within the limits of experimental error, Eq. (6.124) describes all the experimental results on heat transfer in channel 3 (Appendix 4). This means

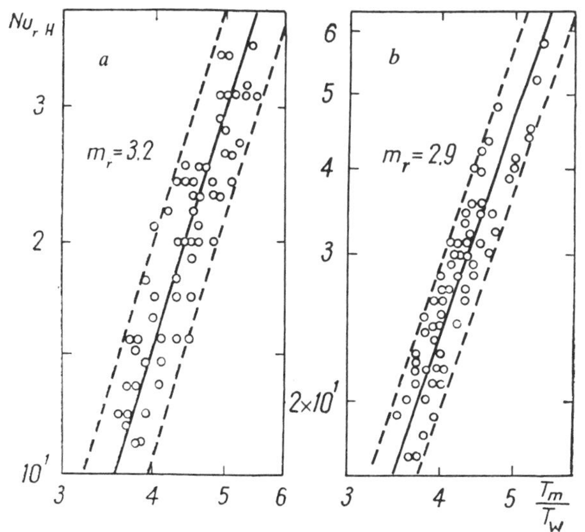


**FIG. 6.12** For determining the power exponent $m_r$ of the temperature dependence of radiative heat transfer. (*a*) Channel 3; (*b*) channel 4.

that the description of experimental results on combined heat transfer in the inlet length of the channel required using quite complex expressions for convective heat transfer in turbulized flows [Eqs. (4.14)–(4.26)], and also for the dependence of $\mathrm{Pr}_T$ on temperature conditions and the free-stream turbulence [Eq. (4.29)].

It is seen from the experiments that the interaction between radiative and convective heat transfer results in some reduction in heat transfer (Fig. 6.13).

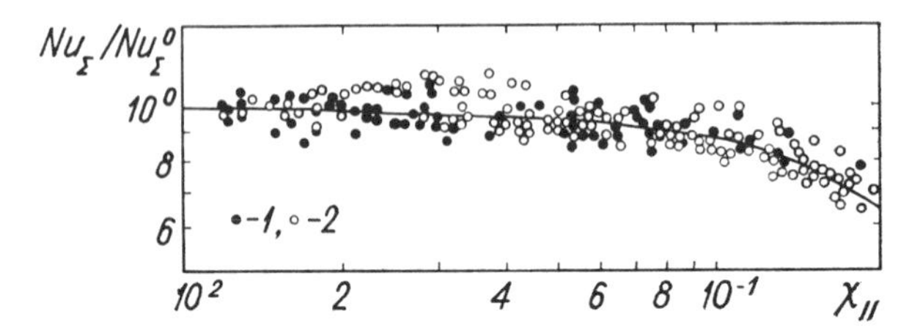


**FIG. 6.13** Relative combined heat transfer of combustion products as a function of $\chi_{II}$. (1) Channel 3; (2) channel 4. The solid curve represents Eq. (6.124).

The interaction between radiative and convective heat transfer in test channel 4 was analyzed in the same manner. Analysis was performed for data on heat transfer over a range of $x/H_h$ from 7 to 24, where convective heat transfer in the absence of radiation is described by Eq. (4.33), according to which it is assumed that for the series of experiments under study $m_c = -0.05$. Correlation was performed for experimental data obtained with both complete and incomplete combustion of the gas. In the latter case the flow was luminant, since it contained soot particles. Here, radiative flux $q_r$ was not determined analytically but experimentally. It was assumed that $q_r$ is independent of the temperature distribution, i.e.,

$$q_r \simeq q_r^0 \tag{6.125}$$

This assumption is based on calculations described in Chapter 5.

Since the bulk temperature in pipe flows does not change uniformly at different $\mathrm{Re}_{f,H}$, the value of $m_r$ is a function of flow conditions (Fig. 6.12). The value of $m_r$ was determined from analysis of experimentally measured radiant heat flux.

Analysis showed that the experimental results on combined heat transfer in channel 4 are satisfactorily described by Eq. (6.124) (Fig. 6.13).

This means that similitude equation (6.124) is applicable for different flow conditions.

The difference between the experimental and analytic results on combined heat transfer can be attributed to the fact that in calculations we compared conditions at the same bulk temperatures, whereas in experiment conditions we compared at the same $\mathrm{Re}_f$.

Analysis of combined heat transfer showed that there exists a close relationship between purely convective heat transfer and radiant heat transfer calculated without allowance for interaction with combined heat transfer. Analysis of this relationship is performed on the basis of the similitude parameter [Eq. (6.123) ]. It can be expected that with the future accumulation of analytic and experimental results, Eq. (6.125), suggested here for correlation of data on combined heat transfer, will be refined, and its limits of application will be widened.

CHAPTER

# SEVEN

# RADIATIVE AND COMBINED HEAT TRANSFER IN APPLIED PROBLEMS

Radiative and combined heat transfer play an important role in processes occurring in many high-temperature power-generating devices and chemical reactors. Not only is it impossible to examine the wide class of practical problems that results from this within the scope of a single book, but it is also unnecessary, since individual engineering problems are analyzed in special studies [2, 3, 48, 200–210]. The purpose of this chapter is to illustrate the capabilities of application of some methods of calculation of radiative and combined heat transfer to the solution of individual practical problems. Recently, allowance has been made in solving engineering problems also for selectiveness of radiation. Naturally, such calculations have as their primary purpose the study of heat transfer at high temperatures, when the devices are subjected to high heat loads. One hopes that in the future methods for calculating radiation with allowance for selectivity will come into extensive use also at low and moderate temperatures, for example, in drying, and they will aid in finding more efficient methods for the performance of a number of industrial processes.

## 7.1. ON THE CALCULATION OF RADIATIVE HEAT TRANSFER IN HEAT ENGINEERING

Heat engineering makes extensive use of various combustion chambers used in boilers and in industrial furnaces.

Calculations of heat transfer in combustion chambers employ the standard method [47, 206], based on certain integral characteristics of the flame of various fuels, and also of water-tube cooled walls of furnace devices and modified in order to make more detailed allowance for radiation [208]. The computational techniques, based on a number of experimentally determined characteristics, cannot always be used in designing new forms of furnaces in which the selectivity of wall radiation sometimes plays a rather important role. Such conditions occur, in part, in furnaces with metal, air-cooled walls, which are promising for use in the drying of products. Selectivity of radiation plays just as an important role in various industrial furnaces.

At present, extensive use is made in various industries of tubular furnaces [203, 207, 209], which are one of the main parts of chemical equipment used in petroleum processing and the petrochemical industry. Owing to the extensive use of furnaces for chemical processing of raw materials, various conditions may come about in the heated coils, which makes their unification on the basis of typical catalogs impossible. The standard method of industrial design of tubular furnaces under development in the All-Union Research Institute of Petroleum Machinery (USSR) will consist not only of thermal design of the furnace chamber, but also of calculation of the entire chemical reactor as a whole [209], with a place of importance assigned to thermal design of tubular furnaces.

An approximate determination of a number of important heat-engineering characteristics can be obtained by methods based on the method of furnace design suggested by Belokon' [207, 209]. However, such an integral approach does not satisfy practical needs. A new stage in thermal design of tubular furnaces is a combined integrodifferential method of calculation, based on describing heat and mass transfer at the differential level using a number of integral characteristics for describing radiation [209]. This technique and high-capacity computers allow us not only to perform a complete thermal design of the tubular furnaces, but also to obtain the distribution of parameters of the product being processed. It is possible, in particular, to determine the temperature distribution of the outer surface of the reaction tube, which, as a rule, operates under very rigid conditions. Given the use of integral characteristics of radiation, this technique makes no allowance for selectivity of radiation of the gases filling the furnace space and the surfaces enclosing the furnace chamber. However, when necessary this can be taken into account using methods for calculating radiation examined in the present study.

Any heat engineering facility consists basically of a number of enclosing surfaces, the volume between which is filled by absorbing and radiating gases. The location of each of these enclosures can be specified by corresponding equations of their boundaries. The spectral optical properties of the enclosures are usually known. The temperatures of the gas space and of the enclosures can be specified, or they may have to be determined for the given situation. In iterative calculation of such devices, one should assume some initial temperature distribution and then refine it on the basis of the material and energy balances of each of the surfaces separately and the entire device as a whole. In such a formulation of the problem, the calculation of any engineering device can be reduced to consecutive calculation of heat transfer between individual surfaces.

Highly nonisothermal surfaces are best subdivided into individual zones, within which the surface temperature can be regarded as constant and each zone can be treated as a separate surface.

In the preceding chapters we examined a technique for calculating radiative heat transfer in a space of arbitrary shape, filled by an isothermal medium. This computational technique can be developed also for more complicated conditions, when the radiative volume is enclosed by any number of surfaces at different temperatures. Here it is possible to incorporate the nonisothermicity of the gas volume, contained between individual surfaces.

Let us consider the calculation of radiative heat flux between two arbitrarily located surfaces, the space between which is filled with a nonisothermal medium with specified temperature distribution (Fig. 7.1). The location of the surfaces is specified by one or several equations of their boundaries in the form $f(x, y, z) = 0$, where $x$, $y$, and $z$ are the current coordinates in the Cartesian system.

If one selects arbitrarily two beams $R_1$ and $R_2$, emanating from surface $F_1$ to surface $F_2$, then the intensity of integral radiation in these directions can be easily calculated. In calculating the intensity of radiation from Eq. (2.144) it is always possible to make allowance for the optical properties of the medum and surfaces. If in general the directions of beams $R_1$ and $R_2$ make angles $\theta_1$ and $\theta_2$ with normal $\vec{n}$ to the surface, then the elementary resultant radiant flux $\Delta q_r$ in these directions can be expressed by the equations

$$\Delta q_r (R_1) = \hat{I}_1 \cdot \cos \theta_1 \tag{7.1}$$

$$\Delta q_r (R_2) = \hat{I}_2 \cdot \cos \theta_2 \tag{7.2}$$

Integration of the elementary resultant flux over the solid angle yields the total radiative heat flux. Hence Eq. (2.146) above can be used also in cases when the selected beams $R_1$ and $R_2$ are not normal to the surface:

$$\hat{I}(R^{ij}) \simeq \hat{I}(R_1^{ij}) \cdot \cos \theta_1 + \\ + [\hat{I}(R_2^{ij}) \cdot \cos \theta_2 - \hat{I}(R_1^{ij}) \cdot \cos \theta_1] \cdot \frac{R^{ij} - R_1^{ij}}{R_2^{ij} - R_1^{ij}} \tag{7.3}$$

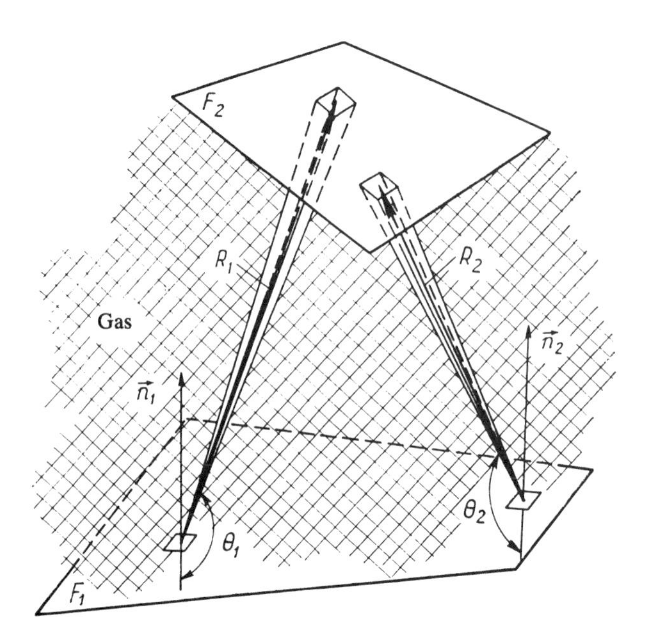


**FIG 7.1** Schematic for calculating the radiation between two artbitrarily spaced surfaces $F_1$ and $F_2$, separated by an absorbing and radiating gas.

With substitution of Eq. (7.3) into Eq. (2.18), we obtain an expression for the radiant heat flux between two surfaces,

$$q_r^{ij} = \pi \frac{\hat{I}(R_1^{ij}) \cdot \cos\theta_1 \cdot (R_2^{ij} - R_e^{ij}) + \hat{I}(R_2^{ij}) \cdot \cos\theta_2 (R_e^{ij} - R_1^{ij})}{R_2^{ij} - R_1^{ij}} \tag{7.4}$$

where $$R_e^{ij} = \frac{1}{\pi F_i} \int_{F_i} \int_0^{2\pi} \int_0^{\frac{\pi}{2}} R^{ij}(\theta, \varphi) \cdot \sin\theta \cdot \cos\theta \, d\theta \, d\varphi \, dF_1 \tag{7.5}$$

It is assumed that quantity $R^{ij}$ in Eqs. (7.4) and (7.5) corresponds to all the possible beams between surfaces $i$ and $j$ (Fig. 7.1). Beams $R_1^{ij}$ and $R_2^{ij}$ can be selected arbitrarily; however, the difference between their magnitudes should amount to from 10 to 30%. In calculating values of $I(R_1^{ij})$ and $I(R_2^{ij})$ from Eq. (7.3), it is possible to make allowance for the real temperature distribution in the gas.

For a radiating system of any configuration, the effective beam lengths $R_e^{ij}$ need be calculated only once.

The total heat flux to any *i*th surface, owing to radiation of the remaining enclosures, can be obtained from the expression

$$q_{r,i} = \sum_{\substack{j=1 \\ j \neq i}}^{N} q_r^{ij} \tag{7.6}$$

The current level of computer technology allows solving such a problem for heat engineering devices of any configuration.

The suggested computational schemes can be used for calculating heat transfer in various heat engineering devices, for example, in furnaces, tubular furnaces, MHD channels, and other objects.

## 7.2. HEAT TRANSFER IN MHD GENERATOR CHANNELS

MHD generators that employ ionized combustion products as the working fluid are one of the promising devices in modern energetics. Such units have already passed pilot-plant tests [176]. Various modifications of channels of industrial MHD generators are being analyzed and refined under the leadership of the High-Temperatures Institute of the USSR Academy of Sciences. The placing on-line of such MHD generators required solving a large number of scientific and engineering problems, of which the more important were thermal protection of the walls of the MHD generators, which are subjected to very difficult conditions. Work on solving this important problem is being performed in a number of studies under way at the High-Temperatures Institute of the USSR Academy of Sciences and in the Engineering Thermophysics Institute of the Ukrainian Academy of Sciences [211–213]. Several studies on this problem also have been performed at our Institute. The accumulated experimental data on heat in channels of pilot-plant units and also in experimental channels require careful analysis, since in designing industrial equipment the existing data must be extrapolated to conditions for which experimental data are not available. This is particularly important because in the future the channels of industrial devices will operate at higher pressures and the role of radiation in them will increase. The existence of strong electromagnetic fields in such channels may in certain cases affect the flow turbulence and also the heat transfer. Ionizing seed, which is present in flow in the form of compounds of alkali metals and products of erosion of channel walls can also affect radiation in the channel. Analysis of these studies is performed together with the High-Temperatures Institute of the USSR Academy of Sciences.* At the

* This work is performed under the leadership of corresponding member of the USSR Academy of Sciences Professor B.S. Petukhov, and Academician of the Lithuanian Academy of Sciences, Professor A.A. Žukauskas.

first stage, heat transfer was calculated for the channel of the U-25 device employing an open-cycle MHD generator. The calculations were performed for a 1D Faraday MHD channel of the U-25, which generated an electrical output of 20.1 MW. Calculations showed that direct use of data on heat transfer in an experimental water-cooled channel may yield erroneous results. Since the channels of MHD generators operate at wall temperatures of about 2000 K, then according to Fig. 4.9 one should expect that under such conditions the value of $Pr_T$ will be higher than under the conditions prevailing in our experiments. The results thus corrected are in rather satisfactory agreement with data on measurement of heat transfer in channels of experimental MHD generators [211]. This additionally confirms the conclusions drawn in Chapter 4 on the significant dependence of $Pr_T$ on the ratio of the wall to flow temperatures.

The above shows that there is a basis for the assumption that the relationships for determining convective heat transfer obtained in this study will be suitable also for thermal design of the channels of industrial MHD generators. The magnitude of heat fluxes in such devices are estimated on the basis of one of the modifications of the channel whose dimensions are shown in Fig. 7.2.

In the first approximation, convective heat transfer was determined from the distributions of pressure, velocity magnetic field induction and temperature as a result of numerical calculations performed in the High-Temperatures Institute of the USSR Academy of Sciences. It was assumed that the flow temperature in individual locations of the channel is constant and equal to the bulk temperature of the flow. The channel wall temperature was also known. Convective heat transfer to the channel walls determined on the basis of these data from Eq. (4.26) was found to be approximately 15% lower than the heat transfer calculated at the High Temperatures Institute of the USSR Academy of Sciences (Fig. 7.3).

A somewhat larger deviation was noted for analytic results on radiant heat transfer, since our calculations of radiant heat flux were performed in the approximation of an infinite plane-parallel layer, whose height is equal to that of the channel. It was assumed that the channel walls are black bodies. Allowance was made for radiation of gaseous $CO_2$, water vapour, CO and NO, corresponding to their equilibrium partial pressure in combustion products. The calculations were performed according to the

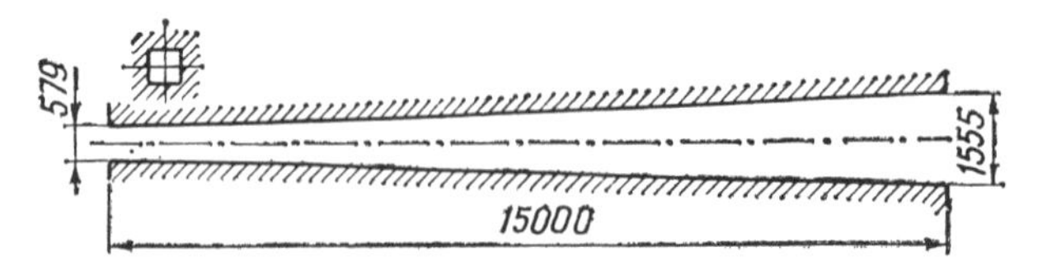


**FIG. 7.2** Schematic of an MHD channel rated at 500 kW.

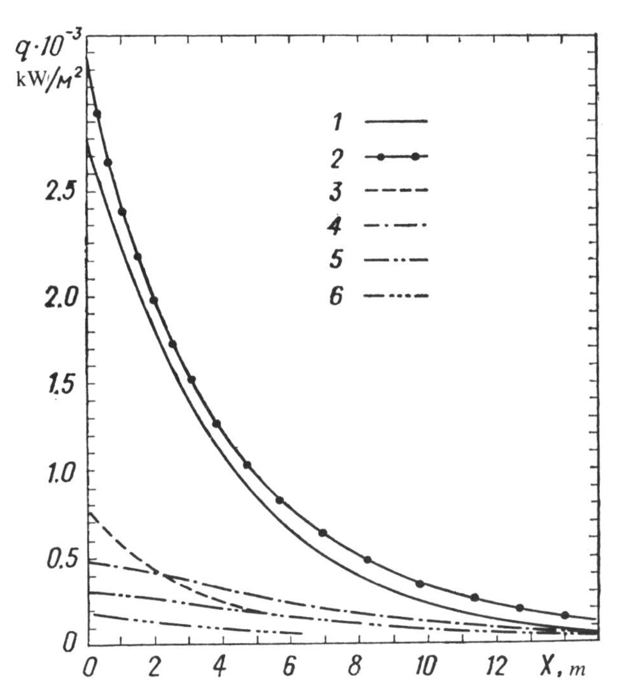


**FIG. 7.3** Comparison of the calculated heat fluxes in MHD channels of various ratings. (1) Convective heat transfer from Eq. (4.26) with allowance for $Pr_T$ (500 kW); (2) convective heat transfer under the same conditions using the technique of the High-Temperatures Institute of the USSR Academy of Sciences (500 kW); (3) convective heat transfer at maximum load of the channel of the U–25 facility; (4) radiative heat transfer from data of the present study for the channel represented in Fig. 7.2; (5) radiative heat transfer using the technique of the High-Temperatures Institute of the USSR Academy of Sciences (500 kW); (6) radiative heat transfer using the technique of the present study, the U–25 facility.

technique described in Chapter 2, and the optical properties of the products of combustion were determined from data listed in Chapter 1. The radiative heat flux determined in this manner was found on the average to be 30–40% higher than the analytic results obtained in the High-Temperatures Institute of the USSR Academy of Sciences. Owing to the presence of a high convective heat flux in the inlet part of the channel, the radiative heat flux amounts to only 10% of the total heat flux, for which reason one should not expect significant interaction between convection and radiation. However, as a result of large optical thicknesses of the layer, the radiative heat flux may affect the temperature distribution across the channel. In these calculations the geometric configuration of the channel has so far been incorporated in a rather rough manner, and allowance for the real optical properties of the walls and a number of other factors have not been taken into account. All these problems will be the subject of our subsequent studies.

## 7.3. HEAT TRANSFER IN INDIRECTLY HEATED FURNACES

High-temperature heating furnaces are subdivided, depending on the mode of heat transfer in their working space, into furnaces with uniformly distributed heating, furnaces with direct directional heating, furnaces with directed indirect heating and furnaces with combined directed heating [205, 214, 215]. A characteristic temperature distribution for such furnaces according to data of Soroka [205] is shown in Fig. 7.4. In many cases, furnaces with directed indirect heating are most effective. Since an improvement of heating effectiveness involves not only fuel savings, but also an increase in furnace productivity, these problems are accorded a great deal of attention; in particular, they were discussed in detail at the first All-Union Conference on Indirect Heating of Materials, which took place in Kiev in 1976. This problem cannot always be resolved using the available techniques of calculation of such furnaces developed by the Gas Institute of the Ukrainian Academy of Sciences, Ural Polytechnic Institute, and other organizations. The main reason limiting a given computational technique may be allowance for radiation in such systems. The above computational techniques incorporate radiation for nongray gases basically by the Hottel model. In conjunction with this, it is interesting to compare such analytic results with data obtained by the technique presented in this study. Such a comparison is performed in Table 7.1 for nine heat-transfer modes presented in Fig. 7.4.

In the simplest approximation one can assume that a furnace is a system of two planar infinite surfaces (radiator and receiver) placed at a distance $H$

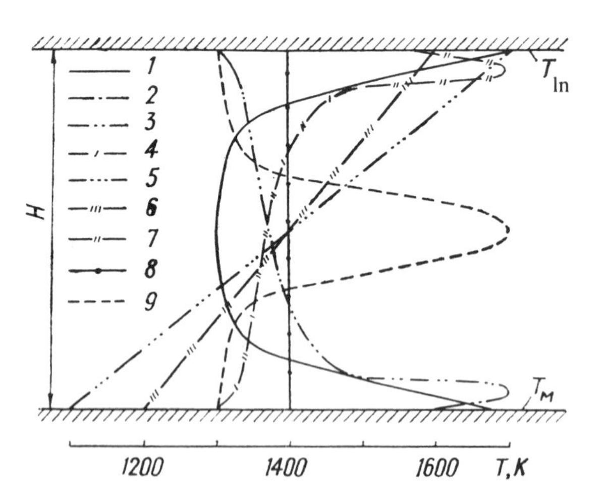


**FIG. 7.4.** Characteristic temperature distribution in heating furnaces at a mean flow temperature of 1400 K and different heat-transfer modes. (1) Directed, direct and indirect; (2) and (3) directed direct; (4)–(7) directed indirect; (8) and (9) uniformly distributed.

**TABLE 7.1** Effectiveness of various heat transfer modes in heating surfaces (the numerator represents results of Soroka [205], the denominator those of the present author).

| $T_м$ K | $\alpha_{кл}$ Вт / W/m²·K | $H$ m | $q_M$, kW/m² or $T_{1n}$, K | Heating modes and temperature distributions by layers according to Fig. 7.4 | | | | | | | | | $q_{M,1}/q_{M,9}$ |
|---|---|---|---|---|---|---|---|---|---|---|---|---|---|
| | | | | 1 | 2 | 3 | 4 | 5 | 6 | 7 | 8 | 9 | |
| 300 | 23.2 | 1.11 | $q_M$ | 180/146 | 166/133 | 160/130 | 154/128 | 146/124 | 144/124 | 150/127 | 147/123 | 138/115 | 1.30/1.27 |
| | | | $T_{1n}$ | 1138/1122 | 1052/1043 | 1052/1044 | 1151/1131 | 1177/1161 | 1144/1138 | 1134/1127 | 1081/1072 | 1063/1051 | |
| 300 | 23.2 | 2.22 | $q_M$ | 214/162 | 196/146 | 188/142 | 177/140 | 167/136 | 165/134 | 172/139 | –/133 | 157/124 | 1.36/1.31 |
| | | | $T_{1n}$ | 1206/1177 | 1103/1086 | 1103/1087 | 1217/1186 | 1247/1220 | 1208/1187 | 1199/1183 | –/1118 | 1113/1094 | |
| 300 | 116 | 1.11 | $q_M$ | 208/175 | 177/145 | 171/141 | 181/156 | 173/152 | 167/145 | 171/149 | 162/138 | 149/126 | 1.40/1.39 |
| | | | $T_{1n}$ | 1242/1235 | 1109/1104 | 1109/1104 | 1250/1240 | 1268/1262 | 1229/1223 | 1217/1215 | 1149/1144 | 1116/1110 | |

**TABLE 7.1** Continued

| $T_M$ K | $\alpha_{ln}$ Bт W/m² K | $H$ m | $q_M$, kW/m² or $T_{ln}$, K | Heating modes and temperature distributions by layers according to Fig. 7.4 | | | | | | | | | $\frac{q_{M,1}}{q_{M,9}}$ |
|---|---|---|---|---|---|---|---|---|---|---|---|---|---|
| | | | | 1 | 2 | 3 | 4 | 5 | 6 | 7 | 8 | 9 | |
| 300 | 116 | 2,22 | $q_M$ | 237/186 | 204/155 | 196/151 | 199/164 | 188/159 | 182/153 | 189/157 | –/145 | 165/132 | 1,43/1,41 |
| | | | $T_{ln}$ | 1288/1272 | 1144/1134 | 1144/1134 | 1296/1277 | 1317/1302 | 1274/1261 | 1264/1254 | –/1177 | 1152/1138 | |
| 1100 | 23,2 | 1,11 | $q_M$ | 113/92 | 98/78 | 92/74 | 87/70 | 80/66 | 77/64 | 83/70 | 80/65 | 71/56 | 1,60/1,64 |
| | | | $T_{ln}$ | 1267/1253 | 1206/1198 | 1206/1198 | 1276/1259 | 1296/1283 | 1271/1260 | 1264/1255 | 1225/1216 | 1213/1202 | |
| 1100 | 23,2 | 2,22 | $q_M$ | 140/106 | 121/90 | 113/85 | 102/80 | –/74 | 90/72 | 98/81 | –/74 | 82/63 | 1,70/1,68 |
| | | | $T_{ln}$ | 1306/1282 | 1227/1212 | 1227/1213 | 1315/1288 | 1339/1317 | 1308/1289 | 1300/1293 | –/1235 | 1234/1217 | |
| 1100 | 116 | 1,11 | $q_M$ | 136/116 | 110/83 | 97/79 | 110/95 | 102/88 | 95/83 | 99/87 | 88/75 | 75/61 | 1,82/1,90 |
| | | | $T_{ln}$ | 1332/1327 | 1222/1215 | 1222/1216 | 1339/1331 | 1354/1347 | 1320/1315 | 1311/1308 | 1255/1250 | 1228/1219 | |
| 1100 | 116 | 2,22 | $q_M$ | 159/127 | 124/94 | 116/87 | 121/100 | 110/93 | 104/88 | 111/93 | –/82 | 85/67 | 1,86/1,89 |
| | | | $T_{ln}$ | 1361/1347 | 1239/1229 | 1239/1229 | 1368/1350 | 1387/1373 | 1349/1337 | 1339/1331 | –/1265 | 1245/1233 | |

apart and separated by a layer of combustion products. The temperature distribution depends basically on the method of fuel combustion in the furnace. The receiver temperature and the convective coefficient of heat transfer about the radiator are specified quantities. It can be assumed that the radiator (lining) is adiabatic. Then the heat energy balance for the radiator is

$$q_{r,\,\mathrm{ln}} + \alpha_{\mathrm{ln}} \cdot (T_g - T_{\mathrm{ln}}) = 0 \tag{7.7}$$

If temperature $T_g$ of the gas and the receiver (heated metal) temperature $T_M$ are specified, and the coefficient of convective heat transfer $\alpha_{\mathrm{ln}}$ between the gas and the lining is known, then Eq. (7.7) contains only one unknown, the lining temperature $T_{\mathrm{ln}}$. In terms of natural radiation of individual zones, such a problem reduces to solving a set of nonlinear algebraic equations [205]. Here the gas layer between the receiver and radiator is subdivided into 20 zones, within each of which the temperature is assumed to be constant.

Using the technique for calculating radiative heat flux described in Chapter 2, Eq. (7.7) can be solved by iterations with respect to $T_{\mathrm{ln}}$, following the determination of which, it is not too difficult to calculate the total heat flux in the receiver. In this way calculations were performed for a number of heating conditions (Table 7.1). There is no principal difference between the values of flux using both methods. However, even an insignificant change in $T_{\mathrm{ln}}$ may in some cases result in a perceptible change in the total heat flux to the receiver. A more detailed allowance for the spectral features of a radiating medium may result in some revision of individual modes both of direct and of indirect heating. Note that in this problem one can easily make allowance for the spectral properties of the radiator and receiver.

When heating materials in furnaces, there always exists some optimum temperature distribution due to the heat sources in individual zones. As of now it is already possible to determine these conditions.

## 7.4. HEAT TRANSFER IN A HYDROGEN PLASMA ARC

An electric arc stabilized by the wall of a channel is extensively used for determining the transfer properties of high-temperature gases, simulating phenomena occurring at high flight velocities in the atmosphere, as a source of radiation in photochemical processes, and also for pumping of gas-dynamic lasers. Implementation of these applications is supported by extensive studies of the arc of various gases, performed in the USSR and in the West. A survey of the main investigations in this field is given by Zhukov and his co-workers [200–202].

Analytic calculations of electric arcs are significantly complicated not only by difficulties in making allowance for the hydrodynamics of laminar and turbulent swirled flows and difficulties in incorporating properties of plasma which are highly temperature dependent, but also by difficulties in the detailed incorporation of radiation, which require a large expenditure of machine time. Additional difficulties arise when allowance must be made for nonequilibrium processes, which may occur in zones with high temperature gradients. Familiar calculations in the simplified formulation of the problem yield, in the majority of cases, results that deviate from experimental data, which may be due not only to absence of thermodynamic equilibrium in the arc, but also to physical simplifications underlying the mathematical model of the arc. Let us consider the elementary problem of the channel model of the arc for the case of a hydrogen plasma arc, with detailed allowance for radiation and other transfer properties of the plasma.

The steady-state cylindrical arc was frequently investigated by many investigators [119, 120, 195, 200–202, 216–220). Some of these studies were concerned with investigation of the hydrogen arc. However, in these studies detailed allowance for radiation was made only for the argon arc [119, 120].

The energy equation for a steady-state cylindrical arc, stabilized by the channel wall, has the form [200]

$$\frac{1}{y}\frac{\partial}{\partial y}\left(\lambda y\frac{\partial T}{\partial y}\right)+\sigma E^{2}-\frac{1}{y}\frac{\partial(yq_r)}{\partial y}=0 \tag{7.8}$$

i.e., it is assumed that the gas is nonmoving, and the magnetic field is moderate.

The energy equation is supplemented by Ohm's law in the form of the requirement that the current along the channel be constant:

$$i=2\pi E\int_0^Y \sigma y\,dy \tag{7.9}$$

The problem is solved with the boundary conditions

$$\frac{\partial T}{\partial y}=0 \quad \text{at} \quad y=0 \tag{7.10}$$

$$T=T_w \quad \text{at} \quad y=Y \tag{7.11}$$

For an arc with cylindrical symmetry in the presence of local thermodynamic equilibrium the radiant heat flux $q_r$ was calculated from Eqs. (2.117)–(2.124). The absorption coefficient of hydrogen plasma was calculated with allowance for the following radiation processes: (1) photoionization of atoms of the ground and excited states; (2) bremsstrahlung radiation in a field of positive ions; (3) bremsstrahlung radiation in a field of neutral particles; (4) photoionization of negative fluxes; (5) bremsstrahlung radiation in a field of negative ions; (6) line radiation of atoms; (7) pressure-induced radiation in fundamental bands [221]. A characteristic absorption spectrum of hydrogen plasma is shown in Fig. 7.5.

In calculating the total radiant heat flux from Eq. (2.107), the entire spectral interval was subdivided into 40 parts, where the width of 20 of these corresponded to the width of atomic lines. The spectral interval within the limits of the lines was subdivided into 10 parts, and the remaining intervals into 5–10 parts.

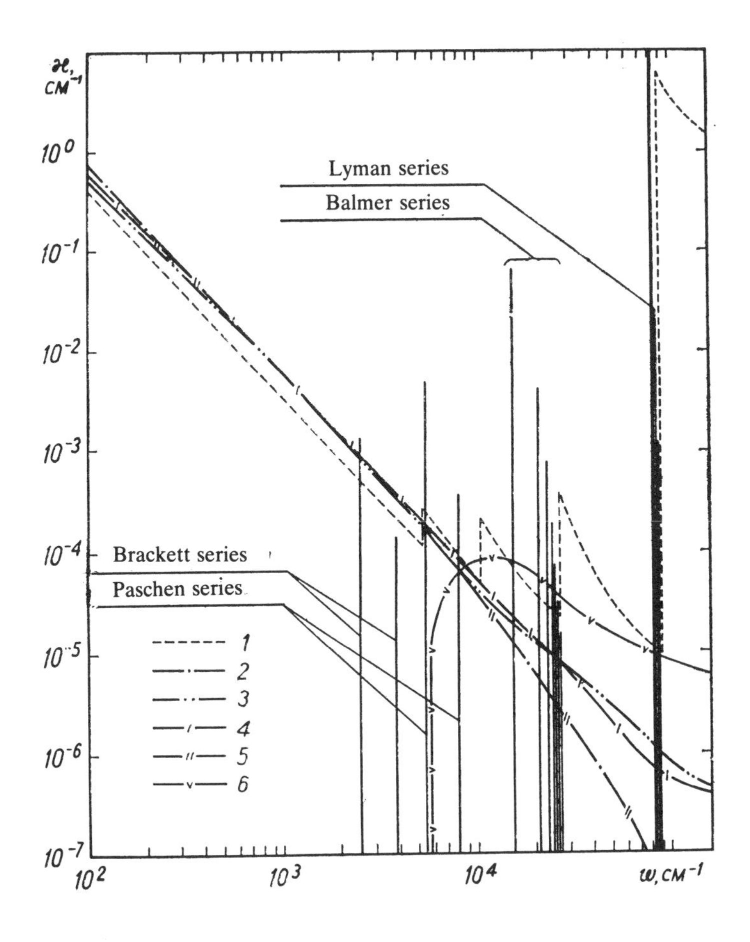


**FIG. 7.5** Absorption coefficient of hydrogen as a function of the wave number at atmospheric pressure and T = 10,000 K for individual processes. (1) Photoionization of atoms from the ground and excited states; (2)–(5) bremsstrahlung radiation in fields of positive ions, atoms, molecules, and negative ions, respectively; (6) photonionzation of negative ions. The vertical lines serve as a nominal designation of the coefficient of the absorption in spectral lines.

The temperature behavior of the thermal conductivity λ and electrical conductivity σ of atomic hydrogen were described as in the paper by Devoto [222], whereas the values of λ for molecular hydrogen were taken from the paper by Vargaftik and Vasilevskaya [223]. Note that dissociation of molecular hydrogen at temperatures to 6000 K makes a significant contribution to the total thermal conductivity coefficient (Fig. 7.6), and neglect of this effect, as this was done by some investigators, may result in a significant distortion of results of calculations.

Equation (7.8) was solved numerically by a simpler iterative computational scheme than for the problem of stabilized flow in a circular pipe analyzed in Chapter 6. After substitution of variables

$$\eta = \frac{y}{Y}, \quad \Theta = \frac{T}{T_0}, \quad \bar{q}_r = \frac{q_r \cdot Y}{\lambda_0 T_0} \tag{7.12}$$

the solution of Eq. (7.8) can be represented in the form

$$\Theta = \Theta_w - \int_1^{\eta} \frac{1}{g_\lambda} \cdot \left\{ \frac{i^2}{\eta \cdot \lambda_0 \cdot T_0 \cdot \left[2\pi Y \cdot \int_0^1 \sigma(\eta_1) \cdot \eta_1 \, d\eta_1\right]^2} \times \int_0^{\eta} \sigma(\eta_1) \cdot \eta_1 \, d\eta_1 - \bar{q}_r \right\} d\eta \tag{7.13}$$

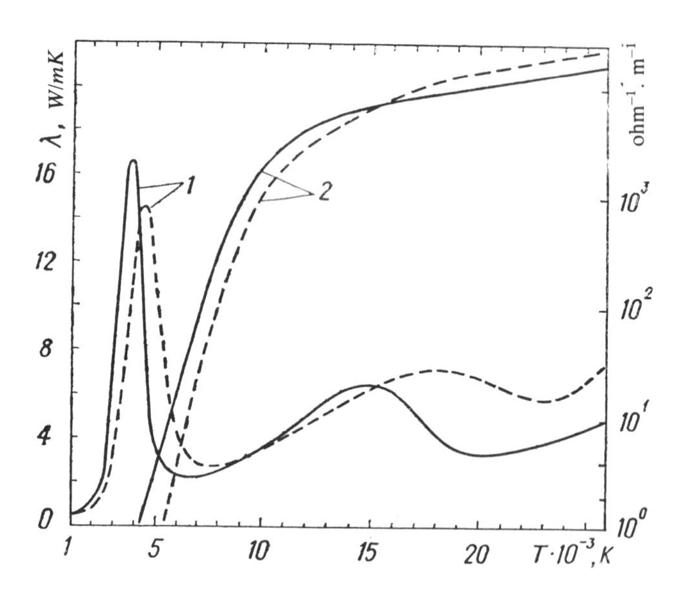


**FIG. 7.6** Thermal conductivity λ and electrical conductivity σ coefficients as a function of temperature of hydrogen plasma at different pressures. (1) 1 atm; (2) 10 atm.

Equation (7.13) incorporates boundary conditions (7.10) and (7.11).

The temperature field was determined from Eq. (7.13) by numerical integration using the trapezoid rule. The zero approximation for the temperature field in calculation of the nonradiating arc was the square temperature distribution, whereas for the radiating arc it was the solution of Eq. (7.13) in the absence of radiation.

For subsequent approximations the temperature field was determined using the damping factor of Eq. (6.60). In numerical integration the nondimensional co-ordinate $\eta$ was subdivided nonuniformly into 200–300 segments; to improve the accuracy, it was necessary here to increase the number of divisions as the wall was approached. For each iteration step the heat flux was calculated in 23 points over coordinate $\eta$, and was determined by linear interpolation for the remining values of $\eta$. The computational time of an individual version for a nonradiating arc on the BESM-6 computer did not exceed 2 min, but when detailed allowance was made for radiation it rose to 2–4 hours.

The temperature profiles of a hydrogen arc, calculated with and without allowance for radiation, have a characteristic inflection in the low-temperature region, which is due to molecular thermal conductivity of the plasma (Fig. 7.7). Radiation has the greatest effect on the temperature distribution in the center of the channel. These effects become significantly amplified with increasing diameter of the arc and current. Temperatures calculated for the central part of the arc are in satisfactory agreement with the not-too-numerous experimental values obtained by Morris and Weiss [220]; a perceptible divergence is observed only in the zone of high temperature gradients, which can be attributed to significant errors of experimental measurement.

The calculated volt–ampere characteristics for the hydrogen arc are in satisfactory agreement with available experimental results (Fig. 7.8 and 7.9). In spite of the fact that radiation effects for 0.1- and 0.15-cm hydrogen arcs are quite insignificant (Fig. 7.8), our experimental data are only insignificantly lower than the experimental results of Maecker, are virtually identical to the measurements of Morris and are approximately 20% higher than the experimental results of Weiss for a 0.15-cm diameter arc. Such an agreement between analytic and experimental results over a wide range of values of the arc current points to satisfactory correspondence of the values of transport properties of hydrogen plasma used in this book. Note that the deviation of results obtained analytically by Incropera, who made no allowance for the thermal conductivity of molecular hydrogen (although he did allow for the nonequilibrium of the hydrogen arc), from the experimental results of the above investigators is about 200%. This allows the assumption that the processes occurring in a hydrogen arc are close to conditions of local thermodynamic equilibrium. Analogous results were obtained also from a 0.5-cm-diameter

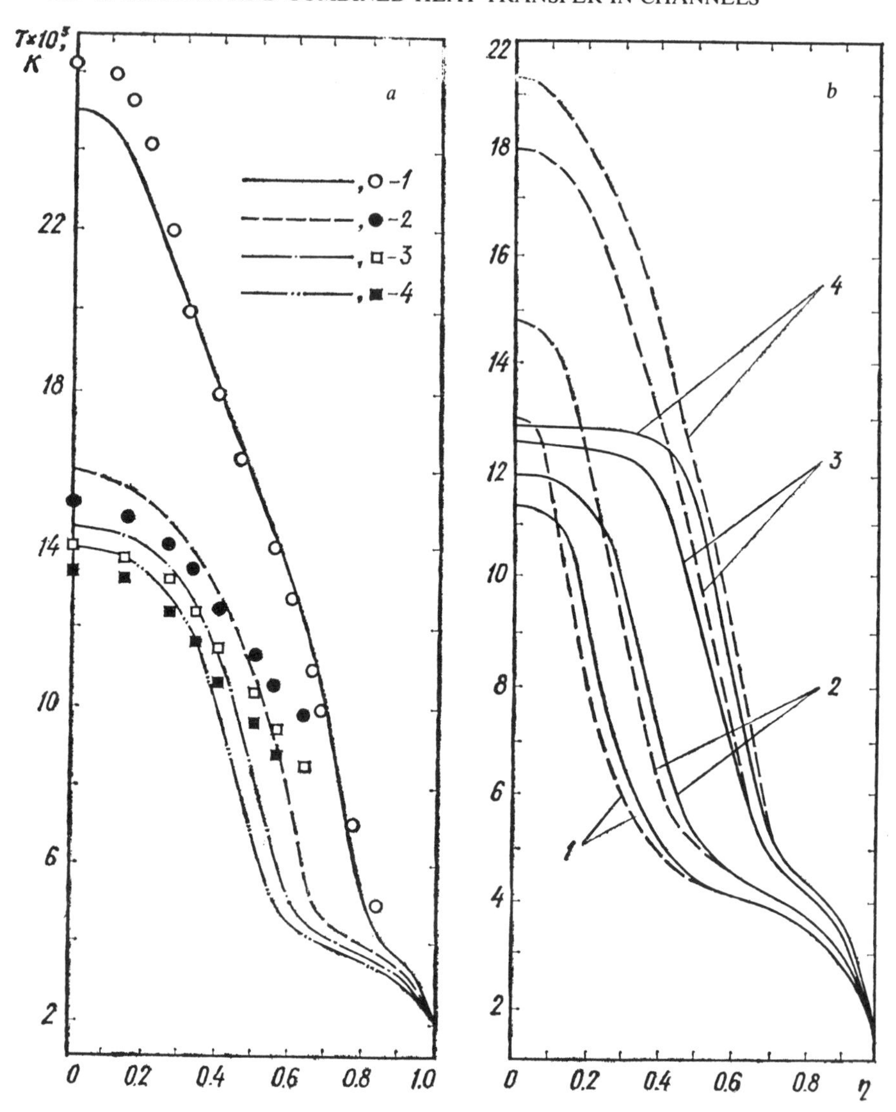


**FIG. 7.7** Temperature distribution in a hydrogen plasma arc stabilized by the channel walls, as a function of current $i$, channel radius $Y$, and pressure $p$. ($a$) $p$ = 1atm, $Y$ = 0.15; the curves represent analytic results and the points represent experimental data, $i$ = 150, 70, 50, and 40 A, respectively; (1) experimental results of Shabasho and Nizovskiy; (2)–(4) of Morris [220], ($b$) $p$ = 10 atm, $Y$ = 0.75; (1–4) analytic results for currents $i$ = 50, 120, 240, and 360 A, with allowance for radiation (solid curves) and without allowance for radiation (dashed curves).

hydrogen arc (Fig. 7.9). Calculations of a nonradiating arc are in satisfactory agreement with the results of Zarudi [219]. In the case of high arc currents, allowance for radiation results in a slight increase in the volt–ampere characteristic. The analytic data of Green and Incropera lie significantly above the data of the present study.

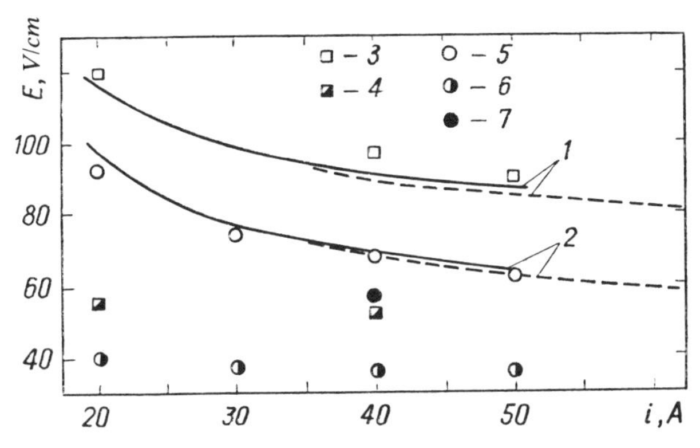


**FIG. 7.8** Analytic volt–ampere characteristics (continuous curves) for a hydrogen arc. (1) Radius of 0.1 cm; (2) radius of 0.15 cm; the dashed segment is without allowance for radiation; (3) experimental data of Mecker for a 0.1-cm-radius arc [220]; (4) analytic results of Incropera for a 0.1-cm-radius arc [220]; (5) experiments by Morris for a 0.15-cm-radius arc [220]; (6) analytic results of Incropera for a 0.15-cm-radius arc; (7) experimental results of Weiss for a 0.15-cm-radius arc.

The satisfactory agreement between analytic and experimental results confirms in a certain sense the reliability of the technique developed here for calculating the hydrogen arc. For this reason this technique was used for calculating the characteristics of the hydrogen arc under conditions not explored in the experiments. However, since such calculations required a large expenditure of computer time, it became necessary to correlate the analytic results in nondimensional form.

In addition to volt–ampere characteristics, in designing arc devices it is important to know the total heat flux in the channel. In the presence of a radiating arc this heat flux is due to interaction between radiation and other

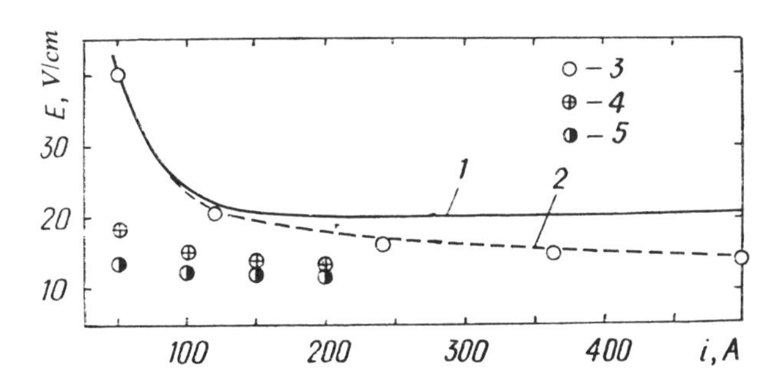


**FIG. 7.9** Analytic volt–ampere characteristics (solid curve) for a 0.5-cm-radius hydrogen arc. (1) With allowance for radiation; (2) without allowance for radiation; (3) according to Zarudi and Edel'baum [219], without allowance for radiation; (4) according to Green [220]; (5) according to Incropera, with allowance for nonequilibrium of the hydrogen plasma.

forms of energy transfer. The governing relationships of combined heat transfer in a hydrogen arc were analyzed using as the reference criterion the parameter $\chi_{\text{III}}$, obtained in Chapter 6 for the problem of combined heat transfer with internal heat sources. If it is remembered that the strength of internal heat sources for a channel arc is controlled by Joule heating,

$$q_b Y = \frac{i \cdot E^0}{2\pi Y} \tag{7.14}$$

the criterion $\chi_{\text{III}}$ can be written as

$$\chi_{\text{III}} = \frac{2\pi q_{\text{p}}^0 Y}{iE^0} \cdot \frac{d \ln q_\Sigma^0}{d \ln (h)} \tag{7.15}$$

Quantities $q_r^0$, $q_\Sigma^0$ and $h$ in Eq. (7.15) were determined from the temperature field calculated without radiation (Table 7.2). To determine the derivative in Eq. (7.15), the calculations were performed for two or three values of arc current for a nonradiating arc, and the derivative was determined by finite-difference methods [115]. Parameter $\chi_{\text{III}}$ determined in this manner is quite general for a wide range of arc conditions (Fig. 7.10).

Combined heat transfer in a hydrogen arc can be described by the expression

$$\frac{q_\Sigma}{q_\Sigma^0} = \frac{1}{1 + 0.229\, \chi_{\text{III}}^{0.736}} \tag{7.16}$$

which is rather similar to results obtained by Biberman for combined heat transfer in the stagnation point of different bodies [182].

The use of Fig. 7.10 may significantly facilitate calculation of combined heat transfer in arc devices, provided that it is found to be universal also for

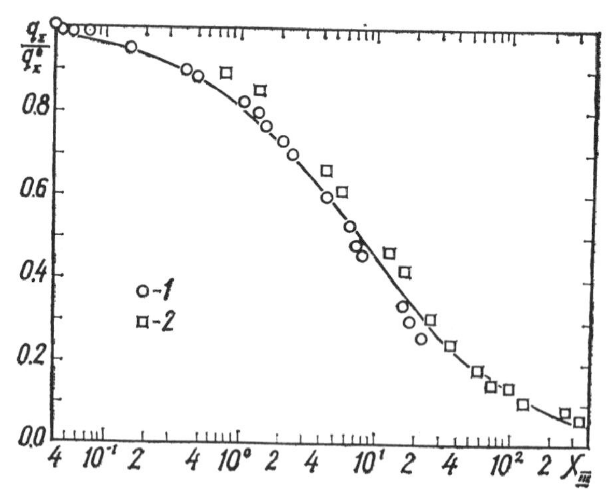


**FIG. 7.10** Combined heat transfer as a function of nondimensional parameter $\chi_{\text{III}}$ for different diameters and currents in a hydrogen plasma arc. (1) $p = 1$ atm; (2) $p = 10$ atm.

**TABLE 7.2** Principal Results on Combined Heat Transfer in a Hydrogen Arc Stabilized by the Channel Walls at Atmospheric Pressure

| $Y$ cm | $i$ $A$ | $E^0$ W/cm | $q^0_{conv} \cdot 10^{-3}$ kW/m² | $q^0_p \cdot 10^{-3}$ kW/m² | $E$ W/cm | $q_{conv} \cdot 10^{-3}$ кВт/м² | $q_p \cdot 10^{-3}$ kW/m² | $\frac{q_\Sigma}{q^0_\Sigma}$ | $\chi_{III}$ |
|---|---|---|---|---|---|---|---|---|---|
| 0,10 | 40 | 88,6 | 56,41 | 1,61 | 93.4 | 55,08 | 4,37 | 1,025 | 0,096 |
| 0.10 | 50 | 83.4 | 66,36 | 2.50 | 85,4 | 61.06 | 6.88 | 0.987 | 0,125 |
| 0,15 | 40 | 67,8 | 28,78 | 0,87 | 69,3 | 28.98 | 0.79 | 1,001 | 0,106 |
| 0.15 | 50 | 62.6 | 33,20 | 1,40 | 64,5 | 33,10 | 1,27 | 0,993 | 0,149 |
| 0,15 | 70 | 57,2 | 42,47 | 2,92 | 59,3 | 41,40 | 2.68 | 0,971 | 0.221 |
| 0,50 | 120 | 21,2 | 8,10 | 1,53 | 22,7 | 7,81 | 0,87 | 0,902 | 0,951 |
| 0,50 | 240 | 16.9 | 12,87 | 6.63 | 21,0 | 12,23 | 3,83 | 0,823 | 2,72 |
| 0,50 | 280 | 16.2 | 14,47 | 8,58 | 20,9 | 13,40 | 5,18 | 0.806 | 2,80 |
| 0,70 | 120 | 17,5 | 4,79 | 0,97 | 18,4 | 4,52 | 0,50 | 0,872 | 1,10 |
| 0,70 | 240 | 13,3 | 7,25 | 3.97 | 15,9 | 6,84 | 1,84 | 0.774 | 3,45 |
| 0,70 | 360 | 11,8 | 9,69 | 9,03 | 15.8 | 8,85 | 4,09 | 0,691 | 5,43 |
| 1.50 | 500 | 6,3 | 3,32 | 5,53 | 8,8 | 3,06 | 1,60 | 0,525 | 14,5 |
| 1,50 | 600 | 5,9 | 3,76 | 8,09 | 8,7 | 3,34 | 2,22 | 0,469 | 18,1 |
| 1,50 | 700 | 5.7 | 4,20 | 10,95 | 8.7 | 3,62 | 2,85 | 0,428 | 19,7 |
| 3,00 | 1200 | 3,0 | 1,88 | 10,54 | 5,1 | 1,50 | 1,72 | 0,259 | 52,6 |
| 3,00 | 1000 | 3,1 | 1,66 | 7,30 | 5,1 | 1,39 | 1,32 | 0,303 | 43,7 |
| 3,00 | 900 | 3,2 | 1,55 | 5,87 | 5,1 | 1,32 | 1,10 | 0,326 | 38.8 |
| 3,00 | 480 | 4,2 | 1,08 | 1,83 | 5.4 | 1,01 | 0,38 | 0,477 | 17.7 |

plasma of other gases. When this relationship is used, it is virtually sufficient to determine the temperature field for nonradiating plasma at two values of the current and also the radiant heat flux in the fundamental formulation of the problem, i.e., to calculate its value at a temperature field obtained in a nonradiating arc. This significantly shortens the computer time needed as compared with solution of the complete problem for a radiating arc. This offers promise for simulation of combined radiation.

# APPENDICES

## 1. NOMOGRAMS OF THE VALUES OF $Nu_T$ IN HIGH-TEMPERATURE TURBULENT FLOWS

These nomograms were compiled for facilitating the calculation of $Nu_{0,T}$ from Eqs. (4.14)–(4.24). The turbulent Nusselt number $Nu_{0,T}$ at constant properties is determined from the nomograms in Figs. A.1–A.5, from which one first determines the values of $Nu_1$, $Nu_2$, $Nu_3$, $Nu_4$, and $\varphi_K$ at specified $Re_f$, Pr, $Pr_T$, and $K_T$. The reference parameters must be calculated from the values of thermophysical properties at the flow temperature. The value of $Nu_{0,T}$ is obtained from the expression

$$Nu_{0,T} = \varphi_K (Nu_1 + Nu_2 + Nu_3 + Nu_4) \qquad 206A$$

Illustrative example: Re = $3.5\times10^6$, Pr = 0.7, $Pr_T = 0.5$, and $K_T = 0.2$. From the nomograms we find that $Nu_1 = 8.0\times10^1$, $Nu_2 = 2.2\times10^3$, $Nu_3 = 3.7\times10^3$, $Nu_4 = 4.6\times10^2$, and $\varphi_K = 0.94$. Then $Nu_{0,T} = 6054$. Exact calculations from Eqs. (4.13)–(4.19) yield the following values: $Nu_1 = 82.8$, $Nu_2 = 2274$, $Nu_3 = 3754.1$, $Nu_4 = 485.1$, $\varphi_K = 0.943$, and $Nu_{0,T} = 6227.8$.

The value of temperature factor $\varphi_T$ for high-temperature flows can be determined from the nomograms in Figs. A.6 and A.7 when the values of Re, Pr, $Pr_T$, $K_T$, and $\varrho_f/\varrho_w$ are known. Then:

$$\phi_T = \phi_1 + \phi_2$$

For the same illustrative example for $\varrho_f/\varrho_w$ we obtain $\varphi_1 = 0.61$ and $\varphi_2 = 0.17$. Accordingly, $\varphi_T = 0.78$ and $Nu_T^* = 4722$. Exact calculations with Eqs. (4.20)–(4.24) yield the following values: $\varphi_1 = 0.616$, $\varphi_2 = 0.174$, $\varphi_T = 0.790$, and $Nu_T^* = 4918.8$.

The error due to the use of nomograms thus does not exceed ±10%.

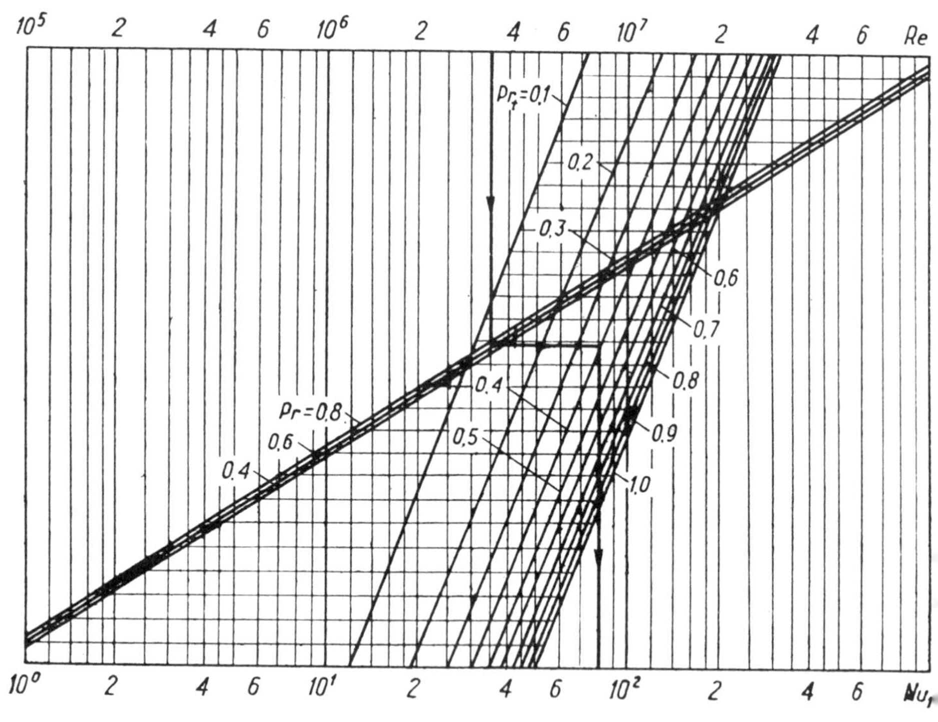


Fig. A.1. Nomogram for determining $\mathrm{Nu}_1 = f(\mathrm{Re}, \mathrm{Pr}, \mathrm{Pr}_T)$.

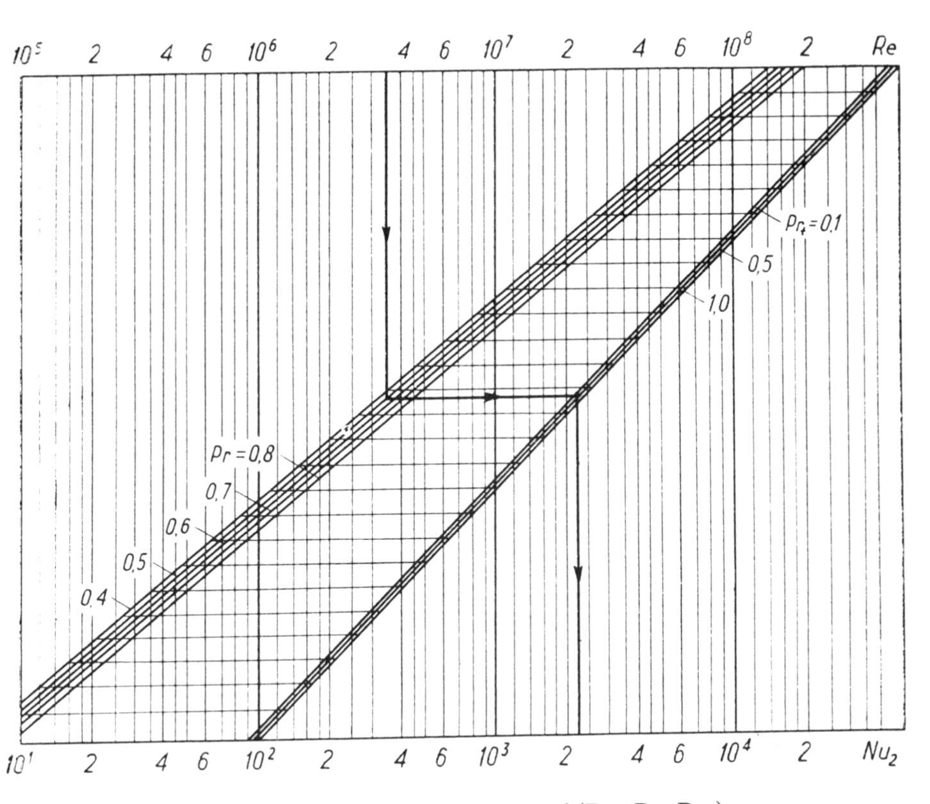


Fig. A.2. Nomogram for determining $Nu_2 = f\,(Re, Pr, Pr_T)$.

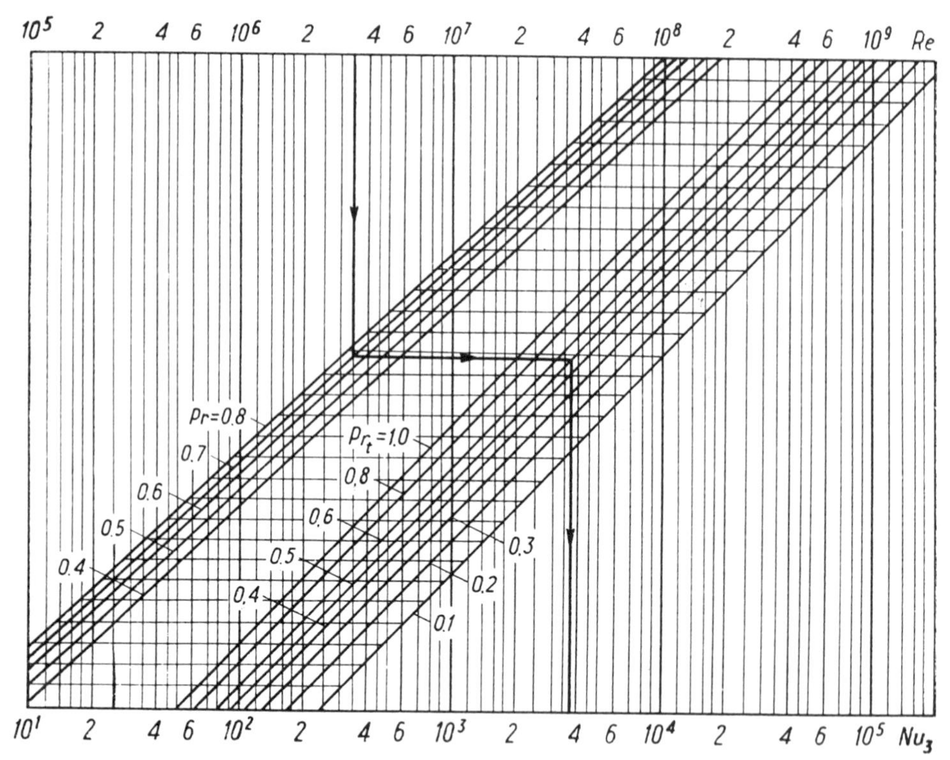


Fig. A.3. Nomogram for determining $Nu_3 = f(Re, Pr, Pr_T)$.

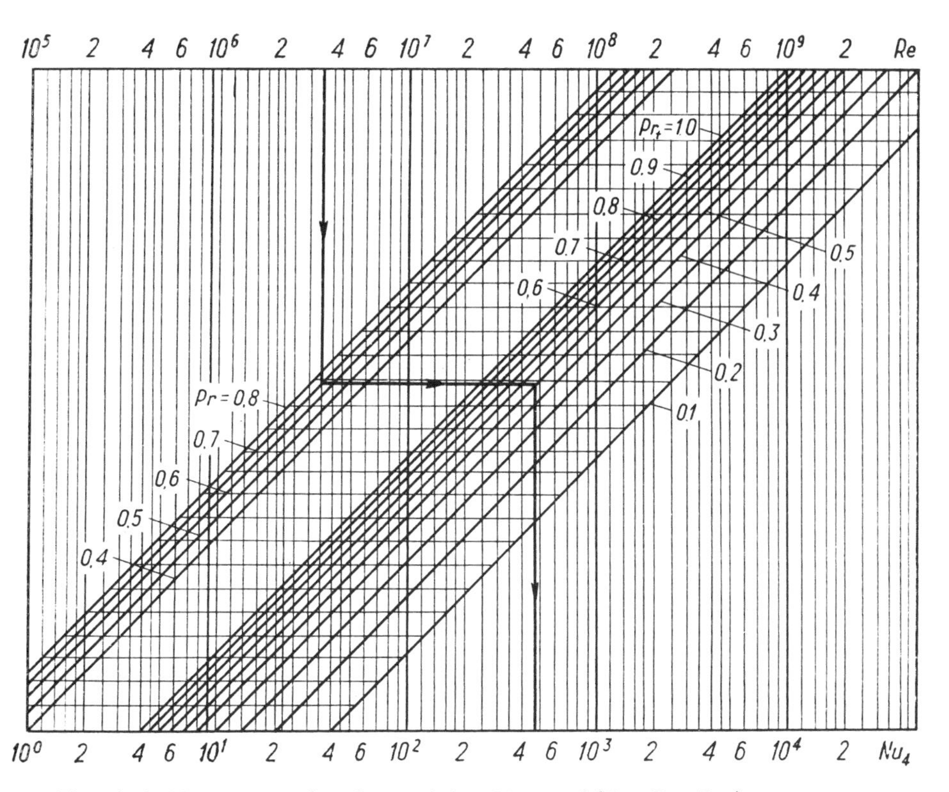


Fig. A.4. Nomogram for determining $Nu_4 = f\,(Re, Pr, Pr_T)$.

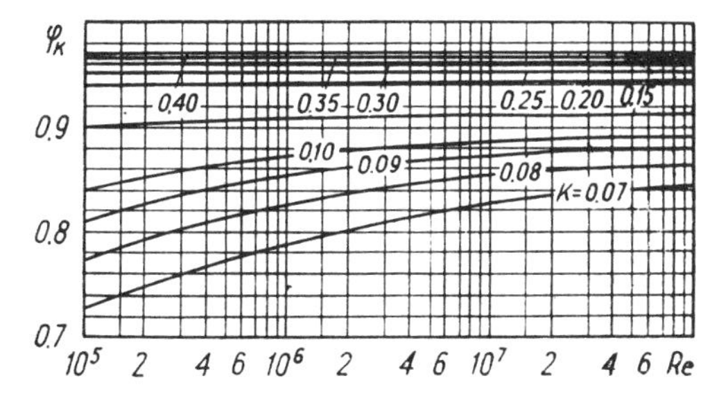


Fig. A.5. Nomogram for determining $\phi_K = f\,(Re, K_T)$.

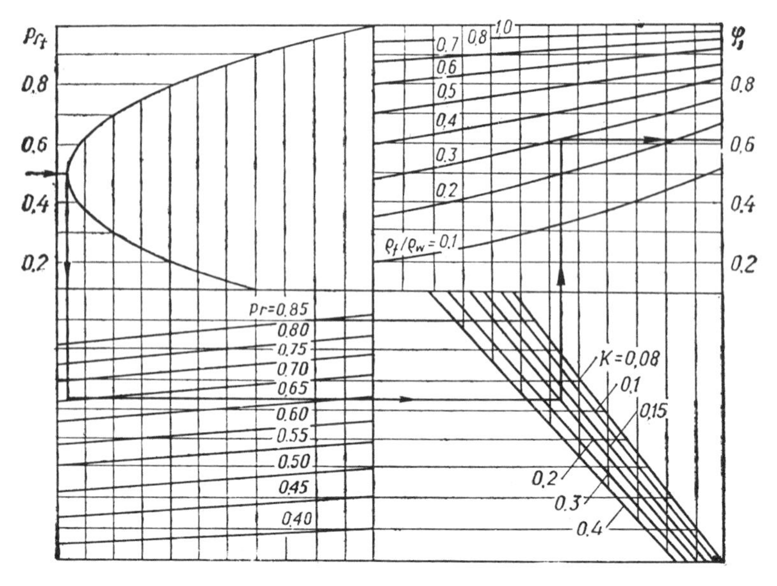


Fig. A.6. Nomogram for determining $\phi_1 = f(\varrho_f / \varrho_w, \mathrm{Pr}_T, K_T, \mathrm{Pr})$.

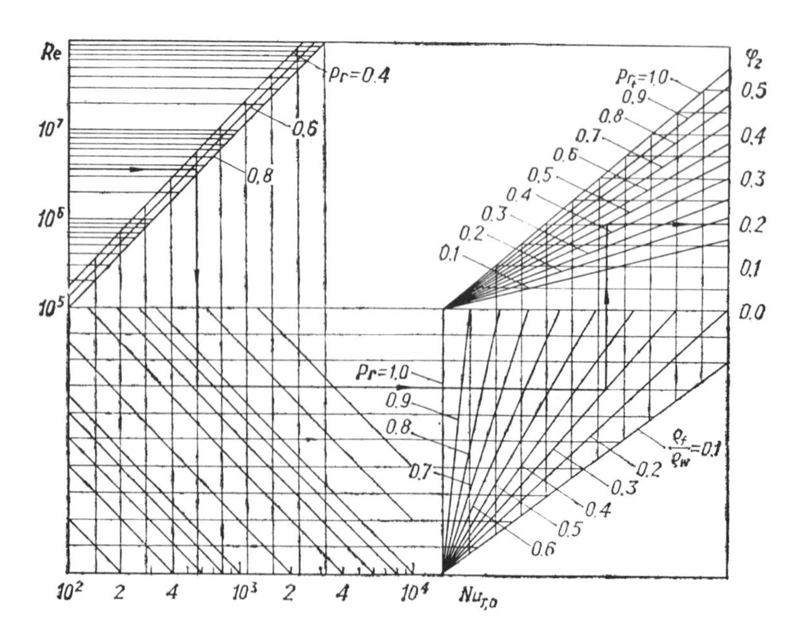


Fig. A.7. Nomogram for determining $\phi_2 = f(\mathrm{Re}, \mathrm{Pr}, \mathrm{Nu}_{T,0}, \varrho_f / \varrho_w, \mathrm{Pr}_T)$.

## 2. EXPERIMENTAL DATA ON CONVECTIVE HEAT TRANSFER IN TEST CHANNELS 1 AND 2

**TABLE A.1** Heat Transfer to a Plate in a Flow of Combustion Products

| $T_f$ K | $\frac{T_f}{T_w}$ | $U_f$ m/sec | $\alpha$ | [O] | $x$ mm | $\frac{\Delta h_{chem}}{\Delta h}$ | $Nu_f^*$ | $\frac{Re_f}{10^3}$ | $Pr_f$ | $Pr_T$ | $\varphi_l$ | $\varphi_T$ |
|---|---|---|---|---|---|---|---|---|---|---|---|---|
| 2240 | 6.9 | 283 | 1.05 | 0.37 | 183 | 0,05 | 267 | 02 | 0,76 | 0.15 | 0.96 | 0.50 |
| 2233 | 6.8 | 175 | 1.07 | 0.37 | 497 | 0.05 | 418 | 172 | 0.76 | 0.15 | 0.96 | 0.52 |
| 2233 | 6.8 | 175 | 1.07 | 0.37 | 487 | 0.05 | 389 | 168 | 0.76 | 0.15 | 0.96 | 0.52 |
| 2233 | 6.6 | 175 | 1.07 | 0.37 | 388 | 0.05 | 301 | 134 | 0.76 | 0.15 | 0.96 | 0.52 |
| 2233 | 6.7 | 175 | 1.07 | 0.37 | 377 | 0.05 | 301 | 130 | 0.76 | 0.15 | 0.96 | 0.52 |
| 2233 | 6.7 | 175 | 1.07 | 0.37 | 290 | 0.05 | 265 | 100 | 0.76 | 0.15 | 0.96 | 0.51 |
| 2650 | 8.1 | 190 | 1.09 | 0.68 | 497 | 0.23 | 275 | 133 | 0.75 | 0.13 | 0.91 | 0.47 |
| 2650 | 8.1 | 190 | 1.09 | 0.68 | 487 | 0.23 | 253 | 130 | 0.75 | 0.13 | 0.91 | 0.47 |
| 2650 | 7.7 | 190 | 1.09 | 0.68 | 388 | 0.23 | 247 | 104 | 0.75 | 0.13 | 0.91 | 0.47 |
| 2553 | 7.9 | 183 | 1.11 | 0.68 | 497 | 0.18 | 318 | 138 | 0.76 | 0.13 | 0.92 | 0.48 |
| 2553 | 7.9 | 183 | 1.11 | 0.68 | 487 | 0.18 | 300 | 135 | 0.76 | 0.13 | 0.92 | 0.48 |
| 2553 | 7.9 | 183 | 1.11 | 0.68 | 443 | 0.18 | 239 | 123 | 0.76 | 0.13 | 0.92 | 0.47 |
| 2553 | 7.5 | 183 | 1.11 | 0.68 | 388 | 0.18 | 255 | 107 | 0.76 | 0.14 | 0.92 | 0.48 |
| 2528 | 7.2 | 311 | 1.10 | 0.56 | 226 | 0.16 | 272 | 110 | 0.76 | 0.14 | 0.93 | 0.49 |
| 2447 | 7.4 | 302 | 1.10 | 0.56 | 497 | 0.12 | 439 | 243 | 0.76 | 0.14 | 0.93 | 0.50 |
| 2447 | 7.4 | 302 | 1.10 | 0.56 | 487 | 0.12 | 439 | 243 | 0.76 | 0.14 | 0.93 | 0.50 |
| 2447 | 7.1 | 302 | 1.10 | 0.56 | 388 | 0.12 | 446 | 194 | 0.76 | 0.14 | 0.93 | 0.51 |
| 2447 | 7.1 | 302 | 1.10 | 0.56 | 377 | 0.12 | 465 | 188 | 0.76 | 0.14 | 0.93 | 0.51 |
| 2447 | 7.1 | 302 | 1.10 | 0.56 | 311 | 0.12 | 288 | 155 | 0.76 | 0.14 | 0.93 | 0.50 |
| 2447 | 7.1 | 302 | 1.10 | 0.56 | 290 | 0.12 | 300 | 145 | 0,76 | 0.14 | 0.93 | 0.50 |
| 2447 | 7.1 | 302 | 1.10 | 0.56 | 258 | 0,12 | 260 | 129 | 0.76 | 0.14 | 0.93 | 0.50 |
| 2447 | 7.1 | 302 | 1.10 | 0.56 | 226 | 0.12 | 225 | 113 | 0.76 | 0.14 | 0.93 | 0.50 |
| 2343 | 7.0 | 169 | 1.00 | 0.39 | 497 | 0,10 | 388 | 151 | 0.76 | 0.15 | 0.95 | 0.51 |
| 2292 | 6.6 | 280 | 1,08 | 0.36 | 388 | 0.06 | 503 | 205 | 0.76 | 0.15 | 0.96 | 0.53 |
| 2553 | 7.5 | 183 | 1.11 | 0.68 | 226 | 0.18 | 161 | 63 | 0.76 | 0.14 | 0.92 | 0.47 |
| 2553 | 8.1 | 183 | 1.11 | 0,68 | 183 | 0.18 | 100 | 51 | 0.76 | 0.13 | 0.92 | 0.45 |
| 2553 | 8.1 | 183 | 1.11 | 0.68 | 97 | 0.18 | 80 | 27 | 0.76 | 0.13 | 0.92 | 0.44 |
| 2553 | 8.1 | 183 | 1.11 | 0.68 | 55 | 0.18 | 32 | 15 | 0.76 | 0.13 | 0.92 | 0.42 |

Table 1 continued

| $T_f$ K | $\frac{T_f}{T_w}$ | $U_f$ m/sec | $\alpha$ | [O] | $x$ mm | $\frac{\Delta h_{chem}}{\Delta h}$ | $Nu_f^*$ | $\frac{Re_f}{10^3}$ | $Pr_f$ | $Pr_T$ | $\varphi_l$ | $\varphi_T$ |
|---|---|---|---|---|---|---|---|---|---|---|---|---|
| 2626 | 7.6 | 190 | 1.08 | 0.69 | 290 | 0.22 | 174 | 79 | 0.75 | 0.13 | 0.91 | 0.47 |
| 2626 | 7.6 | 190 | 1.08 | 0.69 | 258 | 0.22 | 161 | 70 | 0.75 | 0.13 | 0.91 | 0.47 |
| 2086 | 5.8 | 423 | 1.09 | 0.36 | 55 | 0.02 | 136 | 51 | 0.77 | 0.17 | 0.96 | 0.53 |
| 2086 | 5.6 | 423 | 1.09 | 0.36 | 23 | 0.02 | 72 | 21 | 0.77 | 0.17 | 0.96 | 0.52 |
| 2086 | 5.6 | 423 | 1.09 | 0.36 | 12 | 0.02 | 25 | 11 | 0.77 | 0.17 | 0.96 | 0.50 |
| 2071 | 5.6 | 395 | 1.05 | 0.37 | 65 | 0.02 | 177 | 58 | 0.77 | 0.17 | 0.96 | 0.54 |
| 2071 | 5.7 | 395 | 1.05 | 0.37 | 55 | 0.02 | 124 | 49 | 0.77 | 0.17 | 0.96 | 0.53 |
| 2071 | 5.6 | 395 | 1.05 | 0.37 | 23 | 0.02 | 62 | 20 | 0.77 | 0.17 | 0.96 | 0.52 |
| 2308 | 6.9 | 291 | 1.03 | 0.37 | 140 | 0.08 | 210 | 76 | 0.76 | 0.15 | 0.96 | 0.50 |
| 2308 | 6.7 | 291 | 1.03 | 0.37 | 65 | 0.08 | 93 | 35 | 0.76 | 0.15 | 0.96 | 0.49 |
| 2308 | 6.8 | 291 | 1.03 | 0.37 | 55 | 0.08 | 74 | 30 | 0.76 | 0.15 | 0.96 | 0.48 |
| 2308 | 6.7 | 291 | 1.03 | 0.37 | 23 | 0.08 | 43 | 12 | 0.76 | 0.15 | 0.96 | 0.47 |
| 2553 | 7.5 | 183 | 1.11 | 0.68 | 311 | 0.18 | 160 | 86 | 0.76 | 0.14 | 0.92 | 0.48 |
| 2553 | 7.5 | 183 | 1.11 | 0.68 | 290 | 0.18 | 175 | 80 | 0.76 | 0.14 | 0.92 | 0.48 |
| 2553 | 7.5 | 183 | 1.11 | 0.68 | 258 | 0.18 | 170 | 71 | 0.76 | 0.14 | 0.92 | 0.47 |
| 2233 | 6.9 | 175 | 1.07 | 0.37 | 65 | 0.05 | 61 | 22 | 0.76 | 0.15 | 0.96 | 0.47 |
| 2233 | 6.6 | 175 | 1.07 | 0.37 | 258 | 0.05 | 244 | 89 | 0.76 | 0.15 | 0.96 | 0.51 |
| 2435 | 7.1 | 177 | 1.00 | 0.39 | 290 | 0.13 | 233 | 86 | 0.75 | 0.14 | 0.95 | 0.50 |
| 2435 | 7.1 | 177 | 1.00 | 0.39 | 258 | 0.13 | 206 | 77 | 0.75 | 0.14 | 0.95 | 0.49 |
| 2435 | 7.5 | 177 | 1.00 | 0.39 | 226 | 0.13 | 182 | 67 | 0.75 | 0.14 | 0.95 | 0.48 |
| 2435 | 7.5 | 177 | 1.00 | 0.39 | 183 | 0.13 | 152 | 54 | 0.75 | 0.14 | 0.95 | 0.47 |
| 2435 | 7.5 | 177 | 1.00 | 0.39 | 140 | 0.13 | 94 | 41 | 0.75 | 0.14 | 0.95 | 0.46 |
| 2435 | 7.4 | 177 | 1.00 | 0.39 | 65 | 0.13 | 42 | 19 | 0.75 | 0.14 | 0.95 | 0.45 |
| 2435 | 7.5 | 177 | 1.00 | 0.39 | 55 | 0.13 | 41 | 16 | 0.75 | 0.14 | 0.95 | 0.45 |
| 2435 | 7.4 | 177 | 1.00 | 0.39 | 23 | 0.13 | 23 | 7 | 0.75 | 0.14 | 0.95 | 0.43 |
| 2398 | 6.9 | 174 | 1.00 | 0.39 | 258 | 0.12 | 209 | 77 | 0.75 | 0.15 | 0.95 | 0.50 |
| 2398 | 7.4 | 174 | 1.00 | 0.39 | 226 | 0.12 | 181 | 68 | 0.75 | 0.14 | 0.95 | 0.48 |
| 2398 | 7.4 | 174 | 1.00 | 0.39 | 183 | 0.12 | 159 | 55 | 0.75 | 0.14 | 0.95 | 0.47 |
| 2398 | 7.3 | 174 | 1.00 | 0.39 | 65 | 0.12 | 38 | 19 | 0.75 | 0.14 | 0.95 | 0.45 |
| 2398 | 7.4 | 174 | 1.00 | 0.39 | 55 | 0.12 | 41 | 16 | 0.75 | 0.14 | 0.95 | 0.45 |
| 2398 | 7.3 | 174 | 1.00 | 0.39 | 23 | 0.12 | 24 | 7 | 0.75 | 0.14 | 0.95 | 0.43 |

Table 1 continued

| $T_f$ K | $\frac{T_f}{T_w}$ | $U_f$ m/sec | $\alpha$ | [O] | $x$ mm | $\frac{\Delta h_{chem}}{\Delta h}$ | $Nu_f^*$ | $\frac{Re_f}{10^3}$ | $Pr_f$ | $Pr_T$ | $\varphi_l$ | $\varphi_T$ |
|---|---|---|---|---|---|---|---|---|---|---|---|---|
| 2398 | 7.4 | 174 | 1.00 | 0.39 | 140 | 0.12 | 105 | 42 | 0.75 | 0.14 | 0.95 | 0.47 |
| 2447 | 7.5 | 302 | 1.10 | 0.56 | 23 | 0.12 | 24 | 11 | 0.76 | 0.14 | 0.93 | 0.43 |
| 2086 | 5.6 | 423 | 1.09 | 0.36 | 65 | 0.02 | 137 | 61 | 0.77 | 0.17 | 0.96 | 0.54 |
| 2233 | 7.0 | 175 | 1.07 | 0.37 | 226 | 0.05 | 217 | 78 | 0.76 | 0.15 | 0.96 | 0.49 |
| 2233 | 7.0 | 175 | 1.07 | 0.37 | 140 | 0.05 | 129 | 48 | 0.76 | 0.15 | 0.96 | 0.48 |
| 2233 | 7.0 | 175 | 1.07 | 0.37 | 55 | 0.05 | 42 | 19 | 0.76 | 0.15 | 0.96 | 0.47 |
| 2626 | 8.1 | 190 | 1.08 | 0.69 | 487 | 0.22 | 313 | 132 | 0.75 | 0.13 | 0.91 | 0.47 |
| 2626 | 7.6 | 190 | 1.08 | 0.69 | 388 | 0.22 | 202 | 105 | 0.75 | 0.13 | 0.91 | 0.48 |
| 2626 | 7.6 | 190 | 1.08 | 0.69 | 377 | 0.22 | 201 | 102 | 0.75 | 0.13 | 0.91 | 0.48 |
| 2657 | 8.1 | 192 | 1.10 | 0.68 | 487 | 0.23 | 314 | 131 | 0.75 | 0.13 | 0.92 | 0.47 |
| 2657 | 7.7 | 192 | 1.10 | 0.68 | 388 | 0.23 | 252 | 104 | 0.75 | 0.13 | 0.92 | 0.48 |
| 2528 | 7.6 | 311 | 1.10 | 0.56 | 497 | 0.16 | 528 | 242 | 0.76 | 0.14 | 0.93 | 0.50 |
| 2528 | 7.6 | 311 | 1.10 | 0.56 | 487 | 0.16 | 522 | 237 | 0.76 | 0.14 | 0.93 | 0.50 |
| 2528 | 7.6 | 311 | 1.10 | 0.56 | 443 | 0.16 | 506 | 216 | 0.76 | 0.14 | 0.93 | 0.50 |
| 2528 | 7.2 | 311 | 1.10 | 0.56 | 388 | 0.16 | 400 | 189 | 0.76 | 0.14 | 0.93 | 0.51 |
| 2528 | 7.2 | 311 | 1.10 | 0.56 | 290 | 0.16 | 359 | 141 | 0.76 | 0.14 | 0.93 | 0.50 |
| 2528 | 7.2 | 311 | 1.10 | 0.56 | 258 | 0.16 | 314 | 126 | 0.76 | 0.14 | 0.93 | 0.50 |
| 2071 | 5.9 | 395 | 1.05 | 0.37 | 497 | 0.02 | 1052 | 442 | 0.77 | 0.17 | 0.96 | 0.57 |
| 2071 | 5.7 | 395 | 1.05 | 0.37 | 140 | 0.02 | 326 | 124 | 0.77 | 0.17 | 0.96 | 0.55 |
| 2308 | 6.7 | 291 | 1.03 | 0.37 | 388 | 0.08 | 475 | 211 | 0.76 | 0.15 | 0.96 | 0.53 |
| 2308 | 6.7 | 291 | 1.03 | 0.37 | 377 | 0.08 | 514 | 205 | 0.76 | 0.15 | 0.96 | 0.53 |
| 2308 | 6.7 | 291 | 1.03 | 0.37 | 311 | 0.08 | 374 | 170 | 0.76 | 0.15 | 0.96 | 0.52 |
| 2308 | 6.7 | 291 | 1.03 | 0.37 | 290 | 0.08 | 413 | 158 | 0.76 | 0.15 | 0.96 | 0.52 |
| 2308 | 6.7 | 291 | 1.03 | 0.37 | 258 | 0.08 | 344 | 141 | 0.76 | 0.15 | 0.96 | 0.52 |
| 2308 | 6.7 | 291 | 1.03 | 0.37 | 226 | 0.08 | 281 | 123 | 0.76 | 0.15 | 0.96 | 0.52 |
| 2308 | 6.9 | 291 | 1.03 | 0.37 | 183 | 0.08 | 288 | 100 | 0.76 | 0.15 | 0.96 | 0.50 |
| 2240 | 6.5 | 283 | 1.05 | 0.37 | 226 | 0.06 | 331 | 126 | 0.76 | 0.16 | 0.96 | 0.52 |
| 2240 | 6.5 | 283 | 1.05 | 0.37 | 258 | 0.06 | 382 | 144 | 0.76 | 0.16 | 0.96 | 0.53 |
| 2240 | 6.5 | 283 | 1.05 | 0.37 | 290 | 0.06 | 437 | 162 | 0.76 | 0.16 | 0.96 | 0.53 |
| 2240 | 6.5 | 283 | 1.05 | 0.37 | 377 | 0.06 | 539 | 211 | 0.76 | 0.16 | 0.96 | 0.53 |
| 2553 | 7.5 | 183 | 1.11 | 0.68 | 377 | 0.18 | 257 | 104 | 0.76 | 0.14 | 0.92 | 0.48 |

Table 1 continued

| $T_f$ K | $\frac{T_f}{T_w}$ | $U_f$ m/sec | $\alpha$ | [O] | $x$ mm | $\frac{\Delta h_{chem}}{\Delta h}$ | $Nu_f^*$ | $\frac{Re_f}{10^3}$ | $Pr_f$ | $Pr_T$ | $\varphi_l$ | $\varphi_T$ |
|---|---|---|---|---|---|---|---|---|---|---|---|---|
| 2435 | 7.1 | 177 | 1.00 | 0.39 | 388 | 0.13 | 280 | 115 | 0.75 | 0.14 | 0.95 | 0.50 |
| 2086 | 6.0 | 423 | 1.09 | 0.36 | 487 | 0.02 | 1074 | 456 | 0.77 | 0.16 | 0.96 | 0.57 |
| 2398 | 7.1 | 174 | 1.00 | 0.39 | 443 | 0.12 | 311 | 133 | 0.75 | 0.14 | 0.95 | 0.50 |
| 2086 | 6.0 | 423 | 1.09 | 0.36 | 497 | 0.02 | 1078 | 467 | 0.77 | 0.16 | 0.96 | 0.57 |
| 2086 | 5.7 | 423 | 1.09 | 0.36 | 377 | 0.02 | 848 | 354 | 0.77 | 0.17 | 0.96 | 0.58 |
| 2086 | 5.8 | 423 | 1.09 | 0.36 | 183 | 0.02 | 352 | 172 | 0.77 | 0.17 | 0.96 | 0.56 |
| 2086 | 5.7 | 423 | 1.09 | 0.36 | 388 | 0.02 | 852 | 364 | 0.77 | 0.17 | 0.96 | 0.58 |
| 2292 | 6.6 | 280 | 1.08 | 0.36 | 377 | 0.06 | 504 | 199 | 0.76 | 0.15 | 0.96 | 0.53 |
| 2343 | 7.0 | 169 | 1.00 | 0.39 | 487 | 0.10 | 350 | 148 | 0.76 | 0.15 | 0.95 | 0.51 |
| 2343 | 6.9 | 169 | 1.00 | 0.39 | 388 | 0.10 | 252 | 118 | 0.76 | 0.15 | 0.95 | 0.51 |
| 2343 | 6.9 | 169 | 1.00 | 0.39 | 290 | 0.10 | 238 | 88 | 0.76 | 0.15 | 0.95 | 0.50 |
| 2343 | 6.9 | 169 | 1.00 | 0.39 | 258 | 0.10 | 209 | 79 | 0.76 | 0.15 | 0.95 | 0.50 |
| 2343 | 7.3 | 169 | 1.00 | 0.39 | 226 | 0.10 | 175 | 69 | 0.76 | 0.14 | 0.95 | 0.48 |
| 2343 | 7.3 | 169 | 1.00 | 0.39 | 140 | 0.10 | 108 | 42 | 0.76 | 0.14 | 0.95 | 0.47 |
| 2343 | 7.3 | 169 | 1.00 | 0.39 | 65 | 0.10 | 41 | 20 | 0.76 | 0.14 | 0.95 | 0.46 |
| 2343 | 7.3 | 169 | 1.00 | 0.39 | 23 | 0.10 | 25 | 7 | 0.76 | 0.14 | 0.95 | 0.44 |
| 2240 | 6.9 | 283 | 1.05 | 0.37 | 140 | 0.05 | 218 | 78 | 0.76 | 0.15 | 0.96 | 0.50 |
| 2240 | 6.7 | 283 | 1.05 | 0.37 | 55 | 0.05 | 78 | 31 | 0.76 | 0.15 | 0.96 | 0.48 |
| 2240 | 6.6 | 283 | 1.05 | 0.37 | 23 | 0.05 | 37 | 13 | 0.76 | 0.15 | 0.96 | 0.47 |
| 2650 | 7.7 | 190 | 1.09 | 0.68 | 311 | 0.23 | 208 | 83 | 0.75 | 0.13 | 0.91 | 0.47 |
| 2650 | 7.7 | 190 | 1.09 | 0.68 | 290 | 0.23 | 171 | 77 | 0.75 | 0.13 | 0.91 | 0.47 |
| 2650 | 7.7 | 190 | 1.09 | 0.68 | 258 | 0.23 | 144 | 69 | 0.75 | 0.13 | 0.91 | 0.47 |
| 2650 | 7.7 | 190 | 1.09 | 0.68 | 226 | 0.23 | 121 | 60 | 0.75 | 0.13 | 0.91 | 0.46 |
| 2650 | 8.3 | 190 | 1.09 | 0.68 | 183 | 0.23 | 120 | 49 | 0.75 | 0.13 | 0.91 | 0.44 |
| 2650 | 8.3 | 190 | 1.09 | 0.68 | 65 | 0.23 | 35 | 17 | 0.75 | 0.13 | 0.91 | 0.42 |
| 2528 | 7.5 | 311 | 1.10 | 0.56 | 140 | 0.16 | 176 | 68 | 0.76 | 0.14 | 0.93 | 0.47 |
| 2528 | 7.5 | 311 | 1.10 | 0.56 | 97 | 0.16 | 121 | 47 | 0.76 | 0.14 | 0.93 | 0.46 |
| 2528 | 7.6 | 311 | 1.10 | 0.56 | 65 | 0.16 | 74 | 31 | 0.76 | 0.14 | 0.93 | 0.45 |
| 2626 | 7.6 | 190 | 1.08 | 0.69 | 226 | 0.22 | 124 | 61 | 0.75 | 0.13 | 0.91 | 0.47 |

Table 1 continued

| $T_f$ K | $\frac{T_f}{T_w}$ | $U_f$ m/sec | $\alpha$ | [O] | $x$ mm | $\frac{\Delta h_{chem}}{\Delta h}$ | $Nu_f^*$ | $\frac{Re_f}{10^3}$ | $Pr_f$ | $Pr_T$ | $\varphi_l$ | $\varphi_T$ |
|---|---|---|---|---|---|---|---|---|---|---|---|---|
| 2626 | 8.3 | 190 | 1.08 | 0.69 | 183 | 0.22 | 101 | 50 | 0.75 | 0.13 | 0.91 | 0.44 |
| 2626 | 8.3 | 190 | 1.08 | 0.69 | 97 | 0.22 | 70 | 26 | 0.75 | 0.13 | 0.91 | 0.43 |
| 2657 | 7.7 | 192 | 1.10 | 0.68 | 258 | 0.23 | 137 | 69 | 0.75 | 0.13 | 0.92 | 0.47 |
| 2657 | 7.7 | 192 | 1.10 | 0.68 | 290 | 0.23 | 151 | 78 | 0.75 | 0.13 | 0.92 | 0.47 |
| 2657 | 7.7 | 192 | 1.10 | 0.68 | 226 | 0.23 | 124 | 61 | 0.75 | 0.13 | 0.92 | 0.46 |
| 2657 | 8.4 | 192 | 1.10 | 0.68 | 65 | 0.23 | 36 | 17 | 0.75 | 0.13 | 0.92 | 0.42 |
| 2447 | 7.4 | 302 | 1.10 | 0.56 | 183 | 0.12 | 245 | 91 | 0.76 | 0.14 | 0.93 | 0.48 |
| 2447 | 7.4 | 302 | 1.10 | 0.56 | 140 | 0.12 | 172 | 70 | 0.76 | 0.14 | 0.93 | 0.48 |
| 2447 | 7.5 | 302 | 1.10 | 0.56 | 65 | 0.12 | 74 | 32 | 0.76 | 0.14 | 0.93 | 0.46 |
| 2292 | 7.0 | 280 | 1.08 | 0.36 | 140 | 0.06 | 211 | 74 | 0.76 | 0.15 | 0.96 | 0.49 |
| 2292 | 7.0 | 280 | 1.08 | 0.36 | 65 | 0.06 | 97 | 34 | 0.76 | 0.15 | 0.96 | 0.48 |
| 2292 | 6.8 | 280 | 1.08 | 0.36 | 55 | 0.06 | 79 | 29 | 0.76 | 0.15 | 0.96 | 0.48 |
| 2292 | 6.6 | 280 | 1.08 | 0.36 | 23 | 0.06 | 38 | 12 | 0.76 | 0.15 | 0.96 | 0.47 |
| 2292 | 6.6 | 280 | 1.08 | 0.36 | 12 | 0.06 | 20 | 6 | 0.76 | 0.15 | 0.96 | 0.46 |
| 2343 | 6.9 | 169 | 1.00 | 0.39 | 377 | 0.10 | 258 | 115 | 0.76 | 0.15 | 0.95 | 0.51 |
| 2292 | 6.8 | 280 | 1.08 | 0.36 | 497 | 0.06 | 605 | 263 | 0.76 | 0.15 | 0.96 | 0.53 |
| 2292 | 6.8 | 280 | 1.08 | 0.36 | 443 | 0.06 | 494 | 234 | 0.76 | 0.15 | 0.96 | 0.53 |
| 2086 | 6.0 | 423 | 1.09 | 0.36 | 443 | 0.02 | 862 | 415 | 0.77 | 0.16 | 0.96 | 0.57 |
| 2292 | 6.8 | 280 | 1.08 | 0.36 | 487 | 0.06 | 579 | 257 | 0.76 | 0.15 | 0.96 | 0.53 |

*Note*: The first column of the table lists the values of the bulk temperature of the flow at the inlet of the test channel. It was assumed in working up experimental data that it is equal to the flow temperature in the center of the channel, which was taken to be constant over the entire length of the channel. Flow velocity $U_f$ was determined with allowance for the density of combustion products at bulk temperature $T_f$ at the inlet. The composition of combustion products is characterized by the oxidant excess ratio $\alpha$ and the fraction by weight of oxygen [O] in the oxidant. In all cases the oxidant consisted of an oxygen–nitrogen mixture such that [O] + [N] = 1. The value of $Nu_f^*$ was determined from the total enthalpy difference, whereas factors $\varphi_l$ and $\varphi_T$, allowing for variability of thermophysical properties, were determined from Eqs. (4.8) and (4.22), respectively. The table also lists values of $Pr_T$ obtained from Eq. (4.29) at Tu = 9%.

**TABLE A.2** Heat Transfer of the Inlet Region of a Channel with a Flow of Combustion Products

| $T_f$ K | $\frac{T_f}{T_w}$ | $U_f$ m/sec | $\alpha$ | [O] | $x$ mm | $\frac{\Delta h_{chem}}{\Delta h}$ | $Nu_f^*$ | $\frac{Re_f}{10^3}$ | $Pr_f$ | $Pr_T$ | $\varphi_l$ | $\varphi_T$ |
|---|---|---|---|---|---|---|---|---|---|---|---|---|
| 2244 | 4.2 | 400 | 0.99 | 0.40 | 238 | 0.07 | 505 | 219 | 0.73 | 0.21 | 0.95 | 0.66 |
| 2244 | 4.3 | 400 | 0.99 | 0.40 | 273 | 0.06 | 591 | 251 | 0.73 | 0.21 | 0.95 | 0.66 |
| 2244 | 4.3 | 400 | 0.99 | 0.40 | 308 | 0.06 | 631 | 283 | 0.73 | 0.21 | 0.95 | 0.66 |
| 2244 | 4.4 | 400 | 0.99 | 0.40 | 343 | 0.06 | 679 | 316 | 0.73 | 0.21 | 0.95 | 0.66 |
| 2244 | 4.5 | 400 | 0.99 | 0.40 | 378 | 0.06 | 792 | 347 | 0.73 | 0.20 | 0.95 | 0.65 |
| 2062 | 4.2 | 347 | 0.99 | 0.36 | 238 | 0.03 | 540 | 215 | 0.74 | 0.21 | 0.96 | 0.66 |
| 2062 | 4.2 | 347 | 0.99 | 0.36 | 273 | 0.03 | 607 | 246 | 0.74 | 0.21 | 0.96 | 0.66 |
| 2062 | 4.3 | 347 | 0.99 | 0.36 | 308 | 0.03 | 652 | 278 | 0.74 | 0.21 | 0.95 | 0.66 |
| 2062 | 4.4 | 347 | 0.99 | 0.36 | 343 | 0.03 | 689 | 309 | 0.74 | 0.20 | 0.95 | 0.65 |
| 1827 | 4.0 | 309 | 1.03 | 0.31 | 238 | 0.01 | 568 | 232 | 0.74 | 0.22 | 0.96 | 0.67 |
| 1827 | 4.2 | 309 | 1.03 | 0.31 | 308 | 0.01 | 671 | 300 | 0.74 | 0.21 | 0.96 | 0.66 |
| 1827 | 4.3 | 309 | 1.03 | 0.31 | 343 | 0.01 | 836 | 334 | 0.74 | 0.21 | 0.96 | 0.66 |
| 1827 | 4.4 | 309 | 1.03 | 0.31 | 378 | 0.01 | 889 | 368 | 0.74 | 0.21 | 0.96 | 0.66 |
| 1548 | 3.7 | 253 | 0.96 | 0.27 | 238 | 0.01 | 639 | 244 | 0.73 | 0.23 | 0.97 | 0.69 |
| 1548 | 3.8 | 253 | 0.96 | 0.27 | 273 | 0.01 | 733 | 280 | 0.73 | 0.23 | 0.97 | 0.69 |
| 1548 | 3.8 | 253 | 0.96 | 0.27 | 308 | 0.01 | 788 | 316 | 0.73 | 0.23 | 0.97 | 0.69 |
| 1548 | 3.9 | 253 | 0.96 | 0.27 | 343 | 0.01 | 824 | 352 | 0.73 | 0.22 | 0.97 | 0.69 |
| 1548 | 4.0 | 253 | 0.96 | 0.27 | 378 | 0.01 | 907 | 388 | 0.73 | 0.22 | 0.97 | 0.68 |
| 1548 | 4.1 | 253 | 0.96 | 0.27 | 413 | 0.01 | 1026 | 424 | 0.73 | 0.22 | 0.97 | 0.68 |
| 1548 | 4.2 | 253 | 0.96 | 0.27 | 448 | 0.01 | 1088 | 459 | 0.73 | 0.21 | 0.97 | 0.68 |
| 1548 | 4.3 | 253 | 0.96 | 0.27 | 483 | 0.01 | 1074 | 495 | 0.73 | 0.21 | 0.98 | 0.67 |
| 1548 | 4.4 | 253 | 0.96 | 0.27 | 522 | 0.01 | 1067 | 536 | 0.73 | 0.20 | 0.98 | 0.67 |
| 1899 | 4.2 | 330 | 0.95 | 0.34 | 273 | 0.02 | 677 | 262 | 0.73 | 0.21 | 0.97 | 0.67 |
| 1899 | 4.3 | 330 | 0.95 | 0.34 | 308 | 0.02 | 757 | 296 | 0.73 | 0.21 | 0.97 | 0.66 |
| 1899 | 4.4 | 330 | 0.95 | 0.34 | 343 | 0.02 | 801 | 329 | 0.73 | 0.21 | 0.97 | 0.66 |
| 1899 | 4.4 | 330 | 0.95 | 0.34 | 378 | 0.02 | 904 | 363 | 0.73 | 0.20 | 0.97 | 0.66 |
| 2081 | 3.8 | 375 | 0.97 | 0.37 | 203 | 0.02 | 520 | 194 | 0.73 | 0.23 | 0.96 | 0.68 |
| 2081 | 3.9 | 375 | 0.97 | 0.37 | 243 | 0.02 | 521 | 232 | 0.73 | 0.22 | 0.96 | 0.68 |
| 2081 | 4.0 | 375 | 0.97 | 0.37 | 273 | 0.02 | 589 | 261 | 0.73 | 0.22 | 0.96 | 0.68 |
| 2081 | 4.1 | 375 | 0.97 | 0.37 | 308 | 0.02 | 751 | 294 | 0.73 | 0.22 | 0.96 | 0.67 |

Table 2 continued

| $T_f$ K | $\frac{T_f}{T_w}$ | $U_f$ m/sec | α | [O] | $x$ mm | $\frac{\Delta h_{chem}}{\Delta h}$ | $Nu_f^*$ | $\frac{Re_f}{10^3}$ | $Pr_f$ | $Pr_T$ | $\varphi_l$ | $\varphi_T$ |
|---|---|---|---|---|---|---|---|---|---|---|---|---|
| 2081 | 4.2 | 375 | 0.97 | 0.37 | 343 | 0.02 | 698 | 328 | 0.73 | 0.21 | 0.96 | 0.67 |
| 2081 | 4.4 | 375 | 0.97 | 0.37 | 378 | 0.02 | 892 | 361 | 0.73 | 0.21 | 0.96 | 0.66 |
| 2200 | 3.9 | 408 | 1.01 | 0.40 | 243 | 0.06 | 540 | 235 | 0.74 | 0.22 | 0.95 | 0.68 |
| 2200 | 4.0 | 408 | 1.01 | 0.40 | 273 | 0.06 | 652 | 263 | 0.74 | 0.22 | 0.95 | 0.67 |
| 2200 | 4.1 | 408 | 1.01 | 0.40 | 308 | 0.06 | 709 | 298 | 0.64 | 0.21 | 0.95 | 0.67 |
| 2071 | 5.8 | 395 | 1.05 | 0.37 | 236 | 0.02 | 467 | 210 | 0.77 | 0.17 | 0.96 | 0.57 |
| 2130 | 6.1 | 256 | 1.08 | 0.37 | 487 | 0.03 | 561 | 269 | 0.77 | 0.16 | 0.96 | 0.56 |
| 2130 | 6.2 | 256 | 1.08 | 0.37 | 366 | 0.03 | 473 | 203 | 0.77 | 0.16 | 0.96 | 0.55 |
| 1983 | 5.8 | 241 | 1.09 | 0.37 | 487 | 0.01 | 597 | 288 | 0.76 | 0.17 | 0.96 | 0.58 |
| 2017 | 5.9 | 245 | 1.09 | 0.36 | 487 | 0.02 | 612 | 284 | 0.76 | 0.17 | 0.96 | 0.57 |
| 2292 | 6.6 | 280 | 1.08 | 0.36 | 487 | 0.06 | 527 | 257 | 0.76 | 0.15 | 0.96 | 0.54 |
| 2017 | 5.9 | 245 | 1.09 | 0.36 | 366 | 0.02 | 523 | 214 | 0.76 | 0.17 | 0.96 | 0.57 |
| 1999 | 4.6 | 236 | 1.07 | 0.38 | 133 | 0.02 | 185 | 82 | 0.74 | 0.20 | 0.95 | 0.61 |
| 1999 | 4.7 | 236 | 1.07 | 0.38 | 168 | 0.02 | 268 | 104 | 0..74 | 0.20 | 0.95 | 0.61 |
| 1999 | 4.8 | 236 | 1.07 | 0.38 | 203 | 0.02 | 300 | 126 | 0.74 | 0.19 | 0.95 | 0.61 |
| 1999 | 4.8 | 236 | 1.07 | 0.38 | 238 | 0.02 | 360 | 147 | 0.74 | 0.19 | 0.95 | 0.61 |
| 1999 | 4.9 | 236 | 1.07 | 0.38 | 273 | 0.02 | 384 | 169 | 0.74 | 0.19 | 0.95 | 0.61 |
| 2244 | 5.6 | 400 | 0.99 | 0.40 | 522 | 0.06 | 873 | 480 | 0.73 | 0.17 | 0.95 | 0.60 |
| 2173 | 6.1 | 265 | 1.08 | 0.36 | 487 | 0.04 | 678 | 266 | 0.76 | 0.16 | 0.96 | 0.56 |
| 2263 | 6.6 | 276 | 1.08 | 0.36 | 366 | 0.06 | 420 | 195 | 0.76 | 0.15 | 0.96 | 0.54 |
| 2263 | 6.5 | 276 | 1.08 | 0.36 | 487 | 0.06 | 588 | 259 | 0.76 | 0.16 | 0.96 | 0.55 |
| 2240 | 6.6 | 283 | 1.05 | 0.37 | 366 | 0.05 | 493 | 205 | 0.76 | 0.15 | 0.96 | 0.54 |
| 2308 | 6.7 | 291 | 1.03 | 0.37 | 366 | 0.08 | 425 | 200 | 0.76 | 0.15 | 0.96 | 0.53 |
| 2308 | 6.6 | 291 | 1.03 | 0.37 | 487 | 0.08 | 532 | 265 | 0.76 | 0.15 | 0.96 | 0.55 |
| 1983 | 5.8 | 241 | 1.09 | 0.37 | 366 | 0.01 | 525 | 217 | 0.76 | 0.17 | 0.96 | 0.57 |
| 1999 | 4.6 | 236 | 1.07 | 0.38 | 95 | 0.02 | 154 | 59 | 0.74 | 0.20 | 0.95 | 0.61 |
| 1965 | 5.0 | 198 | 0.97 | 0.38 | 343 | 0.01 | 433 | 181 | 0.73 | 0.19 | 0.96 | 0.61 |
| 1965 | 4.9 | 198 | 0.97 | 0.38 | 308 | 0.01 | 411 | 162 | 0.73 | 0.19 | 0.96 | 0.61 |
| 1965 | 4.8 | 198 | 0.97 | 0.38 | 273 | 0.01 | 369 | 144 | 0.73 | 0.19 | 0.96 | 0.62 |
| 1965 | 4.7 | 198 | 0.97 | 0.38 | 238 | 0.01 | 351 | 126 | 0.73 | 0.20 | 0.96 | 0.62 |
| 1965 | 4.6 | 198 | 0.97 | 0.38 | 168 | 0.01 | 254 | 89 | 0.73 | 0.20 | 0.96 | 0.62 |

Table 2 continued

| $T_f$ K | $\frac{T_f}{T_w}$ | $U_f$ m/sec | $\alpha$ | [O] | $x$ mm | $\frac{\Delta h_{chem}}{\Delta h}$ | $Nu_f^*$ | $\frac{Re_f}{10^3}$ | $Pr_f$ | $Pr_T$ | $\varphi_l$ | $\varphi_T$ |
|---|---|---|---|---|---|---|---|---|---|---|---|---|
| 1965 | 4.6 | 198 | 0.97 | 0.38 | 133 | 0.01 | 182 | 70 | 0.73 | 0.20 | 0.96 | 0.61 |
| 1865 | 4.5 | 201 | 0.96 | 0.38 | 343 | 0.01 | 474 | 188 | 0.73 | 0.20 | 0.96 | 0.64 |
| 1865 | 4.5 | 201 | 0.96 | 0.38 | 308 | 0.01 | 435 | 169 | 0.73 | 0.20 | 0.96 | 0.64 |
| 1865 | 4.4 | 201 | 0.96 | 0.38 | 273 | 0.01 | 406 | 150 | 0.73 | 0.20 | 0.96 | 0.64 |
| 1865 | 4.4 | 201 | 0.96 | 0.38 | 238 | 0.01 | 342 | 131 | 0.73 | 0.21 | 0.96 | 0.64 |
| 1865 | 4.3 | 201 | 0.96 | 0.38 | 203 | 0.01 | 317 | 111 | 0.73 | 0.21 | 0.96 | 0.64 |
| 1865 | 4.3 | 201 | 0.96 | 0.38 | 168 | 0.01 | 239 | 92 | 0.73 | 0.21 | 0.96 | 0.64 |
| 1865 | 4.2 | 201 | 0.96 | 0.38 | 133 | 0.01 | 192 | 73 | 0.73 | 0.21 | 0.96 | 0.63 |
| 1705 | 4.7 | 100 | 0.85 | 0.43 | 522 | 0.05 | 339 | 157 | 0.70 | 0.20 | 0.99 | 0.63 |
| 1965 | 5.3 | 198 | 0.97 | 0.38 | 413 | 0.01 | 528 | 218 | 0.73 | 018 | 0.96 | 0.60 |
| 1965 | 5.2 | 198 | 0.97 | 0.38 | 378 | 0.01 | 495 | 197 | 0.73 | 0.18 | 0.96 | 0.61 |
| 1865 | 4.8 | 201 | 0.96 | 0.38 | 522 | 0.01 | 606 | 287 | 0.73 | 0.19 | 0.96 | 0.63 |
| 1865 | 4.8 | 201 | 0.96 | 0.38 | 483 | 0.01 | 677 | 265 | 0.73 | 0.19 | 0.96 | 0.63 |
| 1865 | 4.7 | 201 | 0.96 | 0.38 | 413 | 0.01 | 576 | 227 | 0.73 | 0.20 | 0.96 | 0.63 |
| 1865 | 4.6 | 201 | 0.96 | 0.38 | 378 | 0.01 | 547 | 208 | 0.73 | 0.20 | 0.96 | 0.63 |
| 2071 | 5.4 | 395 | 1.05 | 0.37 | 487 | 0.02 | 811 | 433 | 0.77 | 0.18 | 0.96 | 0.60 |
| 2010 | 5.5 | 277 | 1.18 | 0.35 | 487 | 0.02 | 797 | 322 | 0.77 | 0.18 | 0.96 | 0.59 |
| 2244 | 5.2 | 400 | 0.99 | 0.40 | 448 | 0.06 | 885 | 412 | 0.13 | 0.18 | 0.95 | 0.62 |
| 1999 | 5.4 | 236 | 1.07 | 0.38 | 413 | 0.02 | 559 | 256 | 0.74 | 0.18 | 0.95 | 0.60 |
| 2244 | 5.0 | 400 | 0.99 | 0.40 | 413 | 0.06 | 742 | 380 | 0.73 | 0.19 | 0.95 | 0.63 |
| 2244 | 5.4 | 400 | 0.99 | 0.40 | 483 | 0.06 | 863 | 444 | 0.73 | 0.18 | 0.95 | 0.61 |
| 2062 | 4.6 | 347 | 0.99 | 0.36 | 378 | 0.03 | 774 | 341 | 0.74 | 0.20 | 0.95 | 0.65 |
| 2062 | 4.8 | 347 | 0.99 | 0.36 | 413 | 0.03 | 847 | 372 | 0.74 | 0.19 | 0.95 | 0.64 |
| 2062 | 4.9 | 347 | 0.99 | 0.36 | 448 | 0.03 | 916 | 404 | 0.74 | 0.19 | 0.95 | 0.63 |
| 2062 | 5.1 | 347 | 0.99 | 0.36 | 483 | 0.03 | 905 | 436 | 0.74 | 0.19 | 0.95 | 0.63 |
| 2062 | 5.4 | 347 | 0.99 | 0.36 | 522 | 0.03 | 916 | 471 | 0.74 | 0.18. | 0.95 | 0.62 |
| 1827 | 4.6 | 309 | 1.03 | 0.31 | 413 | 0.01 | 982 | 402 | 0.74 | 0.20 | 0.96 | 0.65 |
| 1827 | 4.7 | 309 | 1.03 | 0.31 | 448 | 0.01 | 1027 | 436 | 0.74 | 0.20 | 0.96 | 0.64 |
| 1827 | 4.8 | 309 | 1.03 | 0.31 | 483 | 0.01 | 1008 | 470 | 0.74 | 0.19 | 0.96 | 0.64 |
| 1827 | 5.0 | 309 | 1.03 | 0.31 | 522 | 0.01 | 1022 | 509 | 0.74 | 0.19 | 0.96 | 0.63 |
| 1899 | 4.5 | 330 | 0.95 | 0.34 | 413 | 0.02 | 907 | 396 | 0.73 | 0.20 | 0.97 | 0.66 |

Table 2 continued

| $T_f$ K | $\frac{T_f}{T_w}$ | $U_f$ m/sec | $\alpha$ | [O] | $x$ mm | $\frac{\Delta h_{chem}}{\Delta h}$ | $Nu_f^*$ | $\frac{Re_f}{10^3}$ | $Pr_f$ | $Pr_T$ | $\varphi_l$ | $\varphi_T$ |
|---|---|---|---|---|---|---|---|---|---|---|---|---|
| 1899 | 4.6 | 330 | 0.95 | 0.34 | 483 | 0.02 | 1046 | 463 | 0.73 | 0.20 | 0.97 | 0.65 |
| 1899 | 4.7 | 330 | 0.95 | 0.34 | 522 | 0.02 | 1049 | 501 | 0.73 | 0.20 | 0.97 | 0.65 |
| 2081 | 4.6 | 375 | 0.97 | 0.37 | 413 | 0.02 | 839 | 395 | 0.73 | 0.20 | 0.96 | 0.65 |
| 2081 | 4.7 | 375 | 0.97 | 0.37 | 448 | 0.02 | 978 | 428 | 0.73 | 0.20 | 0.96 | 0.64 |
| 2081 | 4.9 | 375 | 0.97 | 0.37 | 483 | 0.02 | 999 | 462 | 0.73 | 0.19 | 0.96 | 0.64 |
| 2081 | 5.1 | 375 | 0.97 | 0.37 | 522 | 0.02 | 958 | 499 | 0.73 | 0.19 | 0.96 | 0.63 |
| 2200 | 4.6 | 408 | 1.01 | 0.40 | 378 | 0.06 | 866 | 365 | 0.74 | 0.20 | 0.95 | 0.65 |
| 2200 | 4.8 | 408 | 1.01 | 0.40 | 413 | 0.06 | 943 | 399 | 0.74 | 0.19 | 0.95 | 0.64 |
| 2200 | 5.1 | 408 | 1.01 | 0.40 | 448 | 0.06 | 960 | 433 | 0.74 | 0.19 | 0.95 | 0.63 |
| 2200 | 5.3 | 408 | 1.01 | 0.40 | 483 | 0.06 | 982 | 467 | 0.74 | 0.18 | 0.95 | 0.62 |
| 2200 | 5.7 | 408 | 1.01 | 0.40 | 522 | 0.06 | 929 | 505 | 0.74 | 0.17 | 0.95 | 0.60 |
| 1965 | 5.9 | 198 | 0.97 | 0.38 | 522 | 0.01 | 610 | 276 | 0.73 | 0.17 | 0.96 | 0.58 |
| 1965 | 5.7 | 198 | 0.97 | 0.38 | 483 | 0.01 | 558 | 255 | 0.73 | 0.17 | 0.96 | 0.59 |
| 1705 | 4.6 | 100 | 0.85 | 0.43 | 308 | 0.05 | 213 | 93 | 0.70 | 0.20 | 0.99 | 0.63 |
| 1705 | 4.5 | 100 | 0.85 | 0.43 | 273 | 0.05 | 217 | 82 | 0.70 | 0.20 | 0.99 | 0.63 |
| 1705 | 4.5 | 100 | 0.85 | 0.43 | 238 | 0.05 | 170 | 72 | 0.70 | 0.20 | 0.99 | 0.63 |
| 1705 | 4.5 | 100 | 0.85 | 0.43 | 203 | 0.05 | 163 | 61 | 0.70 | 0.20 | 0.99 | 0.62 |
| 1705 | 4.5 | 100 | 0.85 | 0.43 | 168 | 0.05 | 141 | 51 | 0.70 | 0.20 | 0.99 | 0.62 |
| 1705 | 4.4 | 100 | 0.85 | 0.43 | 133 | 0.05 | 100 | 40 | 0.70 | 0.20 | 0.99 | 0.62 |
| 1705 | 4.4 | 100 | 0.85 | 0.43 | 95 | 0.05 | 85 | 29 | 0.70 | 0.21 | 0.99 | 0.61 |
| 2650 | 7.9 | 191 | 1.09 | 0.68 | 487 | 0.23 | 237 | 130 | 0.75 | 0.13 | 0.92 | 0.48 |
| 2650 | 7.9 | 191 | 1.09 | 0.68 | 302 | 0.23 | 174 | 80 | 0.75 | 0.13 | 0.92 | 0.47 |
| 2650 | 8.2 | 191 | 1.09 | 0.68 | 44 | 0.23 | 24 | 12 | 0.75 | 0.13 | 0.92 | 0.42 |
| 2553 | 7.7 | 183 | 1.11 | 0.68 | 302 | 0.18 | 191 | 84 | 0.76 | 0.13 | 0.92 | 0.48 |
| 2553 | 8.0 | 183 | 1.11 | 0.68 | 44 | 0.18 | 27 | 12 | 0.76 | 0.13 | 0.92 | 0.43 |
| 2528 | 7.4 | 311 | 1.10 | 0.56 | 302 | 0.16 | 299 | 147 | 0.76 | 0.14 | 0.93 | 0.50 |
| 2447 | 7.4 | 302 | 1.10 | 0.56 | 129 | 0.12 | 134 | 64 | 0.76 | 0.14 | 0.93 | 0.48 |
| 2447 | 7.2 | 302 | 1.10 | 0.56 | 44 | 0.12 | 42 | 22 | 0.76 | 0.14 | 0.93 | 0.47 |
| 1921 | 5.6 | 149 | 1.12 | 0.38 | 488 | 0.01 | 423 | 187 | 0.77 | 0.17 | 0.96 | 0.58 |
| 1921 | 5.7 | 149 | 1.12 | 0.38 | 366 | 0.01 | 370 | 140 | 0.77 | 0.17 | 0.96 | 0.56 |
| 1921 | 6.0 | 149 | 1.12 | 0.38 | 130 | 0.01 | 100 | 50 | 0.77 | 0.17 | 0.96 | 0.53 |

Table 2 continued

| $T_f$ K | $\frac{T_f}{T_w}$ | $U_f$ m/sec | α | [O] | $x$ mm | $\frac{\Delta h_{chem}}{\Delta h}$ | $Nu_f^*$ | $\frac{Re_f}{10^3}$ | $Pr_f$ | $Pr_T$ | $\varphi_l$ | $\varphi_T$ |
|---|---|---|---|---|---|---|---|---|---|---|---|---|
| 1921 | 5.9 | 149 | 1.12 | 0.38 | 44 | 0.01 | 43 | 17 | 0.77 | 0.17 | 0.96 | 0.51 |
| 2086 | 5.8 | 423 | 1.09 | 0.36 | 129 | 0.02 | 263 | 121 | 0.77 | 0.17 | 0.96 | 0.56 |
| 2071 | 5.7 | 395 | 1.05 | 0.37 | 44 | 0.02 | 96 | 39 | 0.77 | 0.17 | 0.96 | 0.54 |
| 2130 | 6.2 | 256 | 1.08 | 0.37 | 236 | 0.03 | 350 | 131 | 0.77 | 0.16 | 0.96 | 0.54 |
| 2130 | 6.5 | 256 | 1.08 | 0.37 | 44 | 0.03 | 63 | 24 | 0.77 | 0.16 | 0.96 | 0.49 |
| 2017 | 5.9 | 245 | 1.09 | 0.36 | 236 | 0.02 | 349 | 138 | 0.76 | 0.17 | 0.96 | 0.56 |
| 2017 | 6.1 | 245 | 1.09 | 0.36 | 129 | 0.02 | 167 | 75 | 0.76 | 0.16 | 0.96 | 0.54 |
| 2017 | 6.2 | 245 | 1.09 | 0.36 | 44 | 0.02 | 72 | 26 | 0.76 | 0.16 | 0.96 | 0.51 |
| 2086 | 5.7 | 423 | 1.09 | 0.36 | 44 | 0.02 | 99 | 41 | 0.77 | 0.17 | 0.96 | 0.54 |
| 2240 | 6.3 | 283 | 1.05 | 0.37 | 44 | 0.06 | 67 | 25 | 0.76 | 0.16 | 0.96 | 0.50 |
| 1983 | 6.0 | 241 | 1.09 | 0.37 | 129 | 0.01 | 170 | 76 | 0.76 | 0.16 | 0.96 | 0.54 |
| 1983 | 5.8 | 241 | 1.09 | 0.37 | 236 | 0.01 | 354 | 140 | 0.76 | 0.17 | 0.96 | 0.56 |
| 2292 | 6.8 | 280 | 1.08 | 0.36 | 44 | 0.06 | 58 | 23 | 0.76 | 0.15 | 0.96 | 0.48 |
| 2292 | 6.7 | 280 | 1.08 | 0.36 | 236 | 0.06 | 285 | 125 | 0.76 | 0.15 | 0.96 | 0.52 |
| 2292 | 6.7 | 280 | 1.08 | 0.36 | 366 | 0.06 | 432 | 193 | 0.76 | 0.15 | 0.96 | 0.53 |
| 2343 | 7.4 | 169 | 1.00 | 0.39 | 44 | 0.10 | 28 | 13 | 0.76 | 0.14 | 0.95 | 0.45 |
| 2343 | 6.9 | 169 | 1.00 | 0.39 | 236 | 0.10 | 163 | 72 | 0.76 | 0.15 | 0.95 | 0.51 |
| 1983 | 6.0 | 241 | 1.09 | 0.37 | 44 | 0.01 | 69 | 26 | 0.76 | 0.16 | 0.96 | 0.52 |
| 2308 | 6.7 | 291 | 1.03 | 0.37 | 236 | 0.08 | 285 | 129 | 0.76 | 0.15 | 0.96 | 0.52 |
| 2308 | 6.5 | 291 | 1.03 | 0.37 | 44 | 0.08 | 61 | 24 | 0.76 | 0.16 | 0.96 | 0.50 |
| 2263 | 6.6 | 283 | 1.05 | 0.37 | 236 | 0.06 | 291 | 132 | 0.76 | 0.15 | 0.96 | 0.53 |
| 2010 | 6.0 | 276 | 1.08 | 0.36 | 236 | 0.02 | 319 | 126 | 0.76 | 0.17 | 0.96 | 0.55 |
| 2010 | 6.1 | 277 | 1.18 | 0.35 | 129 | 0.02 | 219 | 85 | 0.77 | 0.16 | 0.97 | 0.54 |
| 2173 | 6.5 | 265 | 1.08 | 0.36 | 236 | 0.04 | 279 | 129 | 0.76 | 0.16 | 0.96 | 0.53 |
| 2263 | 6.6 | 276 | 1.08 | 0.36 | 44 | 0.06 | 64 | 23 | 0.76 | 0.15 | 0.96 | 0.49 |
| 2173 | 6.7 | 265 | 1.08 | 0.36 | 129 | 0.04 | 164 | 71 | 0.76 | 0.15 | 0.96 | 0.51 |
| 2263 | 6.9 | 276 | 1.08 | 0.36 | 129 | 0.06 | 147 | 69 | 0.76 | 0.15 | 0.96 | 0.50 |
| 2010 | 5.8 | 277 | 1.18 | 0.35 | 44 | 0.02 | 96 | 29 | 0.77 | 0.17 | 0.97 | 0.52 |
| 1899 | 3.9 | 330 | 0.95 | 0.34 | 203 | 0.02 | 542 | 195 | 0.73 | 0.22 | 0.97 | 0.68 |
| 1899 | 3.7 | 330 | 0.95 | 0.34 | 133 | 0.02 | 336 | 128 | 0.73 | 0.23 | 0.97 | 0.68 |
| 2062 | 4.0 | 347 | 0.99 | 0.36 | 133 | 0.03 | 288 | 120 | 0.74 | 0.22 | 0.96 | 0.66 |

Table 2 continued

| $T_f$ K | $\frac{T_f}{T_w}$ | $U_f$ m/sec | $\alpha$ | [O] | $x$ mm | $\frac{\Delta h_{chem}}{\Delta h}$ | $Nu_f^*$ | $\frac{Re_f}{10^3}$ | $Pr_f$ | $Pr_T$ | $\varphi_l$ | $\varphi_T$ |
|---|---|---|---|---|---|---|---|---|---|---|---|---|
| 2062 | 4.1 | 347 | 0.99 | 0.36 | 203 | 0.03 | 485 | 183 | 0.74 | 0.22 | 0.96 | 0.66 |
| 2081 | 3.6 | 375 | 0.97 | 0.37 | 133 | 0.02 | 331 | 127 | 0.73 | 0.24 | 0.96 | 0.69 |
| 2081 | 3.7 | 375 | 0.97 | 0.37 | 168 | 0.02 | 459 | 161 | 0.73 | 0.23 | 0.96 | 0.68 |
| 1548 | 3.6 | 253 | 0.96 | 0.27 | 133 | 0.01 | 356 | 136 | 0.73 | 0.24 | 0.97 | 0.69 |
| 2200 | 3.8 | 408 | 1.01 | 0.40 | 203 | 0.06 | 525 | 196 | 0.74 | 0.23 | 0.95 | 0.68 |
| 1827 | 3.8 | 309 | 1.03 | 0.31 | 133 | 0.01 | 336 | 130 | 0.74 | 0.23 | 0.96 | 0.67 |
| 1548 | 3.6 | 253 | 0.96 | 0.27 | 168 | 0.01 | 490 | 172 | 0.73 | 0.24 | 0.97 | 0.69 |
| 2244 | 3.8 | 400 | 0.99 | 0.40 | 95 | 0.07 | 228 | 87 | 0.73 | 0.23 | 0.96 | 0.66 |
| 2244 | 4.0 | 400 | 0.99 | 0.40 | 168 | 0.07 | 389 | 154 | 0.73 | 0.22 | 0.96 | 0.66 |
| 2244 | 4.1 | 400 | 0.99 | 0.40 | 203 | 0.07 | 475 | 187 | 0.73 | 0.22 | 0.95 | 0.66 |
| 2244 | 3.9 | 400 | 0.99 | 0.40 | 133 | 0.07 | 275 | 122 | 0.73 | 0.22 | 0.96 | 0.66 |
| 2010 | 6.0 | 277 | 1.18 | 0.35 | 236 | 0.02 | 406 | 157 | 0.77 | 0.17 | 0.97 | 0.56 |
| 2200 | 3.6 | 408 | 1.01 | 0.40 | 133 | 0.06 | 312 | 129 | 0.74 | 0.24 | 0.96 | 0.68 |
| 2200 | 4.4 | 408 | 1.01 | 0.40 | 343 | 0.06 | 713 | 332 | 0.74 | 0.21 | 0.95 | 0.66 |
| 1999 | 5.3 | 236 | 1.07 | 0.38 | 378 | 0.02 | 533 | 234 | 0.74 | 0.18 | 0.95 | 0.60 |
| 1999 | 5.1 | 236 | 1.07 | 0.38 | 343 | 0.02 | 437 | 212 | 0.74 | 0.18 | 0.95 | 0.61 |
| 1999 | 5.0 | 236 | 1.07 | 0.38 | 308 | 0.02 | 441 | 191 | 0.74 | 0.19 | 0.95 | 0.61 |
| 1965 | 5.5 | 198 | 0.97 | 0.38 | 448 | 0.01 | 591 | 236 | 0.73 | 0.18 | 0.96 | 0.59 |
| 1705 | 4.7 | 100 | 0.85 | 0.43 | 483 | 0.05 | 291 | 145 | 0.70 | 0.20 | 0.99 | 0.63 |
| 1705 | 4.7 | 100 | 0.85 | 0.43 | 448 | 0.05 | 361 | 135 | 0.70 | 0.20 | 0.99 | 0.63 |
| 1705 | 4.6 | 100 | 0.85 | 0.43 | 413 | 0.05 | 297 | 124 | 0.70 | 0.20 | 0.99 | 0.63 |
| 1705 | 4.6 | 100 | 0.85 | 0.43 | 378 | 0.05 | 246 | 114 | 0.70 | 0.20 | 0.99 | 0.63 |
| 1705 | 4.6 | 100 | 0.85 | 0.43 | 343 | 0.05 | 284 | 103 | 0.70 | 0.20 | 0.99 | 0.63 |
| 2343 | 6.9 | 169 | 1.00 | 0.39 | 366 | 0.10 | 241 | 111 | 0.76 | 0.15 | 0.95 | 0.52 |
| 2234 | 7.0 | 175 | 1.07 | 0.37 | 44 | 0.05 | 36 | 15 | 0.76 | 0.15 | 0.96 | 0.47 |
| 2234 | 6.6 | 175 | 1.07 | 0.37 | 236 | 0.05 | 226 | 81 | 0.76 | 0.15 | 0.96 | 0.52 |
| 2234 | 6.6 | 175 | 1.07 | 0.37 | 366 | 0.05 | 316 | 126 | 0.76 | 0.15 | 0.96 | 0.53 |
| 2234 | 6.5 | 175 | 1.07 | 0.37 | 487 | 0.05 | 344 | 168 | 0.76 | 0.15 | 0.96 | 0.54 |
| 2528 | 7.4 | 311 | 1.10 | 0.56 | 487 | 0.16 | 428 | 237 | 0.76 | 0.14 | 0.93 | 0.51 |
| 2086 | 5.7 | 423 | 1.09 | 0.36 | 487 | 0.02 | 850 | 456 | 0.77 | 0.17 | 0.96 | 0.59 |
| 2086 | 5.8 | 423 | 1.09 | 0.36 | 366 | 0.02 | 752 | 344 | 0.77 | 0.17 | 0.96 | 0.58 |
| 2086 | 5.8 | 423 | 1.09 | 0.36 | 236 | 0.02 | 501 | 222 | 0.77 | 0.17 | 0.96 | 0.57 |
| 2071 | 5.8 | 395 | 1.05 | 0.37 | 366 | 0.02 | 754 | 326 | 0.77 | 0.17 | 0.96 | 0.58 |

*Note*: For heat transfer in the inlet part of the channel, there are listed the same data as in Table 1 for flow over a plate.

**TABLE A.3** Heat transfer of the Inlet Region of a Channel with a Flow of Hot Air

| $T_f$ K | $\frac{T_f}{T_w}$ | $U_f$ m/sec | $x$ mm | $Nu_f$ | $\frac{Re_f}{10^3}$ | $Pr_f$ | $Pr_T$ | $\varphi_l$ | $\varphi_T$ |
|---|---|---|---|---|---|---|---|---|---|
| 602 | 1.8 | 81 | 71 | 299 | 110 | 0.68 | 0.39 | 1.00 | 0.90 |
| 440 | 1.4 | 59 | 71 | 379 | 136 | 0.69 | 0.46 | 1.00 | 0.96 |
| 440 | 1.4 | 59 | 127 | 577 | 242 | 0.69 | 0.46 | 1.00 | 0.97 |
| 440 | 1.4 | 59 | 179 | 893 | 342 | 0.69 | 0.46 | 1.00 | 0.97 |
| 440 | 1.4 | 59 | 249 | 1168 | 476 | 0.69 | 0.45 | 1.00 | 0.97 |
| 440 | 1.4 | 59 | 390 | 1724 | 743 | 0.69 | 0.45 | 1.00 | 0.97 |
| 632 | 2.0 | 36 | 215 | 315 | 112 | 0.68 | 0.36 | 1.00 | 0.87 |
| 632 | 2.0 | 36 | 179 | 303 | 134 | 0.68 | 0.36 | 1.00 | 0.87 |
| 632 | 2.0 | 36 | 267 | 408 | 167 | 0.68 | 0.36 | 1.00 | 0.87 |
| 635 | 2.0 | 45 | 127 | 236 | 101 | 0.68 | 0.37 | 1.00 | 0.87 |
| 635 | 2.0 | 45 | 215 | 376 | 170 | 0.68 | 0.36 | 1.00 | 0.87 |
| 680 | 1.9 | 92 | 71 | 291 | 102 | 0.68 | 0.37 | 1.00 | 0.88 |
| 680 | 2.0 | 92 | 215 | 760 | 307 | 0.68 | 0.36 | 1.00 | 0.88 |
| 639 | 1.8 | 86 | 71 | 266 | 106 | 0.68 | 0.38 | 1.00 | 0.89 |
| 639 | 1.9 | 86 | 127 | 466 | 188 | 0.68 | 0.38 | 1.00 | 0.89 |
| 639 | 1.9 | 86 | 179 | 654 | 266 | 0.68 | 0.37 | 1.00 | 0.89 |
| 639 | 1.9 | 86 | 320 | 963 | 474 | 0.68 | 0.37 | 1.00 | 0.89 |
| 639 | 2.0 | 86 | 355 | 1223 | 526 | 0.68 | 0.36 | 1.00 | 0.89 |
| 639 | 2.0 | 86 | 390 | 1251 | 578 | 0.68 | 0.36 | 1.00 | 0.89 |
| 602 | 1.8 | 81 | 127 | 466 | 195 | 0.68 | 0.39 | 1.00 | 0.90 |
| 602 | 1.8 | 81 | 179 | 689 | 276 | 0.68 | 0.38 | 1.00 | 0.90 |
| 602 | 1.9 | 81 | 320 | 1047 | 491 | 0.68 | 0.38 | 1.00 | 0.91 |
| 602 | 1.9 | 81 | 355 | 1289 | 545 | 0.68 | 0.38 | 1.00 | 0.90 |
| 602 | 1.9 | 81 | 390 | 1311 | 599 | 0.68 | 0.38 | 1.00 | 0.91 |
| 602 | 1.9 | 81 | 442 | 1511 | 680 | 0.68 | 0.37 | 1.00 | 0.91 |
| 632 | 2.0 | 36 | 355 | 528 | 222 | 0.68 | 0.36 | 1.00 | 0.87 |
| 632 | 2.0 | 36 | 390 | 584 | 244 | 0.68 | 0.36 | 1.00 | 0.87 |
| 635 | 2.0 | 45 | 355 | 623 | 281 | 0.68 | 0.36 | 1.00 | 0.87 |
| 635 | 2.0 | 45 | 390 | 661 | 309 | 0.68 | 0.35 | 1.00 | 0.87 |
| 635 | 2.1 | 45 | 464 | 827 | 368 | 0.68 | 0.35 | 1.00 | 0.87 |
| 719 | 2.1 | 84 | 127 | 405 | 150 | 0.69 | 0.35 | 1.00 | 0.86 |
| 719 | 2.1 | 84 | 179 | 578 | 213 | 0.69 | 0.35 | 1.00 | 0.86 |
| 719 | 2.1 | 84 | 215 | 631 | 254 | 0.69 | 0.35 | 1.00 | 0.86 |
| 719 | 2.1 | 84 | 267 | 792 | 316 | 0.69 | 0.34 | 1.00 | 0.86 |
| 719 | 2.2 | 84 | 320 | 808 | 378 | 0.69 | 0.34 | 1.00 | 0.86 |
| 719 | 2.2 | 84 | 355 | 968 | 420 | 0.69 | 0.34 | 1.00 | 0.86 |
| 719 | 2.2 | 84 | 390 | 1064 | 461 | 0.69 | 0.34 | 1.00 | 0.86 |

Table 3 continued

| $T_f$ K | $\frac{T_f}{T_w}$ | $U_f$ m/sec | $x$ mm | $Nu_f$ | $\frac{Re_f}{10^3}$ | $Pr_f$ | $Pr_T$ | $\varphi_l$ | $\varphi_T$ |
|---|---|---|---|---|---|---|---|---|---|
| 703 | 2.0 | 87 | 127 | 433 | 163 | 0.69 | 0.36 | 1.00 | 0.86 |
| 703 | 2.1 | 87 | 215 | 676 | 274 | 0.69 | 0.35 | 1.00 | 0.87 |
| 703 | 2.1 | 87 | 267 | 840 | 342 | 0.69 | 0.35 | 1.00 | 0.87 |
| 703 | 2.1 | 87 | 320 | 820 | 409 | 0.69 | 0.35 | 1.00 | 0.87 |
| 703 | 2.1 | 87 | 355 | 1047 | 454 | 0.69 | 0.34 | 1.00 | 0.87 |
| 703 | 2.1 | 87 | 390 | 1139 | 498 | 0.69 | 0.34 | 1.00 | 0.87 |
| 703 | 2.2 | 87 | 442 | 1300 | 566 | 0.69 | 0.34 | 1.00 | 0.87 |
| 703 | 2.2 | 87 | 499 | 1160 | 638 | 0.69 | 0.34 | 1.00 | 0.87 |
| 680 | 2.0 | 92 | 267 | 939 | 383 | 0.68 | 0.36 | 1.00 | 0.88 |
| 680 | 2.0 | 92 | 320 | 916 | 458 | 0.68 | 0.35 | 1.00 | 0.88 |
| 680 | 2.1 | 92 | 355 | 1176 | 508 | 0.68 | 0.35 | 1.00 | 0.88 |
| 680 | 2.1 | 92 | 390 | 1250 | 558 | 0.68 | 0.35 | 1.00 | 0.88 |
| 680 | 2.1 | 92 | 442 | 1452 | 633 | 0.68 | 0.35 | 1.00 | 0.88 |
| 680 | 2.1 | 92 | 499 | 1291 | 715 | 0.68 | 0.35 | 1.00 | 0.88 |
| 791 | 2.4 | 44 | 215 | 292 | 115 | 0.69 | 0.31 | 1.01 | 0.81 |
| 791 | 2.4 | 44 | 267 | 416 | 143 | 0.69 | 0.31 | 1.01 | 0.81 |
| 791 | 2.5 | 44 | 355 | 477 | 190 | 0.69 | 0.31 | 1.01 | 0.81 |
| 719 | 2.2 | 84 | 499 | 1121 | 591 | 0.69 | 0.33 | 1.00 | 0.86 |
| 847 | 2.5 | 61 | 179 | 312 | 118 | 0.70 | 0.31 | 1.01 | 0.79 |
| 847 | 2.5 | 61 | 215 | 395 | 141 | 0.70 | 0.30 | 1.01 | 0.80 |
| 847 | 2.6 | 61 | 249 | 393 | 164 | 0.70 | 0.30 | 1.01 | 0.79 |
| 847 | 2.5 | 61 | 285 | 503 | 187 | 0.70 | 0.30 | 1.01 | 0.80 |
| 847 | 2.6 | 61 | 355 | 604 | 232 | 0.70 | 0.30 | 1.01 | 0.80 |
| 847 | 2.6 | 61 | 390 | 649 | 255 | 0.70 | 0.30 | 1.01 | 0.80 |
| 847 | 2.6 | 61 | 442 | 678 | 290 | 0.70 | 0.30 | 1.01 | 0.80 |
| 632 | 1.9 | 36 | 71 | 115 | 45 | 0.68 | 0.37 | 1.00 | 0.86 |
| 632 | 2.0 | 36 | 127 | 191 | 79 | 0.68 | 0.37 | 1.00 | 0.87 |
| 791 | 2.3 | 44 | 71 | 102 | 38 | 0.69 | 0.32 | 1.01 | 0.80 |
| 791 | 2.4 | 44 | 127 | 186 | 68 | 0.69 | 0.32 | 1.01 | 0.80 |
| 719 | 2.0 | 84 | 71 | 251 | 84 | 0.69 | 0.36 | 1.00 | 0.85 |
| 703 | 2.0 | 87 | 71 | 254 | 91 | 0.69 | 0.36 | 1.00 | 0.86 |
| 847 | 2.4 | 61 | 71 | 147 | 47 | 0.70 | 0.31 | 1.01 | 0.79 |
| 847 | 2.5 | 61 | 127 | 234 | 83 | 0.70 | 0.31 | 1.01 | 0.79 |

*Note*: For heat transfer in the inlet part of the channel with the flow of air, there are listed the same data as in Table 1 for combustion products, with the exception of data on the composition of the flow.

## 3. EXPERIMENTAL DATA ON CONVECTIVE HEAT TRANSFER IN TEST CHANNEL 4

**TABLE A.4** Heat Transfer in a Stabilized Flow of Combustion Products

| $T_w$ K | $\frac{T_f}{T_w}$ | $p$ atm | $p_{O_2}$ atm | $p_{N_2}$ atm | $p_{CO_2}$ atm | $p_{H_2O}$ atm | $x$ m | $Nu_{f,\,exp}$ | $Nu_f$ | $\frac{Re_f}{10^3}$ | $Pr_f$ | $\frac{Nu_{f,\,exp}}{Nu_f}$ |
|---|---|---|---|---|---|---|---|---|---|---|---|---|
| 402 | 3.8 | 1.00 | 0.09 | 0.75 | 0.06 | 0.11 | 0.54 | 107 | 89 | 35 | 0.74 | 0.86 |
| 400 | 3.7 | 1.00 | 0.09 | 0.75 | 0.06 | 0.11 | 0.77 | 112 | 95 | 36 | 0.74 | 0.96 |
| 397 | 3.6 | 1.00 | 0.09 | 0.75 | 0.06 | 0.11 | 1.08 | 114 | 96 | 36 | 0.74 | 1.01 |
| 393 | 3.6 | 1.00 | 0.09 | 0.75 | 0.06 | 0.11 | 1.38 | 121 | 104 | 37 | 0.74 | 1.11 |
| 388 | 3.5 | 1.00 | 0.09 | 0.75 | 0.06 | 0.11 | 1.69 | 119 | 102 | 38 | 0.74 | 1.10 |
| 383 | 3.5 | 1.00 | 0.09 | 0.75 | 0.06 | 0.11 | 1.99 | 108 | 92 | 38 | 0.74 | 1.00 |
| 378 | 3.5 | 1.00 | 0.09 | 0.75 | 0.06 | 0.11 | 2.30 | 125 | 109 | 39 | 0.74 | 1.18 |
| 373 | 3.4 | 1.00 | 0.09 | 0.75 | 0.06 | 0.11 | 2.60 | 110 | 95 | 40 | 0.74 | 1.03 |
| 367 | 3.3 | 1.00 | 0.09 | 0.75 | 0.06 | 0.11 | 3.21 | 116 | 101 | 41 | 0.74 | 1.08 |
| 362 | 3.3 | 1.00 | 0.09 | 0.75 | 0.06 | 0.11 | 3.52 | 112 | 98 | 41 | 0.74 | 1.04 |
| 358 | 3.3 | 1.00 | 0.09 | 0.75 | 0.06 | 0.11 | 3.82 | 109 | 95 | 42 | 0.74 | 1.01 |
| 397 | 3.7 | 1.00 | 0.09 | 0.75 | 0.05 | 0.11 | 0.54 | 125 | 108 | 43 | 0.74 | 0.89 |
| 393 | 3.7 | 1.00 | 0.09 | 0.75 | 0.05 | 0.11 | 0.77 | 126 | 109 | 43 | 0.74 | 0.95 |
| 389 | 3.6 | 1.00 | 0.09 | 0.75 | 0.05 | 0.11 | 1.08 | 118 | 102 | 44 | 0.74 | 0.92 |
| 385 | 3.6 | 1.00 | 0.09 | 0.75 | 0.05 | 0.11 | 1.38 | 142 | 126 | 44 | 0.74 | 1.17 |
| 383 | 3.5 | 1.00 | 0.09 | 0.75 | 0.05 | 0.11 | 1.69 | 135 | 119 | 45 | 0.74 | 1.11 |
| 376 | 3.5 | 1.00 | 0.09 | 0.75 | 0.05 | 0.11 | 1.99 | 125 | 109 | 46 | 0.74 | 1.02 |
| 373 | 3.4 | 1.00 | 0.09 | 0.75 | 0.05 | 0.11 | 2.60 | 132 | 117 | 47 | 0.74 | 1.09 |
| 371 | 3.3 | 1.00 | 0.09 | 0.75 | 0.05 | 0.11 | 2.91 | 143 | 128 | 48 | 0.74 | 1.19 |
| 369 | 3.3 | 1.00 | 0.09 | 0.75 | 0.05 | 0.11 | 3.21 | 134 | 120 | 49 | 0.74 | 1.11 |
| 369 | 3.2 | 1.00 | 0.09 | 0.75 | 0.05 | 0.11 | 3.52 | 128 | 114 | 49 | 0.74 | 1.04 |
| 369 | 3.1 | 1.00 | 0.09 | 0.75 | 0.05 | 0.11 | 3.82 | 127 | 114 | 50 | 0.74 | 1.04 |
| 383 | 3.4 | 0.97 | 0.13 | 0.77 | 0.02 | 0.05 | 0.54 | 128 | 117 | 43 | 0.72 | 0.98 |
| 382 | 3.4 | 0.97 | 0.13 | 0.77 | 0.02 | 0.05 | 0.77 | 121 | 111 | 43 | 0.72 | 0.97 |
| 379 | 3.3 | 0.97 | 0.13 | 0.77 | 0.02 | 0.05 | 1.08 | 118 | 108 | 44 | 0.72 | 0.99 |
| 379 | 3.2 | 0.97 | 0.13 | 0.77 | 0.02 | 0.05 | 1.38 | 123 | 113 | 44 | 0.72 | 1.06 |
| 374 | 3.2 | 0.97 | 0.13 | 0.77 | 0.02 | 0.05 | 1.69 | 126 | 117 | 45 | 0.72 | 1.10 |
| 373 | 3.2 | 0.97 | 0.13 | 0.77 | 0.02 | 0.05 | 1.99 | 106 | 97 | 46 | 0.72 | 0.91 |

Table 4 continued

| $T_w$ K | $\frac{T_f}{T_w}$ | $p$ atm | $p_{O_2}$ atm | $p_{N_2}$ atm | $p_{CO_2}$ atm | $p_{H_2O}$ atm | $x$ m | $Nu_{f,exp}$ | $Nu_f$ | $\frac{Re_f}{10^3}$ | $Pr_f$ | $\frac{Nu_{f,exp}}{Nu_f}$ |
|---|---|---|---|---|---|---|---|---|---|---|---|---|
| 369 | 3.1 | 0.97 | 0.13 | 0.77 | 0.02 | 0.05 | 2.30 | 127 | 118 | 46 | 0.72 | 1.12 |
| 366 | 3.1 | 0.97 | 0.13 | 0.77 | 0.05 | 2.60 | 2.60 | 116 | 107 | 47 | 0.72 | 1.02 |
| 409 | 3.4 | 1.00 | 0.11 | 0.75 | 0.04 | 0.09 | 0.54 | 163 | 148 | 68 | 0.74 | 0.83 |
| 404 | 3.4 | 1.00 | 0.11 | 0.75 | 0.04 | 0.09 | 0.77 | 156 | 141 | 69 | 0.74 | 0.84 |
| 398 | 3.3 | 1.00 | 0.11 | 0.75 | 0.04 | 0.09 | 1.08 | 162 | 148 | 70 | 0.74 | 0.92 |
| 396 | 3.3 | 1.00 | 0.11 | 0.75 | 0.04 | 0.09 | 1.38 | 162 | 148 | 71 | 0.74 | 0.94 |
| 393 | 3.3 | 1.00 | 0.11 | 0.75 | 0.04 | 0.09 | 1.69 | 163 | 149 | 72 | 0.74 | 0.96 |
| 391 | 3.2 | 1.00 | 0.11 | 0.75 | 0.04 | 0.09 | 1.99 | 151 | 137 | 72 | 0.74 | 0.89 |
| 388 | 3.2 | 1.00 | 0.11 | 0.75 | 0.04 | 0.09 | 2.30 | 165 | 151 | 73 | 0.74 | 0.98 |
| 385 | 3.2 | 1.00 | 0.11 | 0.75 | 0.04 | 0.09 | 2.60 | 156 | 143 | 74 | 0.74 | 0.93 |
| 383 | 3.1 | 1.00 | 0.11 | 0.75 | 0.04 | 0.09 | 2.91 | 160 | 147 | 75 | 0.74 | 0.96 |
| 380 | 3.1 | 1.00 | 0.11 | 0.75 | 0.04 | 0.09 | 3.21 | 153 | 140 | 76 | 0.74 | 0.91 |
| 378 | 3.1 | 1.00 | 0.11 | 0.75 | 0.04 | 0.09 | 3.52 | 146 | 133 | 76 | 0.74 | 0.86 |
| 375 | 3.1 | 1.00 | 0.11 | 0.75 | 0.04 | 0.09 | 3.82 | 149 | 137 | 77 | 0.74 | 0.88 |
| 393 | 3.1 | 1.00 | 0.12 | 0.76 | 0.04 | 0.08 | 0.54 | 154 | 142 | 61 | 0.73 | 0.88 |
| 391 | 3.1 | 1.00 | 0.12 | 0.76 | 0.04 | 0.08 | 0.77 | 149 | 137 | 61 | 0.73 | 0.89 |
| 388 | 3.1 | 1.00 | 0.12 | 0.76 | 0.04 | 0.08 | 1.08 | 146 | 134 | 62 | 0.73 | 0.91 |
| 385 | 3.0 | 1.00 | 0.12 | 0.76 | 0.04 | 0.08 | 1.38 | 147 | 135 | 63 | 0.73 | 0.94 |
| 383 | 3.0 | 1.00 | 0.12 | 0.76 | 0.04 | 0.08 | 1.69 | 153 | 141 | 63 | 0.73 | 0.99 |
| 380 | 3.0 | 1.00 | 0.12 | 0.76 | 0.04 | 0.08 | 1.99 | 138 | 127 | 64 | 0.74 | 0.90 |
| 378 | 2.9 | 1.00 | 0.12 | 0.76 | 0.04 | 0.08 | 2.30 | 145 | 133 | 65 | 0.74 | 0.95 |
| 375 | 2.9 | 1.00 | 0.12 | 0.76 | 0.04 | 0.08 | 2.60 | 139 | 128 | 66 | 0.74 | 0.92 |
| 373 | 2.9 | 1.00 | 0.12 | 0.76 | 0.04 | 0.08 | 2.91 | 148 | 137 | 66 | 0.74 | 0.98 |
| 370 | 2.9 | 1.00 | 0.12 | 0.76 | 0.04 | 0.08 | 3.21 | 137 | 127 | 67 | 0.74 | 0.90 |
| 367 | 2.8 | 1.00 | 0.12 | 0.76 | 0.04 | 0.08 | 3.52 | 138 | 128 | 68 | 0.74 | 0.91 |
| 365 | 2.8 | 1.00 | 0.12 | 0.76 | 0.04 | 0.08 | 3.82 | 134 | 123 | 68 | 0.74 | 0.88 |
| 414 | 3.3 | 1.00 | 0.12 | 0.75 | 0.04 | 0.09 | 0.54 | 183 | 168 | 77 | 0.74 | 0.86 |

Table 4 continued

| $T_w$ K | $\frac{T_f}{T_w}$ | $p$ atm | $p_{O_2}$ atm | $p_{N_2}$ atm | $p_{CO_2}$ atm | $p_{H_2O}$ atm | $x$ m | $Nu_{f,\,exp}$ | $Nu_f$ | $\frac{Re_f}{10^3}$ | $Pr_f$ | $\frac{Nu_{f,\,exp}}{Nu_f}$ |
|---|---|---|---|---|---|---|---|---|---|---|---|---|
| 405 | 3.4 | 1.00 | 0.12 | 0.75 | 0.04 | 0.09 | 0.77 | 174 | 159 | 78 | 0.74 | 0.86 |
| 400 | 3.4 | 1.00 | 0.12 | 0.75 | 0.04 | 0.09 | 1.08 | 185 | 171 | 78 | 0.74 | 0.96 |
| 397 | 3.3 | 1.00 | 0.12 | 0.75 | 0.04 | 0.09 | 1.38 | 185 | 171 | 79 | 0.74 | 0.99 |
| 393 | 3.3 | 1.00 | 0.12 | 0.75 | 0.04 | 0.09 | 1.69 | 183 | 169 | 80 | 0.74 | 0.99 |
| 388 | 3.3 | 1.00 | 0.12 | 0.75 | 0.04 | 0.09 | 1.99 | 173 | 159 | 81 | 0.74 | 0.94 |
| 384 | 3.3 | 1.00 | 0.12 | 0.75 | 0.04 | 0.09 | 2.30 | 180 | 167 | 82 | 0.74 | 0.99 |
| 380 | 3.2 | 1.00 | 0.12 | 0.75 | 0.04 | 0.09 | 2.60 | 180 | 167 | 83 | 0.74 | 0.99 |
| 375 | 3.2 | 1.00 | 0.12 | 0.75 | 0.04 | 0.09 | 2.91 | 179 | 166 | 84 | 0.74 | 0.99 |
| 373 | 3.2 | 1.00 | 0.12 | 0.75 | 0.04 | 0.09 | 3.21 | 173 | 160 | 85 | 0.74 | 0.95 |
| 367 | 3.2 | 1.00 | 0.12 | 0.75 | 0.04 | 0.09 | 3.52 | 163 | 150 | 86 | 0.74 | 0.89 |
| 354 | 2.9 | 0.97 | 0.10 | 0.76 | 0.04 | 0.08 | 2.60 | 110 | 99 | 41 | 0.72 | 1.03 |
| 356 | 2.8 | 0.97 | 0.10 | 0.76 | 0.04 | 0.08 | 2.91 | 119 | 109 | 42 | 0.72 | 1.13 |
| 352 | 2.8 | 0.97 | 0.10 | 0.76 | 0.04 | 0.08 | 3.21 | 98 | 88 | 43 | 0.72 | 0.91 |
| 349 | 2.8 | 0.97 | 0.10 | 0.76 | 0.04 | 0.08 | 3.52 | 94 | 84 | 43 | 0.72 | 0.86 |
| 349 | 2.7 | 0.97 | 0.10 | 0.76 | 0.04 | 0.08 | 3.82 | 110 | 100 | 44 | 0.72 | 1.03 |
| 406 | 3.8 | 1.00 | 0.08 | 0.75 | 0.06 | 0.12 | 0.54 | 125 | 107 | 42 | 0.74 | 0.89 |
| 401 | 3.8 | 1.00 | 0.08 | 0.75 | 0.06 | 0.12 | 0.77 | 128 | 110 | 43 | 0.74 | 0.96 |
| 396 | 3.8 | 1.00 | 0.08 | 0.75 | 0.06 | 0.12 | 1.08 | 127 | 109 | 43 | 0.74 | 0.99 |
| 391 | 3.7 | 1.00 | 0.08 | 0.75 | 0.06 | 0.12 | 1.38 | 137 | 119 | 44 | 0.74 | 1.11 |
| 385 | 3.7 | 1.00 | 0.08 | 0.75 | 0.06 | 0.12 | 1.69 | 138 | 121 | 45 | 0.74 | 1.14 |
| 380 | 3.6 | 1.00 | 0.08 | 0.75 | 0.06 | 0.12 | 1.99 | 131 | 114 | 46 | 0.74 | 1.08 |
| 375 | 3.6 | 1.00 | 0.08 | 0.75 | 0.06 | 0.12 | 2.30 | 140 | 124 | 46 | 0.74 | 1.17 |
| 371 | 3.6 | 1.00 | 0.08 | 0.75 | 0.06 | 0.12 | 2.60 | 129 | 113 | 47 | 0.74 | 1.06 |
| 370 | 3.5 | 1.00 | 0.08 | 0.75 | 0.06 | 0.12 | 2.91 | 144 | 144 | 48 | 0.74 | 1.20 |
| 369 | 3.4 | 1.00 | 0.08 | 0.75 | 0.06 | 0.12 | 3.21 | 133 | 118 | 49 | 0.74 | 1.09 |
| 369 | 3.4 | 1.00 | 0.08 | 0.75 | 0.06 | 0.12 | 3.52 | 130 | 115 | 49 | 0.74 | 1.06 |
| 369 | 3.3 | 1.00 | 0.08 | 0.75 | 0.06 | 0.12 | 3.82 | 134 | 119 | 50 | 0.74 | 1.09 |

Table 4 continued

| $T_w$ K | $\frac{T_f}{T_w}$ | $p$ atm | $p_{O_2}$ atm | $p_{N_2}$ atm | $p_{CO_2}$ atm | $p_{H_2O}$ atm | $x$ m | $Nu_{f,exp}$ | $Nu_f$ | $\frac{Re_f}{10^3}$ | $Pr_f$ | $\frac{Nu_{f,exp}}{Nu_f}$ |
|---|---|---|---|---|---|---|---|---|---|---|---|---|
| 366 | 3.0 | 0.97 | 0.12 | 0.77 | 0.03 | 0.06 | 0.54 | 119 | 110 | 44 | 0.72 | 0.88 |
| 364 | 3.0 | 0.97 | 0.12 | 0.77 | 0.03 | 0.06 | 0.77 | 117 | 108 | 44 | 0.72 | 0.92 |
| 364 | 2.9 | 9.97 | 0.12 | 0.77 | 0.03 | 0.06 | 1.08 | 119 | 110 | 45 | 0.72 | 0.97 |
| 354 | 2.9 | 0.97 | 0.12 | 0.77 | 0.03 | 0.06 | 1.38 | 112 | 103 | 46 | 0.72 | 0.93 |
| 349 | 2.9 | 0.97 | 0.12 | 0.77 | 0.03 | 0.06 | 1.69 | 118 | 110 | 46 | 0.72 | 1.01 |
| 348 | 2.9 | 0.97 | 0.12 | 0.77 | 0.03 | 0.06 | 1.99 | 99 | 90 | 47 | 0.72 | 0.83 |
| 345 | 2.9 | 0.97 | 0.12 | 0.77 | 0.03 | 0.06 | 2.30 | 126 | 117 | 47 | 0.72 | 1.09 |
| 344 | 2.8 | 0.97 | 0.12 | 0.77 | 0.03 | 0.06 | 2.60 | 107 | 98 | 48 | 0.72 | 0.91 |
| 342 | 2.8 | 0.97 | 0.12 | 0.77 | 0.03 | 0.06 | 2.91 | 117 | 108 | 48 | 0.72 | 1.01 |
| 340 | 2.8 | 0.97 | 0.12 | 0.77 | 0.03 | 0.06 | 3.21 | 96 | 88 | 49 | 0.72 | 0.81 |
| 340 | 2.7 | 0.97 | 0.12 | 0.77 | 0.03 | 0.06 | 3.52 | 92 | 84 | 49 | 0.72 | 0.78 |
| 340 | 2.7 | 0.97 | 0.12 | 0.77 | 0.03 | 0.06 | 3.82 | 110 | 102 | 50 | 0.72 | 0.94 |
| 334 | 2.7 | 0.99 | 0.13 | 0.77 | 0.03 | 0.06 | 0.54 | 105 | 97 | 37 | 0.73 | 0.88 |
| 332 | 2.7 | 0.99 | 0.13 | 0.77 | 0.03 | 0.06 | 0.77 | 114 | 106 | 38 | 0.73 | 1.02 |
| 332 | 2.6 | 0.99 | 0.13 | 0.77 | 0.03 | 0.06 | 1.08 | 93 | 86 | 38 | 0.73 | 0.85 |
| 330 | 2.6 | 0.99 | 0.13 | 0.77 | 0.03 | 0.06 | 1.38 | 99 | 92 | 39 | 0.73 | 0.93 |
| 329 | 2.5 | 0.99 | 0.13 | 0.77 | 0.03 | 0.06 | 1.69 | 104 | 97 | 39 | 0.73 | 0.99 |
| 327 | 2.5 | 0.99 | 0.13 | 0.77 | 0.03 | 0.06 | 1.99 | 94 | 87 | 40 | 0.73 | 0.90 |
| 364 | 3.1 | 0.97 | 0.13 | 0.77 | 0.02 | 0.05 | 2.91 | 122 | 113 | 47 | 0.72 | 1.07 |
| 361 | 3.0 | 0.97 | 0.13 | 0.77 | 0.02 | 0.05 | 3.21 | 118 | 109 | 48 | 0.72 | 1.03 |
| 361 | 3.0 | 0.97 | 0.13 | 0.77 | 0.02 | 0.05 | 3.52 | 106 | 98 | 48 | 0.72 | 0.92 |
| 361 | 2.9 | 0.97 | 0.13 | 0.77 | 0.02 | 0.05 | 3.82 | 114 | 106 | 49 | 0.72 | 0.99 |
| 361 | 3.4 | 0.97 | 0.10 | 0.76 | 0.04 | 0.07 | 0.54 | 139 | 127 | 52 | 0.72 | 0.90 |
| 360 | 3.3 | 0.97 | 0.10 | 0.76 | 0.04 | 0.07 | 0.77 | 119 | 107 | 52 | 0.72 | 0.81 |
| 357 | 3.3 | 0.97 | 0.10 | 0.76 | 0.04 | 0.07 | 1.08 | 115 | 104 | 53 | 0.72 | 0.81 |
| 357 | 3.2 | 0.97 | 0.10 | 0.76 | 0.04 | 0.07 | 1.38 | 119 | 108 | 53 | 0.72 | 0.86 |
| 357 | 3.2 | 0.97 | 0.10 | 0.76 | 0.04 | 0.07 | 1.69 | 135 | 124 | 54 | 0.72 | 1.01 |

Table 4 continued

| $T_w$ K | $\frac{T_f}{T_w}$ | $p$ atm | $p_{O_2}$ atm | $p_{N_2}$ atm | $p_{CO_2}$ atm | $p_{H_2O}$ atm | $x$ m | $Nu_{f,exp}$ | $Nu_f$ | $\frac{Re_f}{10^3}$ | $Pr_f$ | $\frac{Nu_{f,exp}}{Nu_f}$ |
|---|---|---|---|---|---|---|---|---|---|---|---|---|
| 354 | 3.1 | 0.97 | 0.10 | 0.76 | 0.04 | 0.07 | 1.99 | 116 | 105 | 55 | 0.72 | 0.86 |
| 354 | 3.1 | 0.97 | 0.10 | 0.76 | 0.04 | 0.07 | 2.30 | 136 | 126 | 55 | 0.72 | 1.03 |
| 351 | 3.1 | 0.97 | 0.10 | 0.76 | 0.04 | 0.07 | 2.60 | 125 | 114 | 56 | 0.72 | 0.94 |
| 351 | 3.0 | 0.07 | 0.10 | 0.76 | 0.04 | 0.07 | 2.91 | 132 | 121 | 57 | 0.72 | 0.99 |
| 351 | 3.0 | 0.97 | 0.10 | 0.76 | 0.04 | 0.07 | 3.21 | 122 | 112 | 57 | 0.72 | 0.91 |
| 351 | 2.9 | 0.97 | 0.10 | 0.76 | 0.04 | 0.07 | 3.52 | 115 | 105 | 58 | 0.72 | 0.86 |
| 351 | 2.9 | 0.97 | 0.10 | 0.76 | 0.04 | 0.07 | 3.82 | 124 | 114 | 58 | 0.72 | 0.93 |
| 352 | 3.2 | 0.97 | 0.11 | 0.76 | 0.03 | 0.07 | 0.54 | 135 | 125 | 48 | 0.72 | 0.93 |
| 349 | 3.2 | 0.97 | 0.11 | 0.76 | 0.03 | 0.07 | 0.77 | 127 | 117 | 49 | 0.72 | 0.92 |
| 351 | 3.1 | 0.97 | 0.11 | 0.76 | 0.03 | 0.07 | 1.08 | 122 | 112 | 50 | 0.72 | 0.92 |
| 348 | 3.0 | 0.97 | 0.11 | 0.76 | 0.03 | 0.07 | 1.38 | 125 | 115 | 50 | 0.72 | 0.97 |
| 349 | 3.0 | 0.97 | 0.11 | 0.76 | 0.03 | 0.07 | 1.69 | 130 | 120 | 51 | 0.72 | 1.02 |
| 348 | 2.9 | 0.97 | 0.11 | 0.76 | 0.03 | 0.07 | 1.99 | 128 | 118 | 51 | 0.72 | 1.01 |
| 351 | 2.9 | 0.97 | 0.11 | 0.76 | 0.03 | 0.07 | 2.30 | 123 | 113 | 52 | 0.72 | 0.97 |
| 348 | 2.8 | 0.97 | 0.11 | 0.76 | 0.03 | 0.07 | 2.60 | 117 | 108 | 53 | 0.72 | 0.92 |
| 347 | 2.8 | 0.97 | 0.11 | 0.76 | 0.03 | 0.07 | 2.91 | 125 | 116 | 53 | 0.72 | 0.99 |
| 344 | 2.8 | 0.97 | 0.11 | 0.76 | 0.03 | 0.07 | 3.21 | 108 | 98 | 54 | 0.72 | 0.84 |
| 342 | 2.8 | 0.97 | 0.11 | 0.76 | 0.03 | 0.07 | 3.52 | 103 | 94 | 54 | 0.72 | 0.81 |
| 340 | 2.7 | 0.97 | 0.11 | 0.76 | 0.03 | 0.07 | 3.82 | 118 | 109 | 55 | 0.72 | 0.93 |
| 374 | 3.2 | 0.97 | 0.10 | 0.76 | 0.04 | 0.08 | 0.54 | 116 | 104 | 38 | 0.72 | 0.95 |
| 373 | 3.1 | 0.97 | 0.10 | 0.76 | 0.04 | 0.08 | 0.77 | 114 | 102 | 38 | 0.72 | 0.98 |
| 371 | 3.1 | 0.97 | 0.10 | 0.76 | 0.04 | 0.08 | 1.08 | 110 | 99 | 39 | 0.72 | 0.99 |
| 367 | 3.0 | 0.97 | 0.10 | 0.76 | 0.04 | 0.08 | 1.38 | 111 | 100 | 39 | 0.72 | 1.02 |
| 365 | 3.0 | 0.97 | 0.10 | 0.76 | 0.04 | 0.08 | 1.69 | 120 | 109 | 40 | 0.72 | 1.12 |
| 360 | 3.0 | 0.97 | 0.10 | 0.76 | 0.04 | 0.08 | 1.99 | 102 | 91 | 40 | 0.72 | 0.95 |
| 360 | 2.9 | 0.97 | 0.10 | 0.76 | 0.04 | 0.08 | 2.30 | 122 | 111 | 41 | 0.72 | 1.15 |

**TABLE A.5** Heat Transfer in a Stabilized Flow of Hot Air

| $T_w$ K | $\frac{T_f}{T_w}$ | $\frac{x}{m}$ | $Nu_{f,\,exp}$ | $\frac{Re_f}{10^3}$ | $Pr_f$ | $\frac{Nu_{f,\,exp}}{Nu_f}$ | $T_w$ K | $\frac{T_f}{T_w}$ | $\frac{x}{m}$ | $Nu_{f,\,exp}$ | $\frac{Re_f}{10^3}$ | $Pr_f$ | $\frac{Nu_{f,\,exp}}{Nu_f}$ |
|---|---|---|---|---|---|---|---|---|---|---|---|---|---|
| 296 | 1.3 | 3.52 | 235 | 108 | 0.69 | 1.14 | 296 | 1.2 | 2.60 | 240 | 133 | 0.69 | 0.96 |
| 296 | 1.3 | 0.39 | 264 | 117 | 0.69 | 0.88 | 296 | 1.2 | 3.21 | 204 | 134 | 0.69 | 0.82 |
| 296 | 1.3 | 0.77 | 237 | 118 | 0.69 | 0.91 | 296 | 1.2 | 3.52 | 298 | 135 | 0.69 | 1.20 |
| 296 | 1.3 | 1.08 | 258 | 118 | 0.69 | 1.03 | 296 | 1.2 | 3.82 | 231 | 135 | 0.69 | 0.94 |
| 296 | 1.3 | 1.38 | 211 | 119 | 0.69 | 0.87 | 296 | 1.8 | 0.39 | 134 | 44 | 0.68 | 0.99 |
| 296 | 1.3 | 1.69 | 282 | 119 | 0.69 | 1.19 | 296 | 1.7 | 0.77 | 101 | 45 | 0.68 | 0.85 |
| 296 | 1.3 | 1.99 | 190 | 120 | 0.69 | 0.81 | 296 | 1.7 | 1.08 | 109 | 45 | 0.68 | 0.96 |
| 296 | 1.3 | 2.30 | 239 | 120 | 0.69 | 1.03 | 296 | 1.7 | 1.38 | 100 | 46 | 0.68 | 0.90 |
| 296 | 1.3 | 2.60 | 219 | 121 | 0.69 | 0.95 | 296 | 1.7 | 1.69 | 122 | 46 | 0.68 | 1.12 |
| 296 | 1.3 | 3.52 | 269 | 122 | 0.69 | 1.18 | 296 | 1.5 | 0.77 | 192 | 74 | 0.68 | 1.08 |
| 296 | 1.3 | 3.82 | 209 | 122 | 0.69 | 0.92 | 296 | 1.5 | 1.08 | 159 | 74 | 0.68 | 0.93 |
| 296 | 1.3 | 0.39 | 277 | 119 | 0.69 | 0.91 | 296 | 1.5 | 1.38 | 141 | 75 | 0.68 | 0.85 |
| 296 | 1.3 | 0.54 | 234 | 120 | 0.69 | 0.82 | 296 | 1.4 | 2.60 | 152 | 76 | 0.69 | 0.96 |
| 296 | 1.3 | 0.77 | 259 | 120 | 0.69 | 0.97 | 296 | 1.4 | 2.91 | 191 | 76 | 0.69 | 1.21 |
| 296 | 1.3 | 1.08 | 268 | 121 | 0.69 | 1.05 | 296 | 1.4 | 3.21 | 129 | 77 | 0.69 | 0.82 |
| 296 | 1.3 | 1.38 | 223 | 121 | 0.69 | 0.90 | 296 | 1.4 | 3.52 | 206 | 77 | 0.69 | 1.31 |
| 296 | 1.3 | 1.99 | 229 | 122 | 0.69 | 0.96 | 296 | 1.4 | 3.82 | 144 | 78 | 0.69 | 0.92 |
| 296 | 1.3 | 2.30 | 230 | 123 | 0.69 | 0.97 | 296 | 1.6 | 0.39 | 136 | 48 | 0.68 | 0.93 |
| 296 | 1.3 | 2.60 | 237 | 123 | 0.69 | 1.01 | 296 | 1.6 | 0.77 | 111 | 49 | 0.68 | 0.87 |
| 296 | 1.3 | 3.21 | 191 | 124 | 0.69 | 0.82 | 296 | 1.6 | 1.08 | 123 | 49 | 0.68 | 1.01 |
| 296 | 1.2 | 3.82 | 218 | 125 | 0.69 | 0.94 | 296 | 1.6 | 1.38 | 105 | 50 | 0.68 | 0.88 |
| 296 | 1.3 | 0.39 | 285 | 130 | 0.69 | 0.87 | 296 | 1.6 | 1.69 | 137 | 50 | 0.68 | 1.17 |
| 296 | 1.3 | 0.54 | 251 | 130 | 0.69 | 0.83 | 296 | 1.6 | 1.99 | 110 | 50 | 0.68 | 0.95 |
| 296 | 1.3 | 0.77 | 288 | 130 | 0.69 | 1.01 | 296 | 1.5 | 2.30 | 107 | 51 | 0.68 | 0.93 |
| 296 | 1.3 | 1.08 | 285 | 131 | 0.69 | 1.05 | 296 | 1.5 | 2.60 | 113 | 51 | 0.68 | 0.99 |
| 296 | 1.3 | 1.38 | 232 | 132 | 0.69 | 0.88 | 296 | 1.5 | 2.91 | 137 | 51 | 0.68 | 1.20 |

Table 5 continued

| $T_w$ K | $\frac{T_f}{T_w}$ | $x$ m | $Nu_{f,\,exp}$ | $\frac{Re_f}{10^3}$ | $Pr_f$ | $\frac{Nu_{f,\,exp}}{Nu_f}$ | $T_w$ K | $\frac{T_f}{T_w}$ | $x$ m | $Nu_{f,\,exp}$ | $\frac{Re_f}{10^3}$ | $Pr_f$ | $\frac{Nu_{f,\,exp}}{Nu_f}$ |
|---|---|---|---|---|---|---|---|---|---|---|---|---|---|
| 296 | 1.5 | 3.52 | 137 | 52 | 0.68 | 1.21 | 296 | 1.6 | 1.69 | 154 | 59 | 0.68 | 1.15 |
| 296 | 1.5 | 3.82 | 102 | 52 | 0.68 | 0.90 | 296 | 1.5 | 2.60 | 133 | 60 | 0.68 | 1.02 |
| 296 | 1.4 | 0.77 | 166 | 87 | 0.69 | 0.81 | 296 | 1.5 | 2.91 | 150 | 61 | 0.68 | 1.15 |
| 296 | 1.4 | 1.08 | 174 | 87 | 0.69 | 0.89 | 296 | 1.5 | 9.77 | 154 | 71 | 0.68 | 0.89 |
| 296 | 1.4 | 1.38 | 157 | 87 | 0.69 | 0.83 | 296 | 1.5 | 3.52 | 152 | 62 | 0.68 | 1.16 |
| 296 | 1.4 | 1.69 | 200 | 88 | 0.69 | 1.07 | 296 | 1.5 | 3.82 | 119 | 62 | 0.68 | 0.92 |
| 296 | 1.4 | 2.30 | 168 | 89 | 0.69 | 0.93 | 296 | 1.5 | 0.77 | 154 | 71 | 0.68 | 0.89 |
| 296 | 1.4 | 2.60 | 158 | 89 | 0.69 | 0.88 | 296 | 1.5 | 1.08 | 155 | 72 | 0.68 | 0.94 |
| 296 | 1.4 | 2.91 | 193 | 89 | 0.69 | 1.10 | 296 | 1.5 | 1.38 | 150 | 72 | 0.68 | 0.93 |
| 296 | 1.4 | 3.52 | 197 | 90 | 0.69 | 1.10 | 296 | 1.5 | 1.69 | 180 | 72 | 0.68 | 1.14 |
| 296 | 1.4 | 3.82 | 148 | 91 | 0.69 | 0.83 | 296 | 1.5 | 1.99 | 174 | 73 | 0.68 | 1.11 |
| 296 | 1.4 | 0.77 | 202 | 104 | 0.69 | 0.86 | 296 | 1.5 | 2.60 | 154 | 74 | 0.68 | 1.00 |
| 296 | 1.4 | 1.08 | 225 | 104 | 0.69 | 1.00 | 296 | 1.5 | 2.91 | 175 | 74 | 0.68 | 1.14 |
| 296 | 1.4 | 1.38 | 188 | 105 | 0.69 | 0.86 | 296 | 1.4 | 3.21 | 139 | 76 | 0.67 | 0.90 |
| 296 | 1.3 | 1.69 | 246 | 105 | 0.69 | 1.15 | 296 | 1.4 | 3.52 | 181 | 75 | 0.69 | 1.18 |
| 296 | 1.3 | 1.99 | 217 | 106 | 0.69 | 1.02 | 296 | 1.4 | 3.82 | 147 | 75 | 0.69 | 0.97 |
| 296 | 1.3 | 2.60 | 203 | 106 | 0.69 | 0.97 | 296 | 1.4 | 0.77 | 210 | 95 | 0.69 | 0.95 |
| 296 | 1.3 | 2.91 | 249 | 107 | 0.69 | 1.20 | 296 | 1.4 | 1.08 | 210 | 96 | 0.69 | 1.00 |
| 296 | 1.7 | 1.99 | 112 | 46 | 0.68 | 1.04 | 296 | 1.4 | 1.38 | 201 | 96 | 0.69 | 0.98 |
| 296 | 1.6 | 2.60 | 115 | 42 | 0.68 | 1.08 | 296 | 1.4 | 1.99 | 238 | 97 | 0.69 | 1.20 |
| 296 | 1.6 | 2.91 | 120 | 47 | 0.68 | 1.13 | 296 | 1.4 | 2.60 | 200 | 98 | 0.69 | 1.02 |
| 296 | 1.6 | 2.60 | 115 | 47 | 0.68 | 1.08 | 296 | 1.3 | 3.21 | 171 | 99 | 0.69 | 0.88 |
| 296 | 1.6 | 3.52 | 123 | 48 | 0.68 | 1.16 | 296 | 1.3 | 3.82 | 189 | 100 | 0.69 | 0.98 |
| 296 | 1.6 | 3.82 | 95 | 48 | 0.68 | 0.89 | 296 | 1.3 | 0.77 | 244 | 111 | 0.69 | 0.97 |
| 296 | 1.6 | 0.77 | 127 | 58 | 0.68 | 0.87 | 296 | 1.3 | 1.08 | 265 | 112 | 0.69 | 1.11 |
| 296 | 1.6 | 1.08 | 137 | 58 | 0.68 | 0.98 | 296 | 1.3 | 1.38 | 225 | 113 | 0.69 | 0.97 |

## 4. EXPERIMENTAL DATA ON COMBINED AND RADIATIVE HEAT TRANSFER IN TEST CHANNELS 3 AND 4

**TABLE A.6** Radiative and Combined Heat Transfer in Rectangular Duct 3

| $T_f$ K | $\frac{T_f}{T_w}$ | $T_m$ K | $\varkappa$ | [O] | $\frac{Re_f}{10^3}$ | $Nu_\Sigma$ | $Nu_r$ | $Nu_c^0$ | $m_c \cdot 10$ | $St_r^0 \cdot 10$ | $\chi_{II} \cdot 10$ |
|---|---|---|---|---|---|---|---|---|---|---|---|
| 1582 | 4.35 | 1582 | 1.09 | 0.23 | 65 | 260 | 46 | 231 | −1.00 | 0.65 | 0.23 |
| 1582 | 4.63 | 1485 | 1.09 | 0.23 | 194 | 705 | 116 | 604 | −0.54 | 0.56 | 0.21 |
| 1582 | 4.57 | 1399 | 1.09 | 0.23 | 323 | 1032 | 165 | 895 | −0.29 | 0.50 | 0.18 |
| 1582 | 4.54 | 1325 | 1.09 | 0.23 | 452 | 1242 | 199 | 1156 | −0.11 | 0.44 | 0.16 |
| 1582 | 4.75 | 1248 | 1.09 | 0.23 | 581 | 1647 | 213 | 1409 | −0.02 | 0.38 | 0.12 |
| 1604 | 4.43 | 1604 | 1.06 | 0.23 | 58 | 237 | 47 | 208 | −1.07 | 0.68 | 0.26 |
| 1604 | 4.63 | 1505 | 1.06 | 0.23 | 173 | 635 | 120 | 466 | −0.58 | 0.60 | 0.29 |
| 1604 | 4.64 | 1418 | 1.06 | 0.23 | 288 | 929 | 169 | 805 | −0.34 | 0.53 | 0.22 |
| 1604 | 4.60 | 1343 | 1.06 | 0.23 | 404 | 1100 | 205 | 1040 | −0.33 | 0.47 | 0.18 |
| 1604 | 4.73 | 1262 | 1.06 | 0.23 | 519 | 1539 | 218 | 1268 | −0.06 | 0.40 | 0.15 |
| 1613 | 4.36 | 1613 | 1.07 | 0.23 | 59 | 249 | 47 | 211 | −1.06 | 0.69 | 0.27 |
| 1613 | 4.48 | 1613 | 1.07 | 0.23 | 175 | 656 | 141 | 550 | −0.55 | 0.64 | 0.30 |
| 1613 | 4.81 | 1317 | 1.07 | 0.23 | 408 | 1295 | 187 | 1050 | −0.22 | 0.43 | 0.16 |
| 1646 | 4.34 | 1646 | 1.05 | 0.23 | 65 | 292 | 49 | 231 | −1.00 | 0.72 | 0.26 |
| 1646 | 4.45 | 1537 | 1.05 | 0.23 | 194 | 772 | 121 | 605 | −0.51 | 0.62 | 0.24 |
| 1646 | 4.79 | 1341 | 1.05 | 0.23 | 453 | 1443 | 192 | 1153 | −0.16 | 0.46 | 0.16 |
| 1646 | 4.74 | 1265 | 1.05 | 0.23 | 582 | 1564 | 209 | 1408 | −0.02 | 0.40 | 0.13 |
| 2197 | 4.86 | 2197 | 1.04 | 0.36 | 69 | 320 | 69 | 235 | −1.07 | 1.04 | 0.53 |
| 2197 | 5.02 | 2072 | 1.04 | 0.36 | 205 | 734 | 208 | 617 | −0.59 | 1.23 | 0.75 |
| 2197 | 5.59 | 1967 | 1.04 | 0.36 | 341 | 991 | 301 | 909 | −0.45 | 1.19 | 0.75 |
| 2197 | 5.47 | 1778 | 1.04 | 0.36 | 615 | 1627 | 512 | 1441 | −0.13 | 1.28 | 0.86 |
| 2220 | 5.82 | 2220 | 1.08 | 0.38 | 33 | 167 | 63 | 117 | −1.59 | 1.05 | 0.88 |
| 2220 | 5.98 | 2065 | 1.08 | 0.38 | 99 | 418 | 183 | 309 | −1.06 | 1.25 | 1.17 |
| 1688 | 4.81 | 1498 | 1.01 | 0.23 | 310 | 960 | 179 | 858 | −0.34 | 0.59 | 0.24 |
| 1688 | 4.78 | 1421 | 1.01 | 0.23 | 435 | 1161 | 217 | 1108 | −0.17 | 0.53 | 0.21 |
| 1688 | 4.68 | 1335 | 1.01 | 0.23 | 559 | 1661 | 236 | 1353 | −0.02 | 0.46 | 0.17 |
| 1691 | 4.82 | 1691 | 1.08 | 0.23 | 46 | 174 | 47 | 166 | −1.25 | 0.74 | 0.36 |
| 1691 | 5.11 | 1593 | 1.08 | 0.23 | 137 | 464 | 119 | 435 | −0.76 | 0.66 | 0.33 |

Table 6 continued

| $T_f$ K | $\frac{T_f}{T_w}$ | $T_m$ K | α | [O] | $\frac{Re_f}{10^3}$ | $Nu_\Sigma$ | $Nu_r$ | $Nu_c^0$ | $m_c \cdot 10$ | $St_r^0 \cdot 10$ | $\chi_{II} \cdot 10$ |
|---|---|---|---|---|---|---|---|---|---|---|---|
| 1691 | 5.03 | 1505 | 1.08 | 0.23 | 228 | 683 | 172 | 647 | −0.50 | 0.59 | 0.30 |
| 2220 | 6.43 | 1930 | 1,08 | 0.38 | 165 | 591 | 259 | 457 | −0.89 | 1.19 | 1.11 |
| 2220 | 6.43 | 1801 | 1.08 | 0.38 | 231 | 780 | 369 | 594 | —0.71 | 1.31 | 1.30 |
| 1628 | 5.02 | 1337 | 1.05 | 0.23 | 310 | 1016 | 187 | 813 | −0.34 | 0.45 | 0.20 |
| 1628 | 4.97 | 1265 | 1,05 | 0.23 | 398 | 1020 | 205 | 993 | −0.21 | 0.39 | 0.17 |
| 1658 | 4.66 | 1658 | 1,01 | 0.23 | 43 | 178 | 48 | 156 | −1.26 | 0.74 | 0.38 |
| 1658 | 4.81 | 1560 | 1.01 | 0,23 | 128 | 446 | 120 | 351 | −0.74 | 0.65 | 0.39 |
| 1658 | 5.10 | 1346 | 1.01 | 0.23 | 298 | 951 | 187 | 782 | —0.37 | 0.46 | 0.22 |
| 1658 | 5.04 | 1270 | 1.01 | 0.23 | 383 | 1014 | 204 | 955 | −0.24 | 0.40 | 0.18 |
| 1674 | 4.44 | 1674 | 1.16 | 0.26 | 62 | 280 | 49 | 220 | −1.05 | 0.74 | 0.29 |
| 1674 | 4.56 | 1568 | 1.16 | 0.26 | 184 | 709 | 87 | 576 | −0,55 | 0,46 | 0,13 |
| 1674 | 4.91 | 1474 | 1.16 | 0.26 | 307 | 1003 | 174 | 851 | −0.37 | 0.56 | 0.23 |
| 1674 | 4.91 | 1386 | 1.16 | 0.26 | 430 | 1304 | 206 | 1100 | —0.20 | 0.49 | 0.19 |
| 1674 | 4.86 | 1311 | 1.16 | 0.26 | 553 | 1460 | 227 | 1343 | −0.07 | 0.44 | 0.16 |
| 1688 | 4.58 | 1688 | 1.01 | 0.23 | 62 | 246 | 49 | 221 | −1.06 | 0.76 | 0.29 |
| 1688 | 4.86 | 1588 | 1.01 | 0.23 | 186 | 647 | 124 | 578 | −0.59 | 0.66 | 0.27 |
| 1742 | 4.70 | 1742 | 1.02 | 0.23 | 61 | 243 | 50 | 217 | −1.08 | 0.80 | 0.32 |
| 1742 | 4.80 | 1645 | 1,02 | 0.23 | 183 | 597 | 129 | 568 | —0.59 | 0.72 | 0.31 |
| 1742 | 5.12 | 1561 | 1.02 | 0.23 | 629 | 842 | 193 | 840 | −0.40 | 0.66 | 0.30 |
| 1742 | 5.08 | 1428 | 1.02 | 0.23 | 304 | 1849 | 394 | 1506 | −0.05 | 0.53 | 0.28 |
| 1971 | 5.29 | 1585 | 0,96 | 0.40 | 260 | 951 | 243 | 679 | −0.47 | 0.83 | 0.54 |
| 1971 | 4.99 | 1474 | 0.96 | 0.40 | 334 | 1087 | 261 | 832 | —0.29 | 0.72 | 0.43 |
| 2126 | 5.27 | 2126 | 1.03 | 0.32 | 61 | 258 | 62 | 209 | —1.19 | 1.03 | 0.53 |
| 2126 | 5.67 | 1877 | 1.03 | 0.32 | 302 | 951 | 256 | 816 | −0.51 | 1.09 | 0.65 |
| 2126 | 5.75 | 1775 | 1.03 | 0.32 | 423 | 1194 | 314 | 1056 | −0.36 | 1.00 | 0.59 |
| 2126 | 5.63 | 1679 | 1.03 | 0.32 | 544 | 1464 | 353 | 1292 | −0.21 | 0.92 | 0.51 |
| 2129 | 4.63 | 1748 | 1.05 | 0.39 | 454 | 1197 | 312 | 1134 | −0.15 | 1.01 | 0.54 |

Table 6 continued

| $T_f$ K | $\frac{T_f}{T_w}$ | $T_m$ K | α | [O] | $\frac{Re_f}{10^3}$ | $Nu_\Sigma$ | $Nu_r$ | $Nu_C^0$ | $m_C \cdot 10$ | $St_r^0 \cdot 10$ | $\chi_{II} \cdot 10$ |
|---|---|---|---|---|---|---|---|---|---|---|---|
| 2129 | 4.32 | 1647 | 1.05 | 0.39 | 584 | 1748 | 351 | 1389 | —0.03 | 0.93 | 0.46 |
| 2147 | 5.31 | 2001 | 1.08 | 0.38 | 155 | 643 | 172 | 474 | −0.76 | 1.16 | 0.75 |
| 2147 | 5.87 | 1869 | 1.08 | 0.38 | 258 | 929 | 240 | 700 | −0.61 | 1.07 | 0.68 |
| 2147 | 5.78 | 1679 | 1.08 | 0.38 | 465 | 1224 | 333 | 1110 | −0.30 | 0.90 | 0.53 |
| 2150 | 5.28 | 2150 | 1.06 | 0.33 | 61 | 250 | 62 | 211 | −1.19 | 1.05 | 0.54 |
| 2150 | 5.26 | 2016 | 1.06 | 0.33 | 184 | 709 | 180 | 558 | —0.68 | 1.18 | 0.70 |
| 2150 | 5.67 | 1906 | 1.06 | 0.33 | 306 | 932 | 260 | 760 | −0.51 | 1.11 | 0.71 |
| 2150 | 5.81 | 1807 | 1.06 | 0.33 | 429 | 1162 | 320 | 1068 | −0.36 | 1.03 | 0.61 |
| 2150 | 6.08 | 1710 | 1.06 | 0.33 | 551 | 1453 | 356 | 1303 | −0.28 | 0.93 | 0.52 |
| 1619 | 4.57 | 1619 | 1.06 | 0.23 | 45 | 176 | 46 | 166 | −1.21 | 0.69 | 0.33 |
| 1619 | 5.00 | 1318 | 1.06 | 0.23 | 316 | 1040 | 182 | 830 | —0.33 | 0.43 | 0.19 |
| 1619 | 4.94 | 1243 | 1.06 | 0.23 | 407 | 1077 | 197 | 1013 | −0.20 | 0.37 | 0.15 |
| 1626 | 4.43 | 1433 | 0.92 | 0.23 | 460 | 1445 | 177 | 1247 | −0.10 | 0.51 | 0.15 |
| 1626 | 4.42 | 1354 | 0.92 | 0.23 | 645 | 1840 | 211 | 1608 | −0.07 | 0.45 | 0.13 |
| 2245 | 4.92 | 1807 | 1.05 | 0.43 | 393 | 1292 | 323 | 984 | −0.25 | 1.13 | 0.70 |
| 2245 | 4.55 | 1696 | 1.05 | 0.43 | 505 | 1583 | 361 | 1208 | —0.06 | 1.05 | 0.60 |
| 2340 | 5.40 | 2340 | 1.01 | 0.38 | 62 | 274 | 70 | 211 | −1.19 | 0.94 | 0.54 |
| 2340 | 5.32 | 2192 | 1.01 | 0.38 | 186 | 760 | 208 | 559 | −0.67 | 1.14 | 0.76 |
| 2340 | 5.76 | 2066 | 1.01 | 0.38 | 310 | 1030 | 301 | 825 | −0.51 | 1.15 | 0.78 |
| 2340 | 6.33 | 1937 | 1.01 | 0.38 | 435 | 1454 | 360 | 1062 | —0.43 | 1.11 | 0.73 |
| 2340 | 5.75 | 1837 | 1.01 | 0.38 | 559 | 1485 | 408 | 1307 | —0.21 | 1.07 | 0.66 |
| 2382 | 5.87 | 2382 | 0.95 | 0.60 | 18 | 119 | 69 | 68 | −1.92 | 1.07 | 1.35 |
| 2382 | 6.00 | 2009 | 0.95 | 0.60 | 91 | 427 | 251 | 264 | −1.06 | 1.49 | 1.87 |
| 2382 | 6.50 | 1646 | 0.95 | 0.60 | 165 | 651 | 276 | 416 | —0.83 | 1.05 | 1.08 |
| 2421 | 5.72 | 2421 | 0.97 | 0.37 | 31 | 151 | 63 | 112 | −1.56 | 0.92 | 0.79 |
| 2490 | 5.94 | 1721 | 0.86 | 0.64 | 225 | 756 | 273 | 573 | —0.54 | 1.20 | 0.99 |
| 2533 | 5.78 | 2215 | 0.91 | 0.68 | 112 | 457 | 271 | 322 | −0.88 | 1.57 | 1.83 |

**TABLE A.7** Radiative and Combined Heat Transfer in Circular Channel 4

| $T_f$ K | $\frac{T_f}{T_w}$ | $G$ kg/hr | $p_{CO_2}$ atm | $p_{H_2O}$ atm | $x$ m | $\frac{Re_f}{10^3}$ | $Nu_\Sigma$ | $Nu_r$ | $Nu_C^0$ | $St_r^0 \cdot 10^2$ | $\chi_{II} \cdot 10^2$ |
|---|---|---|---|---|---|---|---|---|---|---|---|
| 1487 | 4.42 | 511 | 0.09 | 0.17 | 3.52 | 23 | 78 | 29 | 58 | 4.80 | 4.45 |
| 1446 | 4.32 | 511 | 0.09 | 0.17 | 3.82 | 23 | 77 | 27 | 58 | 4.43 | 3.90 |
| 1567 | 4.45 | 473 | 0.05 | 0.11 | 0.54 | 19 | 81 | 36 | 64 | 4.79 | 4.84 |
| 1535 | 4.34 | 473 | 0.05 | 0.11 | 0.77 | 19 | 100 | 33 | 60 | 4.30 | 4.27 |
| 1488 | 4.23 | 473 | 0.05 | 0.11 | 1.08 | 19 | 98 | 29 | 57 | 3.88 | 3.66 |
| 1444 | 3.96 | 473 | 0.05 | 0.11 | 1.38 | 19 | 83 | 27 | 55 | 3.55 | 3.20 |
| 1403 | 4.14 | 473 | 0.05 | 0.11 | 1.69 | 22 | 103 | 29 | 59 | 3.25 | 2.97 |
| 1363 | 4.04 | 473 | 0.05 | 0.11 | 1.99 | 22 | 93 | 27 | 59 | 2.99 | 2.60 |
| 1326 | 3.95 | 473 | 0.05 | 0.11 | 2.30 | 22 | 92 | 25 | 59 | 2.77 | 2.29 |
| 1291 | 3.78 | 473 | 0.05 | 0.11 | 2.60 | 23 | 88 | 24 | 59 | 2.55 | 2.01 |
| 1257 | 3.82 | 473 | 0.05 | 0.11 | 2.91 | 23 | 89 | 22 | 60 | 2.37 | 1.77 |
| 1225 | 3.73 | 473 | 0.05 | 0.11 | 3.21 | 24 | 78 | 21 | 60 | 2.20 | 1.55 |
| 1194 | 3.61 | 473 | 0.05 | 0.11 | 3.52 | 24 | 82 | 20 | 61 | 2.05 | 1.37 |
| 1165 | 3.47 | 473 | 0.05 | 0.11 | 3.82 | 24 | 78 | 19 | 61 | 1.90 | 1.21 |
| 1550 | 4.59 | 614 | 0.07 | 0.13 | 1.69 | 26 | 118 | 32 | 69 | 4.31 | 3.77 |
| 1508 | 4.40 | 614 | 0.07 | 0.13 | 1.99 | 27 | 107 | 30 | 69 | 3.98 | 3.32 |
| 1469 | 4.31 | 614 | 0.07 | 0.13 | 2.30 | 27 | 107 | 28 | 69 | 3.69 | 2.93 |
| 1431 | 4.08 | 614 | 0.07 | 0.13 | 2.60 | 28 | 102 | 26 | 69 | 3.41 | 2.59 |
| 1395 | 4.21 | 614 | 0.07 | 0.13 | 2.91 | 28 | 103 | 25 | 70 | 3.18 | 2.28 |
| 1360 | 4.07 | 614 | 0.07 | 0.13 | 3.21 | 29 | 91 | 23 | 70 | 2.96 | 2.02 |
| 1328 | 3.96 | 614 | 0.07 | 0.13 | 3.52 | 29 | 92 | 22 | 71 | 2.76 | 1.79 |
| 1297 | 3.87 | 614 | 0.07 | 0.13 | 3.82 | 30 | 92 | 21 | 71 | 2.58 | 1.59 |
| 1400 | 4.06 | 1050 | 0.05 | 0.10 | 1.99 | 48 | 144 | 22 | 110 | 2.44 | 1.06 |
| 1374 | 3.99 | 1050 | 0.05 | 0.10 | 2.30 | 48 | 148 | 21 | 109 | 2.31 | 0.96 |
| 1347 | 3.83 | 1050 | 0.05 | 0.10 | 2.60 | 49 | 142 | 20 | 110 | 2.18 | 0.87 |
| 1322 | 3.92 | 1050 | 0.05 | 0.10 | 2.91 | 50 | 142 | 19 | 110 | 2.07 | 0.79 |
| 1298 | 3.82 | 1050 | 0.05 | 0.10 | 3.21 | 50 | 125 | 18 | 110 | 1.96 | 0.72 |
| 1275 | 3.72 | 1050 | 0.05 | 0.10 | 3.52 | 51 | 136 | 17 | 110 | 1.86 | 0.65 |
| 1253 | 3.66 | 1050 | 0.05 | 0.10 | 3.82 | 52 | 126 | 17 | 111 | 1.77 | 0.59 |
| 1538 | 4.11 | 915 | 0.06 | 0.12 | 1.38 | 37 | 134 | 35 | 92 | 4.93 | 3.70 |
| 1503 | 4.38 | 915 | 0.06 | 0.12 | 1.69 | 40 | 152 | 34 | 96 | 4.27 | 3.06 |
| 1468 | 4.24 | 915 | 0.06 | 0.12 | 1.99 | 41 | 122 | 31 | 96 | 3.74 | 2.47 |
| 1436 | 4.16 | 915 | 0.06 | 0.12 | 2.30 | 41 | 122 | 31 | 96 | 3.74 | 2.47 |
| 1408 | 3.97 | 915 | 0.06 | 0.12 | 2.60 | 42 | 132 | 29 | 96 | 3.53 | 2.24 |
| 1378 | 4.09 | 915 | 0.06 | 0.12 | 2.91 | 42 | 132 | 28 | 96 | 3.33 | 2.01 |
| 1983 | 4.69 | 627 | 0.09 | 0.18 | 0.54 | 23 | 88 | 61 | 73 | 17.34 | 22.30 |
| 1934 | 4.69 | 627 | 0.09 | 0.18 | 0.77 | 24 | 88 | 60 | 70 | 16.17 | 21.08 |

Table 7 continued

| $T_f$ K | $\frac{T_f}{T_w}$ | $G$ kg/hr | $p_{CO_2}$ atm | $p_{H_2O}$ atm | $x$ $m$ | $\frac{Re_f}{10^3}$ | $Nu_\Sigma$ | $Nu_r$ | $Nu^0_c$ | $St^0_r \cdot 10^2$ | $\chi_{II} \cdot 10^2$ |
|---|---|---|---|---|---|---|---|---|---|---|---|
| 1872 | 4.65 | 627 | 0.09 | 0.18 | 1.08 | 24 | 91 | 49 | 67 | 12.59 | 14.96 |
| 1811 | 4.61 | 627 | 0.09 | 0.18 | 1.38 | 25 | 94 | 47 | 66 | 11.40 | 13.32 |
| 1753 | 4.56 | 627 | 0.09 | 0.18 | 1.69 | 25 | 94 | 46 | 66 | 10.38 | 11.89 |
| 1699 | 4.50 | 627 | 0.09 | 0.18 | 1.99 | 26 | 89 | 42 | 66 | 9.14 | 10.02 |
| 1647 | 4.42 | 627 | 0.09 | 0.18 | 2.30 | 26 | 97 | 40 | 66 | 8.07 | 8.45 |
| 1598 | 4.33 | 627 | 0.09 | 0.18 | 2.60 | 27 | 92 | 38 | 67 | 7.37 | 7.51 |
| 1551 | 4.24 | 627 | 0.09 | 0.18 | 2.91 | 27 | 97 | 36 | 68 | 6.46 | 6.22 |
| 1507 | 4.12 | 627 | 0.09 | 0.18 | 3.21 | 28 | 92 | 35 | 68 | 5.92 | 5.54 |
| 1466 | 4.00 | 627 | 0.09 | 0.18 | 3.52 | 28 | 93 | 33 | 69 | 5.36 | 4.84 |
| 1427 | 3.93 | 627 | 0.09 | 0.18 | 3.82 | 29 | 92 | 33 | 70 | 5.01 | 4.44 |
| 2010 | 4.68 | 733 | 0.09 | 0.18 | 0.54 | 27 | 98 | 48 | 83 | 13.75 | 14.09 |
| 1963 | 4.59 | 733 | 0.09 | 0.18 | 0.77 | 27 | 95 | 48 | 78 | 13.25 | 14.12 |
| 1905 | 4.52 | 733 | 0.09 | 0.18 | 1.08 | 28 | 95 | 45 | 76 | 11.69 | 12.16 |
| 1848 | 4.44 | 733 | 0.09 | 0.18 | 1.38 | 28 | 103 | 43 | 75 | 10.66 | 10.93 |
| 1792 | 4.36 | 733 | 0.09 | 0.18 | 1.69 | 29 | 101 | 41 | 74 | 9.62 | 9.60 |
| 1741 | 4.26 | 733 | 0.09 | 0.18 | 1.99 | 30 | 99 | 39 | 74 | 8.55 | 8.18 |
| 1691 | 4.19 | 733 | 0.09 | 0.18 | 2.30 | 30 | 107 | 37 | 74 | 7.68 | 7.06 |
| 1642 | 4.12 | 733 | 0.09 | 0.18 | 2.60 | 31 | 102 | 35 | 75 | 6.99 | 6.22 |
| 1597 | 4.06 | 733 | 0.09 | 0.18 | 2.91 | 31 | 105 | 33 | 75 | 6.29 | 5.35 |
| 1554 | 4.00 | 733 | 0.09 | 0.18 | 3.21 | 32 | 102 | 32 | 76 | 5.77 | 4.76 |
| 1514 | 3.95 | 733 | 0.09 | 0.18 | 3.52 | 33 | 100 | 31 | 77 | 5.25 | 4.15 |
| 1476 | 3.91 | 733 | 0.09 | 0.18 | 3.82 | 33 | 102 | 31 | 77 | 5.11 | 4.07 |
| 2030 | 4.59 | 839 | 0.09 | 0.18 | 0.54 | 30 | 107 | 49 | 92 | 14.17 | 13.70 |
| 1986 | 4.53 | 839 | 0.09 | 0.18 | 0.77 | 31 | 101 | 48 | 87 | 13.32 | 13.17 |
| 1930 | 4.44 | 839 | 0.09 | 0.18 | 1.08 | 32 | 104 | 47 | 84 | 12.48 | 12.53 |
| 1875 | 4.37 | 839 | 0.09 | 0.18 | 1.38 | 32 | 111 | 43 | 83 | 10.71 | 10.16 |
| 1822 | 4.29 | 839 | 0.09 | 0.18 | 1.69 | 33 | 107 | 43 | 82 | 10.24 | 9.81 |
| 1773 | 4.23 | 839 | 0.09 | 0.18 | 1.99 | 33 | 105 | 41 | 82 | 9.38 | 8.77 |
| 1725 | 4.17 | 839 | 0.09 | 0.18 | 2.30 | 34 | 112 | 40 | 82 | 8.71 | 7.98 |
| 1679 | 4.11 | 839 | 0.09 | 0.18 | 2.60 | 35 | 110 | 40 | 82 | 8.21 | 7.46 |
| 1634 | 4.05 | 839 | 0.09 | 0.18 | 2.91 | 35 | 114 | 36 | 83 | 6.95 | 5.78 |
| 1592 | 3.97 | 839 | 0.09 | 0.18 | 3.21 | 36 | 112 | 34 | 84 | 6.24 | 4.94 |
| 1552 | 3.91 | 839 | 0.09 | 0.18 | 3.52 | 37 | 108 | 32 | 84 | 5.71 | 4.36 |
| 1514 | 3.84 | 839 | 0.09 | 0.18 | 3.82 | 37 | 113 | 28 | 85 | 4.79 | 3.28 |
| 1659 | 4.64 | 570 | 0.06 | 0.12 | 0.54 | 22 | 94 | 43 | 72 | 6.30 | 6.61 |
| 1626 | 4.53 | 570 | 0.06 | 0.12 | 0.77 | 22 | 107 | 39 | 67 | 5.69 | 5.88 |
| 1579 | 4.43 | 570 | 0.06 | 0.12 | 1.08 | 22 | 105 | 35 | 64 | 5.18 | 5.11 |

Table 7 continued

| $T_f$ K | $\frac{T_f}{T_w}$ | $G$ kg/hr | $p_{CO_2}$ atm | $p_{H_2O}$ atm | $x$ m | $\frac{Re_f}{10^3}$ | $Nu_\Sigma$ | $Nu_r$ | $Nu_c^0$ | $St_r^0 \cdot 10^2$ | $\chi_{II} \cdot 10^2$ |
|---|---|---|---|---|---|---|---|---|---|---|---|
| 1533 | 4.15 | 570 | 0.06 | 0.12 | 1.38 | 22 | 103 | 32 | 62 | 4.74 | 4.48 |
| 1489 | 4.46 | 570 | 0.06 | 0.12 | 1.69 | 25 | 111 | 35 | 66 | 4.40 | 4.23 |
| 1449 | 4.29 | 570 | 0.06 | 0.12 | 1.99 | 26 | 103 | 33 | 66 | 4.06 | 3.74 |
| 1410 | 4.18 | 570 | 0.06 | 0.12 | 2.30 | 26 | 104 | 31 | 66 | 3.76 | 3.30 |
| 1373 | 3.99 | 570 | 0.06 | 0.12 | 2.60 | 26 | 100 | 29 | 67 | 3.48 | 2.91 |
| 1337 | 3.93 | 570 | 0.06 | 0.12 | 2.91 | 27 | 101 | 27 | 67 | 3.22 | 2.56 |
| 1304 | 3.95 | 560 | 0.06 | 0.12 | 3.21 | 27 | 88 | 26 | 67 | 3.00 | 2.27 |
| 1272 | 3.83 | 570 | 0.06 | 0.12 | 3.52 | 28 | 93 | 24 | 68 | 2.79 | 2.01 |
| 1242 | 3.75 | 570 | 0.06 | 0.12 | 3.82 | 28 | 84 | 23 | 69 | 2.61 | 1.78 |
| 1587 | 4.41 | 720 | 0.06 | 0.11 | 0.54 | 29 | 116 | 36 | 89 | 4.78 | 3.80 |
| 1558 | 4.29 | 720 | 0.06 | 0.11 | 0.77 | 29 | 125 | 35 | 84 | 4.53 | 3.68 |
| 1520 | 4.11 | 720 | 0.06 | 0.11 | 1.08 | 29 | 123 | 32 | 80 | 4.21 | 3.31 |
| 1481 | 3.99 | 720 | 0.06 | 0.11 | 1.38 | 29 | 119 | 29 | 77 | 3.90 | 2.94 |
| 1444 | 4.29 | 720 | 0.06 | 0.11 | 1.69 | 32 | 132 | 31 | 81 | 3.66 | 2.81 |
| 1409 | 4.15 | 720 | 0.06 | 0.11 | 1.99 | 33 | 117 | 30 | 81 | 3.41 | 2.51 |
| 1377 | 4.05 | 720 | 0.06 | 0.11 | 2.30 | 33 | 118 | 28 | 81 | 3.19 | 2.25 |
| 1347 | 3.89 | 720 | 0.06 | 0.11 | 2.60 | 34 | 104 | 27 | 81 | 2.99 | 2.03 |
| 1318 | 3.98 | 720 | 0.06 | 0.11 | 2.91 | 34 | 116 | 25 | 81 | 2.82 | 1.81 |
| 1289 | 3.90 | 720 | 0.06 | 0.11 | 3.21 | 35 | 102 | 24 | 82 | 2.64 | 1.62 |
| 1262 | 3.77 | 720 | 0.06 | 0.11 | 3.52 | 35 | 106 | 23 | 82 | 2.48 | 1.46 |
| 1236 | 3.71 | 720 | 0.06 | 0.11 | 3.82 | 36 | 102 | 22 | 83 | 2.33 | 1.31 |
| 2035 | 5.42 | 511 | 0.09 | 0.17 | 0.54 | 18 | 90 | 57 | 59 | 12.59 | 17.56 |
| 1986 | 5.36 | 511 | 0.09 | 0.17 | 0.77 | 18 | 93 | 52 | 55 | 11.19 | 15.26 |
| 1920 | 5.29 | 511 | 0.09 | 0.17 | 1.08 | 18 | 91 | 45 | 53 | 10.04 | 13.08 |
| 1854 | 4.98 | 511 | 0.09 | 0.17 | 1.38 | 18 | 91 | 40 | 51 | 9.06 | 11.26 |
| 1790 | 5.27 | 511 | 0.09 | 0.17 | 1.69 | 20 | 99 | 44 | 55 | 8.17 | 10.25 |
| 1735 | 5.09 | 511 | 0.09 | 0.17 | 1.99 | 21 | 86 | 41 | 55 | 7.46 | 8.95 |
| 1680 | 4.89 | 511 | 0.09 | 0.17 | 2.30 | 21 | 100 | 38 | 55 | 6.80 | 7.78 |
| 1626 | 4.65 | 511 | 0.09 | 0.17 | 2.60 | 22 | 88 | 35 | 56 | 6.18 | 6.72 |
| 1576 | 4.74 | 511 | 0.09 | 0.17 | 2.91 | 22 | 88 | 33 | 56 | 5.66 | 5.82 |
| 1530 | 4.63 | 511 | 0.09 | 0.17 | 3.21 | 22 | 77 | 31 | 57 | 5.20 | 5.07 |
| 1350 | 3.98 | 915 | 0.06 | 0.12 | 3.21 | 43 | 119 | 26 | 97 | 3.14 | 1.82 |
| 1325 | 4.27 | 915 | 0.06 | 0.12 | 3.52 | 44 | 114 | 25 | 97 | 2.98 | 1.64 |
| 1298 | 4.71 | 915 | 0.06 | 0.12 | 3.82 | 44 | 128 | 23 | 97 | 2.83 | 1.47 |

# REFERENCES


1. Penner, S. S. 1959. *Quantitative Molecular Spectroscopy and Gas Emmissivities*. Reading, Mass.: Addison-Wesley. [Russ. transl. 1963].
2. Blokh, A. G. 1967. *Teplovoye Izlucheniye v Kotel'nykh Ustanovkakh* (*Thermal Radiation in Furnaces*). Leningrad: Energiya.
3. Nevskiy, A. S. 1971. *Luchistyy Teploobmen v Pechakh i Topkakh* (*Radiant Heat Transfer in Ovens and Furnaces*). Moscow: Metallurgiya.
4. Siegel, R., and Howell, J. R. 1972. *Thermal Radiation Heat Transfer*. New York: McGraw-Hill. [Russ. transl. 1975].
5. Kamenshchikov, V. A., et al. 1972. *Radiatsionnyye Svoystva Gazov pri Vysokikh Temperaturakh* (*High-Temperature Radiative Properties of Gases*). Moscow: Mashinostroyeniye.
6. Goody, R. M. 1964. *Atmospheric Radiation, Theoretical Basis, 1*. Oxford: Clarendon. [Russ. transl. 1966].
7. Hottel, H. C. 1954. Radiant-heat transmission. In *Heat Transmission*, ed. W. H. McAdams, pp. 55–125. New York: McGraw Hill. [Russ. transl. 1961].
8. Schack, A. 1929. *Industrielle Wärmeübergang* (*Industrial Heat Transfer*). Dusseldorf.
9. Tien, C. L. 1968. Thermal radiation properties of gases. In *Advances in Heat Transfer*, vol. 5, pp. 253–324. Academic.
10. Ludwig, C. B., Malkmus, W., Rearden, J. E., and Thomson, J. A. L. 1973. *Handbook of Infrared Radiation from Combustion Gases*, p. 486. NASA SP-3080. Washington, D.C.
11. Zuyev, V. E. 1970. *Rasprostraneniye Vidimykh i Infrakrasnykh Voln v Atmosfere* (*Propagation of Visible and Infrared Waves in the Atmosphere*). Moscow: Sovetskoye Radio Press.
12. Bond, J. W., Jr., Watson, K. M., and Welch, J. A., Jr. 1965. *Atomic Theory of Gas Physics*. Reading, Mass.: Addison-Wesley. [Russ. transl. 1968].
13. Golovnev, I. F., Sevast'yanenko, V. G., and Soloukhin, R. I. 1979. Mathematical modelling the optical properties of gaseous carbon dioxide. *Inzh. Fiz. Zhurn.* 36(2):197–203.

14. Sampson, D. H. 1965. *Radiative Contributions to Energy and Momentum Transport in a Gas*. New York: Wiley. [Russ. transl. 1969].
15. Hirschfelder, J. O., Curtiss, C. F., and Bird, R. B. 1954. *Molecular Theory of Gases and Liquids*. New York: Wiley. [Russ. transl. 1961].
16. Yamamoto, G., Tonaka, M., and Aoki, T. 1969. Estimation of rotational line widths of carbon dioxide bands. *J. Quant. Spectrosc. Radiat. Transfer* 9(3): 371.
17. Popov, Yu. A., and Shvartsblat, R. L. 1973. Radiative properties of water vapor and gaseous carbon dioxide. *Teplofiz. Vys. Temp.* 11(4):741–749.
18. Golden, S. A. 1969. The Voigt analog of an Elsasser band. *J. Quant. Spectrosc. Radiat. Transfer* 9(8):1067–1081.
19. King, J. S. F. Band absorption model for arbitrary line variance. *J. Quant. Spectrosc. Radiat. Transfer* 4(5):705–711.
20. Detkov, S. P. 1972. On the calculation of absorption in the vibrational-rotational band. *Prikl. Mat. Tekhn. Fiz.* 2:15–22.
21. Edwards, D. K. and Menard, W.A. 1964. Comparison of models for correlation of total band absorption. *Appl. Optics*. 3:621.
22. Tien, C. L., and Lowder, J. E. 1966. A correlation for the total band absorptance of radiating gases. *Int. J. Heat Mass Transfer* 9:698.
23. Detkov, S. P., Tokmakov, V. N., and Beregovoy, A. N. 1975. Emissivities of gaseous components in the combustion of organic fuel and their utilization in calculation of radiant fluxes. In *Teplotekhnicheskiye Problemy Pryamogo Preobrazovaniya Energii* (*Heat-Engineering Problems of Direct Energy Conversion*) vol. 7, pp. 40–46. Kiev.
24. Lukash, V. P. 1971. Calculation of the emissivity of the products of combustion of hydrocarbon fules ($CO_2$ and $H_2O$) at high temperatures and pressures. *Teplofiz Vys. Temp.* 9(4):708–716.
25. Hsieh, T. C., and Greif, R. 1972. Theoretical determination of the absorption coefficient and the total band absorptance including a specific application to carbon monoxide. *Int. J. Heat Mass Transfer*. 15:1477.
26. Detkov, S. P., and Tokmakov, V. N. 1976. A new wide-band model in the calculation of radiation of gases. VINITI Deposition number 2741–75, abstr. *Zhurn. Prikl. Spektrosk* 1:170–171.
27. Malkmus, W., Thomson. A. 1962. Infrared emissivity of gases for the anharmonic vibrating–rotator model. *J. Quant. Spectrosc. Radiat. Transfer* 2(2):17.
28. Edwards, D. K., Glasen, L. K., Hauser, W. C., and Tuchscher J. S. 1967. Radiation heat transfer in nonisothermal nongray gases. *Int. J. Heat Transfer* 86:219.
29. Edwards, D. K., and Balakrishnan, A. 1973. Thermal radiation by combustion gases. *Int. J. Heat Mass Transfer* 16:25.
30. Balakrishnan, A., and Edwards, D. K. 1974. Established laminar and turbulent channel flow of radiating molecular gases. *Heat Transfer — 1974*. Proceedings of the Fifth International Heat Transfer Conference, Tokyo, vol. 1, pp. 93–97.
31. Wassel, A. T., and Edwards, D. K. 1976. Molecular gas radiation in a laminar or turbulent pipe flow. *J. Heat Transfer* 98(1):101–107.
32. Edwards, D. K., and Balakrishnan, A. 1973. Nongray radiative transfer in a turbulent gas layer. *Int J. Heat. Mass Transfer* 16:1003–1015.
33. Bratis J. C., and Novotny, J. L. 1972. Radiation–convection interaction in real gases. AIAA 7th Thermophysics Conference, San Antonio, Texas, April 10-12, 1972; AIAA paper 72-278.
34. Edwards, D. K., and Balakrishnan, A. 1973. Self-absorption of radiation in turbulent molecular gases. *Combustion and Flame* 20:401–517.
35. Edwards, D. K., and Menard, W. A. 1964. Correlations for absorption by methane and carbon dioxide gases. *Appl. Optics* 3(7):847–852.
36. Edwards, D. K., and Sun W. 1964. Correlations for absorption by the 9.4 and 10.4 micron $CO_2$ bands. *Appl. Optics* 3:1501–1502.

37. Abu-Romia, M. M., and Tien, C. L. 1966. Measurements and correlations of infrared radiation of carbon monoxide at elevated temperatures. *J. Quant. Spectrosc. Radiat. Transfer* 6(2):143–167.
38. Edwards, D. K. 1965. Absorption of radiation by carbon monoxide gas according to the exponential wide-band model. *Appl. Optics* 4(10):1352–1353.
39. Chan, S. H., and Tien, C. L. 1971. Infrared radiation properties of sulphur dioxide. *J. Heat Transfer* 93(2):172–177.
40. Balakrishnan, A., and Edwards, D. K. 1971. Discussion of paper by S. H. Chan and C. L. Tien "Infrared radiation properties of sulphur dioxide." *J. Heat Transfer* 93(2):177–178.
41. Leckner, B. 1972. Spectral and total emissivity of water vapor and carbon dioxide. *Combustion and Flame* 19(1): 33–48.
42. Leckner. B. 1971. The spectral and total emissivity of carbon dioxide. *Combustion and Flame* 17(7):37–44.
43. Detkov, S. P., and Volkov, V. V. 1976. Allowance for the structure of the 4.3 micron band of $CO_2$ in heat-engineering calculations. *Teplofiz. Vys. Temp.* 14(6):1254–1260.
44. Tamonis, M. M., Sinkyvičius, E. E., and Tutljte, O. L. 1978. Radiant energy transport in combustion products of hydrocarbon fuel. I. Allowance for the rotational structure of bands in isothermal gas mixtures. *Trudy Akad. Nauk LitSSR, Ser. B,* 5(108):69–78.
45. Edwards, D. K., Flornes, B. J., Glassen, L. K., and Sun, W. 1965. Correlation of absorption by water vapor at temperatures from 3000 to 1100°K. *Appl. Optics* 4:715–721.
46. Timofeyev, V. N., and Karasina, E. S. 1948. Nomograms for determining the emissivities of carbon dioxide and water vapor. *Izv. VTI [Dzherzhinskiy Heat Engineering Institute]*, numbers 9–10.
47. *Normy Teplovogo Rascheta Kotel'nogo Agregata VIT (Standards of Thermal Calculations of the Dzherzhinskiy Heat Engineering Institute Boiler)*. 1952. Moscow: Gosenergoizdat.
48. Neviskiy, A. S. 1963. *Teploperedacha v Maternovskikh Pechakh (Heat Transfer in Open Hearth Furnaces)*. Moscow: Metallurgizdat.
49. Detkov, S. P. 1969. On the calculation of radiation of carbon dioxide and water vapor at small optical thicknesses. *Izv. Vuz. Chernaya Metallurgiya* 8:153–157.
50. Detkov, S. P., and Vinogradov, A. V. 1969. Emissivity of water vapor. *Teploenergetika* 11:75–76.
51. Penner, S. S., and Varanasi, P. 1965. Approximate band absorption and total emissivity calculations for $H_2O$. *J. Quant. Spectrosc. Radiat. Transfer* 5(2):391–401.
52. Ferriso, C. C., Ludwig, C. B., and Boynton, F. P. 1966. Total emissivity of hot water vapour. I. High pressure limit. *Int. J. Heat Mass Transfer* 9:853–864.
53. Boynton, F. P., and Ludwig, C. B. 1971. Total emissivity of hot water vapour. II. Semi-empirical charts deduced from long-path spectral data. *Int. J. Heat Mass Transfer* 14:963–973.
54. Khakimov, E. A., and Panfilovich, K. B. 1968. Experimental study of the radiation of high-temperature and -pressure steam. *Trudy KKhTI [Kazan' Institute of Chemical Technology]*, number 43.
55. Panfilovich, K. B., and Usmanov, A. G. 1968. Radiation of gaseous $CO_2$ at elevated pressures. In: *Teplo- i Massoperenos (Heat and Mass Transfer)*, vol. 1, pp. 779–782. Moscow.
56. Akhunov, N. Kh., Khekimor, N. Kh., Vasliyera, M. Yu., Panfilovich, K. B., and Usmanor, A. G. 1976. Investigation of the pressure dependence of radiative properties. In *Teplomassoobmen-V (Heat and Mass Transfer, 5*, Proc. 5th All-Union Conference on Heat and Mass Transfer), vol. 8, pp. 186–190. Minsk. [Engl. transl. *Heat Transfer—Soviet Res.* (9)5: 83–86. 1977].
57. Chukanova, L. A., and Nevskiy, A. S. 1965. Experimental study of the radiation of gaseous $CO_2$ at nonuniform temperatures. *Teplofiz. Vys. Temp.* 3(4):577–586.
58. Detkov, S. P., and Girs, V. N. 1967. On the mechanization of calculations of radiation of gaseous $CO_2$. *Izv. Vuz. Chernaya Metallurgiya* 2:162-167.

59. Edwards, D. K. 1962. Radiation interchange in a nongray enclosure containing an isothermal carbon-dioxide–nitrogen gas mixture. *J. Heat Transfer* 84:(1):1.
60. Penner, S. S., and Varanasi, P. 1964. Approximate band absorption and total emissivity calculations for $CO_2$. *J. Quant. Spectrosc. Radiat. Transfer* 4:799–806.
61. Panfilovich, K. B., Akhunov, N. Kh., and Usmanov, A. G. 1973. Radiation of gaseous $CO_2$ at elevated pressures. In *Teplo- i Massoobmen v Khimicheskoy Tekhnologii* (*Heat and Mass Transfer in Chemical Technology*), vol. 1, pp. 26–31. Kazan'.
62. Akhunov, N. Kh., Panfilovich, K. B., and Usmanov, A. G. 1971. An experimental study of the emissive power of $CO_2$ at elevated pressures. *Teplofiz. Vys. Temp.* 9(4):703-707.
63. Chukanova, L. A., and Nevskiy, A. S. 1967. Experimental study of the radiation of water vapor at nonequilibrium temperatures and a method for calculating the radiation of gases. *Teplofiz. Vys. Temp.* 5(5):827.
64. Detkov, S. P., Beregovoy, A. N., and Tokmakov, V. N. 1976. The emissivities of gaseous $SO_2$ and CO. VINITI Deposition number 297–76, abstr. *Inzh. Fiz. Zhurn.* 30(5):933–934.
65. Detkov, S. P., and Tokmakov, V. N. 1976. Emissivities of CO. *Inzh. Fiz. Zhurn.* 30(4):632–639.
66. Green, R. M., and Thien, C. L. 1970. Infrared radiation properties of nitric oxide at elevated temperatures. *J. Quant. Spectrosc. Radiat. Transfer* 16(7):805–817.
67. Plass, G. N. 1967. Radiation from nonisothermal gases. *Appl. Optics* 6(11):1995–1999.
68. Weiner, M. M., and Edwards, D. K. 1968. Non-isothermal gas radiation in superposed vibration–radiation bands. *J. Quant. Spectrosc. Radiat. Transfer* 8(5):1171–1183.
69. Edwards, D. K., and Morizumi, S. J. 1970. Scaling of vibration–rotation band parameters for nonhomogeneous gas radiations. *J. Quant. Spectrosc. Radiat. Transfer* 10(3):175.
70. Cess, R. D., and Wang, L. S. 1970. A band absorptance formulation for nonisothermal gaseous radiation. *Int. J. Heat Mass Transfer* 13(3):547–555.
71. Chan, S. H., and Tien, C. L. 1971. Infrared radiative heat transfer in nongray nonisothermal gases. *Int. J. Heat Mass Transfer* 14:19–26.
72. Lin, J. C., and Greif, R. 1974. Total band absorptance of nonisothermal radiating gases. In *Heat Transfer — 1974*. Proceedings of the Fifth International Heat Transfer Conference, Tokyo, vol. 1, pp.1–5.
73. Popov, Yu. A. 1974. Application of the statistical model of bands to calculation of radiative properties of a nonhomogeneous and nonisothermal gas. *Teplofiz. Vys. Temp.* 12(4):790–796.
74. Detkov, S. P. 1972. Comparison of elementary approximations in calculations of radiation of a nonisothermal gas. *Teplofiz. Vys. Temp.* 10(3):609–614.
75. Detkov S. P. 1974. Method of approximate calculation of radiant heat transfer between gas and surfaces. In *Heat Transfer in Flames*, eds. N. H. Afgan and J. B. Beer, pp. 219–227. Washington, D.C: Scripta.
76. Khodyko, Yu. V., Vitkin, E. I., and Kabashnikov, V. P. 1979. Methods of calculation of radiation of molecular gases on the basis of modelling of the spectral composition. *Inzh. Fiz. Zhurn.* 36(3):204–217.
77. Tamonis, M. M. 1979. Radiative heat transfer between two parallel walls at different temperatures with allowance for the spectral features of a nonhomogenous layer of combustion products. In *Protsessy Napravlennogo Teploobmena* (*Processes of Directed Heat Transfer*), pp. 86–91. Kiev: Naukova Dumka.
78. Schlichting, H. 1968. *Boundry Layer Theory*. New York: McGraw-Hill. [Russ. transl. from German, 1974].
79. Loytyanskiy, L. G. 1970. *Mekhanika Zhidkosti i Gaza* (*Mechanics of Liquids and Gases*). Moscow: Nauka.
80. Pai, Shih-I. 1966. *Radiation Gas Dynamics*. Berlin: Springer-Verlag. [Russ. transl. 1968].
81. Mellor, G. L., and Herring, H. J. 1973. A survey of the mean turbulent field closure models. *AIAA J.* 11(5):590.

82. Žukauskas, A. A., and Šlančiauskas, A. A. 1973. *Teplootdacha v Turbulentnom Potoke Zhidkosti* (*Heat Transfer in Turbulent Flows*). Vilnius: Mintis. [Engl. transl. Hemisphere Publishing Corp., in press].
83. Maksin, P. L., Petukhov, B. S., and Polyakov, A. F. 1976. Calculation of turbulent [eddy] heat transfer in stabilized pipe flow. In *Teplomassoobmen-V* (*Heat and Mass Transfer, 5,* Proc. 5th All-Union Conference on Heat and Mass Transfer) vol. 1, pt. 1, pp. 14–23. Minsk. [Engl. transl. *Heat Transfer, Soviet Res.* 9(4):1–10. 1977].
84. Cebeci, T. 1973. A model of eddy conductivity and turbulent Prandtl number. *J. Heat Transfer* 95(2):227–234.
85. Adzerikho, K. S. 1975. *Lektsii po Teorii Perenosa Luchistoy Energii* (*Lectures on the Theory of Radiant Energy Transport*). Minsk: Belorussian State University Press.
86. Adrianov, V. N. 1972. *Osnovy Radiatsionnogo i Slozhnogo Teploobmena* (*Principles of Radiative and Combined Heat Transfer*). Moscow: Energiya.
87. Özisik, M. N. 1973. *Radiative Transfer and Interactions with Conduction and Convection.* Interscience. [Russ. transl. 1976].
88. Viskanta, R. 1966. Radiation transfer and interaction of convection with radiation heat transfer. In *Advances in Heat Transfer*, eds. T.F. Irvine, Jr. and J.P. Hartnett, vol. 3, pp. 175–251. New York: Academic.
89. Polyak, G. L. 1976. Phenomenology, statistics and analogy in the theory of molecular and radiative transfer. In *Teploobmen-V*, vol. 8: *Teploobmen Izlucheniyem i Slozhnyy Teploobmen* (*Radiative and Combination Heat Transfer*), pp. 41–50. Minsk.
90. Surinov, Yu. A. 1970. Concerning certain principal problems of the theory of radiant heat transfer. (In *Teploobmen-V, vol. 8: Teploobmen Izlucheniyem i Slozhnyy Teploobmen (Radiative and Combined Heat Transfer*), pp. 70–79. Minsk.
91. Motulevich, V. P. 1962. A system of equations of the laminar boundary layer with allowance for chemical reactions and various forms of diffusion. In *Fizicheskaya Gazodinamika, Teploobmen i Termodinamika Gazov Vysokikh Temperatur* (*Physical Gas Dynamics, Heat Transfer and Thermodynamics of High-Temperature Gases*), pp. 159–170. Moscow: USSR Academy of Sciences.
92. Motulevich, V. P. 1962. Heat and mass transfer in a flow of an incompressible fluid with heterogeneous chemical reactions. In *Fizicheskaya Gazodinamite, Teploobmen i Termodinamika Gazov Vysokikh Temperatur* (*Physical Gas Dynamics, Heat Transfer and Thermodynamics of High-Temperature Gases*), pp. 171–179. Moscow: USSR Academy of Sciences.
93. Sheyndlin, A. Ye., ed. 1974. *Izluchatel'nyye Svoystva Tverdykh Materialov* (*Radiative Properties of Solids*). Moscow: Energiya.
94. Khrustalev, B. A. 1970. Radiative properties of solids. *Inzh. Fiz. Zhurn.* 18(4):740–762. [Engl. transl. *Heat Transfer*, *Soviet Res.* 2(5):149–170. 1970].
95. Mikheyev, M. A. 1956. *Osnovy Teploperedachi* (*Principles of Heat Transfer*). Moscow: Gosenergoizdat.
96. Mikheyev, M. A., and Mikheyeva, I.M. 1973. *Osnovy Teploperedachi.* Moscow: Energiya.
97. Kutateladze, S. S., and Leont'yev, A. I. 1962. *The Turbulent Boundary Layer in Compressible Gases*. Novosibirsk. [Engl. transl. Academic. 1964].
98. Shorin, S. N. 1964. *Teploperedacha* (*Heat Transfer*). Moscow: Vysshaya Shkola.
99. Petukhov, B. S. 1967. *Teploobmen i Soprotivleniye pri Laminarnom Techenii Zhidkosti v Trubakh* (*Heat Transfer and Drag in Laminar Pipe Flows*). Moscow: Energiya.
100. Makarevičius, V. J. 1978. *Teploobmen pri Fiziko-khimicheskikh Izmeneniyakh* (*Heat Transfer Attending Physico-Chemical Transformations*). Vilnius: Mokslas.
101. McDonald, H., and Kreskovsky, J. P. 1974. Effect of free stream turbulence on the turbulent boundary layer. *Int. J. Heat Mass Transfer* 17(8):705–716.
102. Dyban, Ye. P., and Epik, E. Ya. 1976. Calculation of convective heat transfer from statistical characteristics of turbulence. In *Teplomassoobmen-V*, vol. 1, pp. 24–34. Minsk. [Engl. transl. *Heat Transfer*, *Soviet Res.* 9(4):11–19. 1977].

103. Sukomel, A. S., Velichko, V. I., Abrosimov, Yu. G., and Erera, O. 1971. Experimental study of local heat transfer in the inlet region of a pipe in the case of transition flow in the boundary layer at different inlet turbulence levels. *Trudy MEI* [Moscow Energetics Institute] 81:3–10.
104. Pedišius, A. A., Kažimekas, P. -V. A. , and Šlančiauskas, A. A. 1978. Heat transfer from a plate to a high-turbulence air flow. *Trudy Akad. Nauk. LisSSR, Ser. B*, 5(108):91–99. [Engl. transl. *Heat Transfer, Soviet Res.* 11(5):125–134. 1979].
105. Tamonis, M. M., Dagys, L. J., and Žukauskas, A. A. 1975. Analysis of the turbulent boundary layer in a flow with variable physical properties (1. Analytic study.). *Trudy Akad. Nauk. LitSSR, Ser.B* 6(91):105–113. 1975.
106. Tamonis, M. M., Dagys, L. J., and Žukauskas, A. A. 1976. Analysis of the turbulent boundary layer in a flow with variable physical properties (2. Results of numerical calculation for a flow of air with combustion gases.). *Trudy Akad. Nauk. LitSSR, Ser. B.* 5(96):105–113.
107. Žukauskas, A. A., Dagys, L. J., and Tamonis, M. M. 1976. Analysis of turbulence characteristics of the boundary layer on the basis of measured temperatures and velocity fields. *Abstracts of the 2nd Interindustry Conference on Theoretical and Applied Aspects of Turbulent Flows*, pp. 6–7. Tallin.
108. Tamonis, M. M. 1965. The laminar boundary layer on a wedge-shaped surface with a stepwise velocity distribution on the outer edge of the layer. (1. The temperature boundary layer). *Trudy Akad. Nauk. LitSSR, Ser. B* 3(42):169–177.
109. Švenčianas, P. P., Makarevičius, V. J., and Tamonis, M. M. 1969. Effect of physical properties of gases on hydrodynamics and heat transfer in flows over bodies. *Trudy Akad. Nauk. LitSSR, Ser. B* 3(58):169–191.
110. Švenčianas, P. P., Makarevičius, V. J., Tamonis, M. M., and Žukauskas, A.A. 1969. Effect of physical properties of liquids on the hydrodynamics and heat transfer of a plate in longitudinal flow. *Trudy Akad. Nauk. LitSSR, Ser. B* 4(59):149–162.
111. Gydraitis, A. I., Makarevičius, V. J., Šuks, B. I., and Tamonis, M. M. 1971. Effect of physical properties of gases and liquids on mass transfer of a plate in longitudinal flow. *Trudy Akad. Nauk. LitSSR, Ser. B*, 2(65):153–160.
112. Popov, V. N. 1970. Heat transfer and frictional drag in longitudinal turbulent flow of a gas with variable physical properties along a plate. *Teplofiz. Vys. Temp.* 8(2):333–345.
113. Clauser, F. H. 1956. The turbulent boundary layer. In *Advances in Applied Mechanics*, vol. 4. New York: Academic.
114. Jaffe, N. A., and Thomas, J. 1970. Application of quasilinearization and Chebyshev series to the numerical analysis of the laminar boundary-layer equation. *AIAA J.* 8(3):483-490.
115. Korn, G. A., and Korn, T. M. 1967. *Mathematical Handbook for Scientists and Engineers*. New York: McGraw-Hill. [Russ transl. 1968].
116. Nevskiy, A. S. 1972. Heat transfer in a layer of selectively radiating gas. In *Metallurgicheskaya Teplotekhnika* (*Metallurgical Heat Engineering*), number 1, pp. 15–22. Moscow.
117. Howell, J. R. 1968. Application of Monte Carlo to heat transfer problems. In *Advances in Heat Transfer*, vol. 5, pp. 2–54. New York: Academic. [Russ. transl. 1971].
118. Shmiglevskiy, Yu. D. 1973. Calculation of radiant energy transfer by the Galerkin method. *Zhurn. Vychisl. Mat. Matemat. Fiz.* 13(2):388–407.
119. Gianaris, R. J. 1972. Calculations for coupled radiative and collisional effects in a cylindrically confined plasma. In Technical Report no. HTGOL-6, Purdue University, West Lafayette, Ind.
120. Lee, J. B. Wai-Yuen. 1973. Spectral distribution of radiation from a constricted arc plasma. Thesis, Purdue University, West Lafayette, Ind.
121. Sparrow, E. H., and Cess, R. D. 1966. *Radiation Heat Transfer*. Brooks-Cole. [Russ. transl. 1971].

122. Kesten A. S. 1968. Radiant heat flux distribution in a cylindrically symmetric nonisothermal gas with temperature-dependent absorption coefficient. *J. Quant. Spectrosc. Radiat. Transfer* 8, 419–434.
123. Habib, I. S., and Greif, R. 1970. Nongray radiative transport in a cylindrical medium. *J. Heat Transfer* 92(1):28.
124. Edwards, D. K., and Wassel, A. T. 1973. The radial radiative heat flux in a cylinder. *J. Heat Transfer* 95(2):276–277.
125. Detkov, S. P., and Khalevich, O. A. 1979. Mathematical simulation of combined heat transfer in a slotted channel. *Inzh. Fiz. Zhurn.* 26(2):270–278.
126. Abramowitch, M., and Stegun, T. A. 1964. *Handbook of Mathematical Functions*, p. 231. Washington, D.C.: U.S. Government Printing Office.
127. Šidlauskas, V. A., and Tamonis, M. M. 1979. Radiant energy transfer in a layer of hydrogen plasma. 2. Radiation of a planar and cylindrical layer of plasma with specified temperatue distribution. *Trudy Akad. Nauk. LitSSR, Ser. B* 3(112):75–83.
128. Detkov, S. P., Ponmarev, N. N., and Petrak, L. V. 19 . Rationalization of Calculations of Radiant Fluxes in Elementary Systems of Bodies. VINITI Deposition number 784–77.
129. Zachor, A. S. 1968. A General approximation for gaseous absorption. *J. Quant. Spectrosc. Radiat. Transfer* 8(2):771–781.
130. Berezin, I. S., and Zhidkov, N. P. 1959. *Metod Vychisleniy* (*Method of Calculations*), vols. 1 and 2. Moscow: Fizmatgiz.
131. Kuprys, A. J., Sazhima, S. A., Sinkevičius, E. Y., and Tamonis, M. M. 1980. Measurement of the radiative heat flux from combustion products in configured channels. In *Metody i Sredstva Mashinnoy Diagnostiki Gazoturbinnykh Dvigateley i ikh Elementov* (*Methods and Means of Machine Diagnostics of Gas Turbine Engines and their Components*), vol. 2, pp. 259–260. Khar'kov.
132. Giedraitis, A. I., Gimbutis, G. I., Žukauskas, A. A., Makarevičius, V. J., Tamonis, M. M. and Tamulionis, A. P. 1966. Heat transfer in the crossflow of a free jet of a thermally dissociated gas over a cylinder. *Trudy Akad. Nauk. LitSSR, Ser. B* 2(45):135–141.
133. Gimbutis, G. I., Žukauskas, A. A., and Tamulionis, A. P. 1966. Local heat transfer in a pipe with subsonic flow of thermally partially dissociated gas. *Trudy Akad. Nauk. LitSSR, Ser. B* 4(47):119–126.
134. Žukauskas, A. A., Tamulionis, A. P., Makarevičius, V. I., and Tamonis, M. M. 1972. Effect of thermal dissociation of the combustion products of a propane–oxygen mixture on heat transfer and frictional drag in pipe flow. In *Teploobmen v Vysokotemperaturnom Potoke Gaza* (*Heat Transfer in High-Temperature Gas Flow*), pp. 144–154. Vilnius: Mintis.
135. *Teplotekhnicheskiy Spravochnik* (*Heat Engineering Handbook*), vol. 1, 1957. Moscow–Leningrad: Gosenergoizdat.
136. Makarevičius, V. J., Dagys, L. J., Tamonis, M. M., and Žukauskas, A. A. 1972. Investigation of heat transfer from a plate in the flow of dissociating combustion products. In *Teploobmen v Vysokotemperaturnom Potoke Gaza* (*Heat Transfer in High-Temperature Gas Flow*), pp. 154–164. Vilnius: Mintis.
137. Bannikov, A. I., D'yachenko, S. G., and Zavolovich, A. L. 1977. Technique for measuring high local temperatures in supersonic gas flow. In *Abstracts of Reports of the All-Union Scientific Conference on Methods and Means of Machine Diagnostics of the State of Gas-Turbine Engines and their Components*, pp. 276–278. Khar'kov.
138. Mikhaylov, A. I., Gorbunov, G. M., Borisov, V. V., Krasnikov, L. A., and Markov, N. I. 1959. *Rabochiy Protsess i Raschet Kamer Sgoraniya Gazoturbinnykh Dvigateley* (*Working Process and Design of Combustion Chambers of Gas-Turbine Engines*). Moscow: Oborongiz.
139. *Luchistyy Teploobmen.* (*Metody i Pribory Issledovaniya Luchistogo Teploobmena*). 1974. Mezhvuzovskiy Sbornik [*Radiant Heat Transfer.* (*Methods and Instruments for the Study of Radiant Heat Transfer*). Intercollegiate collection]. Kaliningrad.

140. Kremenchugskiy, L. S. 1967. Comparison of thermal radiation sensors and methods for measuring their principal parameters. In *Teplovyye Priyemniki Izlucheniya* (*Thermal Radiation Sensors*), pp. 3–20. Kiev.
141. Gerashchenko, O. A. 1971. *Osnovy Teplometrii* (*Fundamentals of Calorimetry*). Kiev: Naukova Dumka.
142. Gerashchenko, O. A., Kuprys, A. J., Sazhina, S. A., Sinkevičius, J. E., and Tamonis, M. M. 1977. Investigation of thermoelectric sensors for the study of radiative properties of the flow of combustion products in channels. In *Abstracts of Reports of the All-Union Scientific Conference on Methods and Means of Machine Diagnostics of the State of Gas-Turbine Engines and their Components*, pp. 248–250. Khar'kov.
143. Kuprys, A. J., and Sinkeivičius, J. E. 1978. Instruments for investigating the radiative properties of combustion products in power-equipment passages. In *Abstracts of the 4th All-Union Conference on Radiant Heat Transfer*, pp. 47–48. Kiev.
144. Gerashchenko, O. A., Kuprys, A. J., Sazhina, S. A., Sinkeivičius, J. E., and Tamonis, M. M. 1980. Radiant energy transfer in combustion products of hydrocarbon fuels. 2. Instruments for measuring the radiant heat flux from combustion products to the channel walls. *Trudy Akad. Nauk. LitSSR, Ser. B* 6(121):49–57.
145. Blokh, A. G. 1962. *Osnovy Teploobmena Izlucheniyem* (*Principles of Radiant Heat Transfer*). Moscow: Gosenergoizdat.
146. Žukauskas, A. A., and Žiugžda, V. J. 1969. *Teplootdacha v Laminarnom Potoke Zhidkosti* (*Heat Transfer in Laminar Flows*). Vilnius: Minits.
147. Petukhov, B. S., and Vilenskiy, V. P. 1972. Heat transfer in the boundary layer on a flat surface at equilibrium dissociation of the gas. In *Teploobmen v Vysokotemperaturnom Potoke Gaza* (*Heat Transfer in High-Temperature Gas Flow*), pp. 108–121. Vilnius: Mintis.
148. Žukauskas, A. A., Makarevičius, V. J., and Šakmanas, A. T. 1972. Heat transfer cross-flow of dissociated combustion products of a propane-oxygen mixture over a cylinder. In *Teplo- i Massoperenos*, vol. 1, pp. 271–276. Minsk.
149. Ambrazevičius, A. B., and Žukauskas, A. A. 1959. Investigation of heat transfer from a plate in the flow of a dropwise liquid. *Trudy Akad. Nauk. LitSSR, Ser. B* 3(19):111–122.
150. Fay, J. A., and Riddel, F. R. 1958. Theory of stagnation point heat transfer in dissociated air. *J. Aeronaut. Sci.* 25:73–85.
151. Probstein, R. F., Adams, Mac C., and Rose, P. H. 1958. On turbulent heat transfer through a highly cooled partially dissociated boundary layer. *Jet Propulsion* 28(1):56–58.
152. Dagys, L. J., and Tamonis, M. M. 1974. Heat transfer and frictional drag of a plate in laminar flow of an equilibrium gas mixture. *Trudy Akad. Nauk. LitSSR, Ser. B* 1(80):107–115.
153. Sutkaitte, I. B., Makarevičius, V. J., and Tamonis, M. M. 1973. A simplified technique for determining the thermal conductivity and viscosity of high-temperature products of combustion of hydrocarbon fuels. *Trudy Akad. Nauk. LitSSR, Ser. B* 6(79):135–142.
154. Dagys, L. J., Tamonis, M. M., and Makarevičius, V. J. 1974. Investigation of the effect of the lewis number on heat transfer in thermodynamically equilibrium combustion products. In *Mekhanika—V. Materialy Konferentsii Razvitiye Tekhnicheskikh Nauk v Respublike i Ispol'zovaniye Ikh Rezul'tatov* (*Mechanics—5. Proc. Conf. Scientific Advances in the [Lithuanian] Republic and Their Practical Applications*), pp.192–195. Kaunas.
155. Popov, V. N. 1970. Heat transfer and drag in longitudinal turbulent flow of air over a plate. *Teplofiz. Vys. Temp.* 8(5):1035–1042.
156. Van Driest, E.R. 1956. On turbulent flow near a wall. *J. Aeronaut. Sci.* 23(11):1007–1011.
157. Šlančiauskias, A. A., Vaitiekunas, P. A., and Žukauskas, A. A. 1971. A method for calculating friction and heat transfer in a turbulent layer with variable physical properties. *Trudy Akad. Nauk. LitSSR, Ser. B* 4(67):85–100.

158. Vilemas, J. V., Česna, B. A., and Survila, V. J. 1977. *Teplootdacha v Gazookhlazhdayemykh Kol'tsevykh Kanalakh* (*Heat Transfer in Gas-Cooled Annuli*). Vilnius: Mosklas. [Engl. transl. Hemisphere Publishing Corp, in press].
159. Velichko, V. I. 1969. Experimental study of local heat transfer in the inlet length of a circular tube. Author's abstract of candidate thesis. Moscow: Moscow Energetics Institute.
160. Chu Tz'u-hesiang. 1965. Experimental study of heat transfer in turbulent pipe flow of gases at high-temperature differentials. Author's abstract of candidate thesis. Moscow: Moscow Energetics Institute.
161. Ambrazevičius, A., Žukauskas, A., Valatkevičius, P., and Keželis, R. 1974. Plasma heat transfer during turbulent gas flow in the entrance region of a circular tube. AIAA/ASME, Thermophysics and Heat Transfer Conference, Boston, pp. 1–4.
162. Shorin, S. N., and Pechurkin, V. A. 1969. Heat transfer from a high-temperature gas jet to a plane. In *Teplofizicheskiye Svoystva Zhidkostey i Gazov pri Vysokikh Temperaturakh Plazmy* (*Thermophysical Properties of Liquids and Gases at High Plasma Temperatures*), pp. 272–280. Moscow: Standartizdat.
163. Dagys, L. J., Tamonis, M. M., and Žukauskas, A. A. 1970. Analysis of the turbulent boundary layer with variable physical properties of the flow. 4. Determination of characheristcs of turbulence from measurement of the velocity and temperature profiles in the boundry layer. *Trudy Akad. Nauk. LitSSR, Ser. B* 4(101):39–46.
164. Dagys, L. J., Tamonis, M. M., and Žukauskas, A. A. 1976. Analysis of the turbulent boundary layer with variable physical properties of the flow. 3. Experimental study of heat transfer in the flow of combustion products in the inlet region of the channel. *Trudy Akad. Nauk. LitSSR, Ser. B* 5(96):115–123.
165. Petukhov, B. S., and Popov, V. N. 1963. Analytic calculation of heat transfer and frictional drag in turbulent pipes flows of incompressible fluids with variable physical properties. *Teplofiz. Vys. Temp.* 1(1).
166. Vinogradov, A. V., and Detkov, S. P. 1975. The mean emissivities of volumes of different shapes. *Inzh. Fiz. Zhurn.* 29(2):313–317.
167. Mikk, I. R. 1976. On the calculation of nonisothermal bulk radiation. *Teplofiz. Vys. Temp.* 14(3):586–592.
168. Mikk, I. R. 1976. On the calculation of nonisothermal bulk raiation. *Izv. Akad. Nuak. EstSSR Fizika Matematika* (3):296–303.
169. Vares, V. A., Mikk, I. R., and Tijkma, T. B. 1976. Tables of coefficients for calculating radiant fluxes of the intrinsic radiation of a nonisothermal volume. *Trudy Tall. Politekh. Inst. peploenergetika* 392:31–44.
170. Kavaderov, A. V. 1952. *Teplovaya Rabota Plamennykh Pechey* (*Thermal Performance of Flame Furnaces*). Moscow: Metallurgizdat.
171. Nevskiy, A. V., and Chukhanova, L. A. 1965. Concerning the physical substance of various methods for calculating the radiation of a electively emitting variable temperature medium. *Teplofiz. Vys. Temp.* 3(1):124–133.
172. Detkov, S. P. 1970. Absorption of thermal radiation by a medium with a band spectrum. *Teplofiz. Vys. Temp.* 8(4):840–846.
173. Detkov, S. P., Girs, V.N., and Luk'yanets, E.A. 1967. The temperature field within gaseous carbon dioxide. *Teplofiz. Vys. Temp.* 5(5):837–840.
174. Detkov, S. P., and Vinogradov, A.V. 1969. Radiant heat transfer in a layer of gaseous $CO_2$ and $H_2O$ and of their mixture. *Izv. Akad. Nauk. SSSR Energetika i Transport* 3:139–144. [Engl. transl. *Heat Transfer, Sov. Res.* 2(2):54–62.
175. Detkov, S. P., and Vinogradov, A. V. 1968. Heat transfer in a layer of gaseous carbon dioxide. In *Teplomassoperenos*, vol. 1, pp. 770–774. Minsk.
176. MHD Electrical Power Generation, 1976 Status Report, April 1977, Nuclear Energy Agency Organization for Economic Cooperation and Development.

177. Viskanta, R., and Grosh, R. J. 1966. Recent advances in radiant heat transfer. In *Applied Mechanics Surveys*. eds H. N. Abramson, H. N. Abramson, H. Leibowitz, T. M. Crowlex, and S. Tuhaz, pp. 1113–1125. Washington, D.C.: Scripta.
178. Cess, R. D. 1964. The interaction of thermal radiation with conductive and convective heat transfer. In *Advances in Heat Transfer*, vol. 1, pp. 1–51. New York: Academic.
179. Filimonov, S. S., and Khrustalev, B. A. 1968. Combined heat transfer calculations. In *Teploobmen, Gidrodinamika i Teplofizicheskiye Svoystva Veshchestv* (*Heat Transfer, Hydrodynamics and Thermophysical Properties of Substances*), pp. 107–122. [Engl. transl. *Heat Transfer, Sov. Res.* 1(4):96–110. 1969].
180. Cess, R. D., and Tiwari, S. N. 1972. Infrared radiative energy transfer in gases. In *Advances in Heat Transfer*, vol. 8, pp. 369. New York: Academic.
181. Biberman, L. M. 1970. Radiant heat transfer at high temperatures. *Izv. Akad. Nauk. SSSR Energetika transport* 3:120–128. [Engl. transl. *Heat Transfer, Sov. Res.* 2(5):88–100. 1970].
182. Biberman, L. M. 1974. Radiant heat transfer at high temperatures. In Heat Transfer, 1974, Proc. Fifth International Heat Transfer Conference, Tokyo, vol. 6, p. 105.
183. Goryainov, L. A. 1965. Methods of decomposition of combined heat transfer. In *Issledovaniye Teploobmena v Teploenergetike* (*Heat Transfer in Heat Power Engineering*), pp. 59–67. Moscow.
184. Konakov, P. K. 1965. On the absorption coefficient of a gray medium. *Trudy Mosk. In-ta Inzh. Zh.-d. Transporta* 224:18–28.
185. Goryainov, L. A. 1958. Investigation of combined heat transfer in a cooled passage. *Sb. Leningrad. Ord. Lenina In-ta Inzh Zh.-d. Transporta* 160:241–250.
186. Adrianov, V. N., and Shorin, S. N. 1957. Heat transfer in channel flow of radiating combustion products. *Teploenergetika* 3:50–55.
187. Adrianov, V. N., Khrustalev, B. A., and Kolchenogova, I. P. 1966. Radiative–convective heat transfer in channel flow of high-temperature gas. In *Teploobmen v Elementakh Energeticheskikh Ustanovok* (*Heat Transfer in Components of Power Plants*), pp. 134–150.
188. Timofeyev, V. N., Bokovikova, A. Kh., Shklyar, B. R., and Denisov, M. A. 1972. Governing relationships for combined heat transfer in turbulent flow in a slotted channel. In *Metallurgicheskaya Teplotekhnika* (*Metallurgical Heat Engineering*), number 1, pp. 3–9. Moscow.
189. Shcherbinin, V. I., and Bokovikova, A. Kh. 1974. Investigation of combined heat transfer in a short cylindrical pipe. In *Metallurgicheskaya Teplotekhnika* (*Metallurgical Heat Engineering*), number 2, pp. 128–133.
190. Sinkevičius, J. E., and Tamonis, M. M. 1978. Governing relationships for combined heat transfer of combustion products in a channel. In *Mekhanika—8. Materialy Konferentsii 1977: Razvitiye Tekhnicheskikh Nauk v Respublike i Ispol'zovaniye ikh Rezul'tatov* (*Mechanics 8. Proc. Conf. Scientific Advances in the [Lithuanian] Republic and Their Practical Applications*). Vilnius.
191. Žukauskas, A. A., Sinkevičius, K. I. E., and Tamonis, M. M. 1978. Governing relationships for radiative and combined heat transfer of combustion products in a cooled square duct. In *IV Vsesoyuznaya Konferentsiya po Radiatsionnomy Teploobmenu* (*Fourth All-Union Conference on Radiative Heat Transfer*). Abstracts of reports, pp. 111–112. Kiev.
192. Wang, L. S., and Tien, C. L. 1967. A study of various limits in radiation heat-transfer problems. *Int. J. Heat Mass Transfer* 10:1327–1338.
193. Greif, R. 1964. Energy transfer by radiation and conduction with variable gas properties. *Int. J. Heat Mass Transfer* 7(8):891–900.
194. Rubtsov, N. A. 1974. Certain aspects of the study of radiative–conductive heat transfer. In *Problemy Teplofiziki i Fizicheskoy Gidrodinamiki* (*Problems of Thermophysics and Physical Hydrodynamics*), pp. 246–261. Novosibirsk.

195. Kutateladze, S. S., ed. 1977. *Teploobmen Izlucheniyem* (*Radiant Heat Transfer*). Novosibirsk.
196. Cess, R. D., and Tiwari S. N. 1969. The interaction of thermal conduction and infrared gaseous radiation. *Appl. Sci. Res.* 20(1): 25–39.
197. Viskanta, R. 1963. Interaction of heat transfer by conduction, convection and radiation in a radiating fluid. *J. Heat Transfer* 85:318–328.
198. Hunn, B. D., and Maffat, R. J. 1974. Radiative Heat transfer from a plasma in tube flow. *Int. J. Heat Mass Transfer* 17:1319–1328.
199. Sheremet'yev, S. V., and Filimonov, S. S. 1978. Radiative–convective heat transfer in the inlet length of a channel. In *Radiatsionnyy i Slozhnyy Teploobmen* (*Radiative and Combined Heat Transfer*), number 67, pp. 63–79. Moscow.
200. Zhukov, M. F., ed. 1977. *Teoriya Elektricheskoy Dugi v Usloviyakh Vynuzhdennogo Teploobmena* (*Theory of the Electric Arc Under Forced Heat Transfer Conditions*). Novosibirsk.
201. Zhukov, M. F., Smolyakov, V. Ya., and Uryukov, B. A. 1973. *Elecktodugovyye Nagrevateli Gaza* (*Plasmotrony*) [Electric-Arc Gas Heaters (*Plasmotrons*)]. Moscow: Nauka.
202. Zhukov, M. F., Koroteyev, A. S., and Uryukov, B. A. 1975. *Prikladnaya Dinamika Termicheskoy Plazmy* (*Applied Dynamics of Thermal Plasma*). Novosibirsk: Nauka.
203. Bakhshiyin, Ts. A. 1960. *Trubchatyye Pechi s Izluchayushchimi Stenami Topki* (*Tubular Furnaces with Radiating Surfaces*). Moscow: Gosinti.
204. Glinkov, M. A. 1962. *Osnovy Obshchey Teorii Pechey* (*Fundamentals of the General Theory of Furnaces*). Moscow: Metallurgizdat.
205. Soroka, B. S. 1977. *Processy Perenosa v Pechakh Kosvennogo Radiatsionnogo Nagreva* (*Transfer Processes in Indirectly Radiatively Heated Furnaces*). Kiev: Znaniye.
206. Gurvich, A. M., and Kuznetosv, N. V., eds. 1957. *Teplovoy Raschet Kotel'nykh Agregatov* (*Normativnyy Metod*) [*Thermal Design of Boiler Units (Standard Method)*]. Moscow–Leningrad: GEI.
207. Bakhshiyan, Ts. A., ed. 1969. Trubchatyye pechi (Tubular furnaces). *Trudy GIPRO–EFTMASH* [State Research and Planning Institute of Petroleum Machinery Manufacture], number 5(15). Moscow: Khimiya.
208. Filimonov, S. S., Adrianov, V. N., and Khrustalev, B. A. 1975. Calculation of heat transfer in furnaces. *Soviet Contributions to the Tokyo International Heat Transfer Conference, Tokyo.*
209. Mathematical simulation of reactor process in the design of petroleum-processing and petrochemical industries. 1976. In *Neftepererabotka i Neftekhimiya* (*Petroleum Processing and Petrochemistry*), number 12. Moscow.
210. Zuber, I., and Konečny, V. 1973. Mathematical model of combustion chambers for technical applications *J. Inst. Fuel* 285–294.
211. Kirillov, V. V., and Semenov, V.D. 1973. Investigation of heat transfer in the channel of an MHD generator. *Teplofiz. Vys. Temp.* 2(5):1092–1099.
212. Kirillov, V. V., and Semenov, V. D. 1975. Investigation of heat transfer in an open-cycle MDH generator channel. In *Teplotekhnicheskiye Problemy Pryamogo Preobrazovaniya Energii* (*Heat Engineering Aspects of Direct Energy Conversion*), number 6, pp. 45–49. Kiev: Naukova Dumka.
213. Shchegolev, G. M. 1975. Comparative analysis of heat transfer and drag in the channel of an MHD generator. In *Teplotekhnicheskiye Problemy Pryamogo Preobrazovaniya Energii* (*Heat Engineering Aspects of Direct Energy Conversion*), number 7. Kiev: Naukova Dumka.
214. Soroka, B. S. 1976. *Gazovyye Promyshlennyye Pechi i Kosvennyy Radiatsionnyy Nagrev Metalla* (*Industrial Gas-Fired Furnaces and Indirect Radiative Heating of Metals*). Moscow: Vniiegazprom.
215. Soroka, B. S., Yerinow, A. Ye., and Petishkin, S. A. 1975. Enhancement of heat transfer in heating furnaces. *Grazovaya Promyshlennost'* 3:34–39.

216. Pyare, R., and Abu-Romia, M. M. 1973. Coupled radiation with turbulent convection in electric arcs. *AIAA J.* 11(8):1188.
217. Pustogarov, A. V. 1965. Calculation of the parameters of an arc column in an argon atmosphere. *Teplofiz. Vys Temp.* 3(1):28–32.
218. Onufriyev, A. T., and Sevast'yenenko, V. G. 1968. Calculation of a cylindrical electrical arc with allowance for radiant heat transfer. Arc in hydrogen at 100 atm. *Prikl. Mekh. Tekhn. Fiz.* 2:17–22.
219. Zarudi, M. Ye., and Edel'baum, I. S. 1967. Calculation of the temperature profiles and parameters of channel arcs for argon, nitrogen and hydrogen by stepwise approximation. *Izv. SO Akad. Nauk. SSSR Ser. Tekhn. Nauk.* 1(3)(123):3–7.
220. Scott, R. K., and Incropera, F. P. 1973. Nonequilbrium flow calculations for the hydrogen constricted arc. *AIAA J.* 11(12):1714.
221. Sidlauskas, V. A., and Tamonis, M. M. 1977. Radiant energy transfer in a layer of hydrogen plasma. 1. Total radiation of a semispherical layer. *Trudy Akad. Nauk. LitSSR, Ser B* 4(101):81–90.
222. Devoto, R. S. 1968. Transport coefficients of partially ionized hydrogen. *J. Plasma Physics* 2(4):617–631.
223. Vargaftik, N.B., and Vasilevskaya, Yu.D. 1969. Transport coefficients for dissociating hydrogen. *Teplofiz. Vys. Temp.* 7(5):913–917.

# INDEX


Absorption, 8
  band, 12–15
  coefficient, 32
  experimental result, 21
Absorptivity, 48

Beer's law, 10
Bernoulli equation, 45
Boltzmann equations, 43
Boundary layer:
  equations, 49
  planar and laminar, 50
  turbulent, 54, 100
Broadening of spectral line:
  Doppler, 3–6, 10, 11
  Lorentz, 3–7, 10
  Stark, 3–4
  Voigt, 3–6, 10, 41

Carbon dioxide, 4, 18, 116, 134, 149–151, 157
Carbon monoxide, 18
Channel:
  cylindrical, 121, 125, 160
  flow temperature measurement, 83
  heat transfer, 50, 57, 82, 103, 166, 191
    radiative, 110
  MHD, 178–179
Combustion products, 75, 92, 105, 148, 168, 170
Continuity equation, 44, 56, 59, 62
Curtis-Godson approximation, 29–31, 34, 41, 126

Doppler spectral line broadening, 3–5, 9, 10–11

Edwards wide-band model, 27
Einstein coefficients, 3
Elsasser model, 8, 13–14, 16
Emissivity:
  of carbon dioxide, 26, 28–29
  of carbon monoxide, 27
  of gases, 22
  of water vapor, 25–26
Energy transport:
  in radiating medium, 43
  transport coefficients, 43
Enthalpy, 139

Equation:
of motion, 44
of state, 44
Equivalent width, 6

Flows:
in channels, 56, 59, 62
constant velocity, 51
convective, 91
cylindrical channel, 121
high-temperature, 52, 92, 193
laminar, 50, 91–93, 140
multicomponent reactive, 94
pipe, 110, 170
planar duct, 159
of radiating and absorbing media, 58
of real gas, 93
stabilized in channels, 62, 107
turbulent, 53, 62, 92, 98, 142, 167, 193
of water vapor, 158
Fresnel reflection laws, 48, 95, 99
Frictional drag, 56
Furnaces, 180

Gas:
emissivity, 11, 22, 133
gray, 113
heated flows, 91
heat transfer, 91
nonisothermal, 42, 115
nonuniform, 29, 34
optical properties, 1
polyatomic, 2
radiative heat transfer, 111
spectral absorption and transmission, 23
Golden model, 8
Grashof number, 50
Growth curve for spectral lines, 6

Harmonic oscillator, 16
Heat flux:
for arbitrary surfaces, 175
convective, 103
finite-difference method used, 64
radiative, 63, 85, 110, 129–132
sensors, 84
Heat transfer:
boundary conditions, 47
in channel, 50, 57
combined, 43, 49, 57, 80, 91, 138, 167, 173, 190
combustion chambers, 174
from combustion products, 75
conductive-diffusive, 94
convective, 91, 103, 107, 169, 171
convective-radiative, 147, 159, 144, 160
cylinder with radiating medium, 67, 112, 114, 121, 160
experimental techniques, 75
in furnaces, 180
in gaseous media, 111
and geometry of systems, 112
in hemispheric layer, 73, 113
in hydrogen arc, 183
in hydrogen plasma, 183
in MHD generator channels, 177
in planar radiating duct, 153, 159
in planar radiating layer, 64, 119–120, 121, 144, 152
in radiating gases, 50, 113–114, 119
radiative, 169, 175
radiative in closed space, 72, 111
in rectangular channels, 82, 136
in turbulent flow, 95, 96, 108, 168
Hottel's nomograms, 24
Hydrogen:
absorption coefficient, 185
arc, 188–191
conductivities, 196

Infrared radiation from hot gas, 1, 3

von Karman constant, 45, 101
King model, 8

Ladenburg-Reiche function, 6
Lagrange equation, 61
Line-width equivalent in spectral lines, 6
Lorentz spectral line broadening, 3–9

Mayer-Goody model, 13–15, 24, 30, 41
Missile engineering, 1
Mixing length, 102
Models of line groups, 9
Molecular spectra, 2
Multiplicity, 7

Nusselt number, 92, 97

Optical properties:
- of gases, 1–2
- of media generally, 111

Oscillator:
- anharmonic, 2
- harmonic, 16

Photon scattering, 2
Planck function, 47, 66, 73
Prandtl:
- hypothesis, 52, 54, 62, 163
- number, 50, 52, 92, 97
- theory of turbulence, 44

Radiometer:
- calibration, 88
- double beam, 88
- narrow angle, 85, 86, 87
- sensors compared, 87

Reflectivity, 48
Reynolds number, 50, 97
Rotator:
- nonrigid, 13
- rigid, 2, 14, 16

Schack triangular model, 15
Spectra:
- banded, 15
- infrared, 3
- molecular, 2
- rotational structure, 15

Spectral absorption coefficient, 15
Spectral bands:
- correlation constants, 16
- envelope, 15
- models compared, 12, 13, 14, 15
- rotational and vibrational, 2, 3, 29
- Schack model, 15

Spectral lines:
- absorption, 8
- bands, 11
- broadening, 3–5, 10
- center, 5
- contour, 3–6
- curve of growth, 6
- Elsasser model, 8
- Golden model, 8
- groups, 8–9, 20, 33
- half-width, 3–4, 36
- hydrogen, 185
- King model, 8
- models compared, 9
- shape, 3, 9, 10
- wings, 5

Stanton number, 168
Stark spectral line broadening, 3–4

Temperature:
- functions, 19
- measurement, 83

Transfer equations, 3
Transitions, bound-bound, 2
Transmissivity, 10–11, 48
Turbulence:
- free-stream, 52, 102
- heat transfer, 95
- high-temperature, 193
  - parameters measured, 99
- in plane layer, 96
- Prandtl theory, 44
- in water vapor, 167

Viscosity, 44
- eddy, 56, 60, 163

Voigt spectral line broadening, 3–6, 9

Water vapor, 24, 114, 150, 157, 167